STUDY GUIDE

Joseph P. Chinnici
Virginia Commonwealth University

Susan M. Wadkowski
Lakeland Community College

Fifth Edition

BIOLOGY
Life on Earth

Teresa Audesirk

Gerald Audesirk

PRENTICE HALL, Upper Saddle River, NJ 07458

Senior Editor: *Teresa Ryu*
Associate Editor: *Mary Hornby*
Special Projects Manager: *Barbara A. Murray*
Production Editor: *James Buckley*
Manufacturing Buyer: *Ben Smith*
Supplement Cover Manager: *Paul Gourhan*
Supplement Cover Designer: *PM Workshop Inc.*
Copy Editor: *Jane Loftus*

 © 1999 by **PRENTICE-HALL, INC.**
Simon & Schuster/A Viacom Company
Upper Saddle River, NJ 07458

Printed in the United States of America

10 9 8 7 6 5 4 3

Reprinted with Corrections June 1999.

ISBN 0-13-081039-8

Prentice-Hall International (UK) Limited, *London*
Prentice-Hall of Australia Pty. Limited, *Sydney*
Prentice-Hall Canada, Inc., *Toronto*
Prentice-Hall Hispanoamericana, S.A., *Mexico*
Prentice-Hall of India Private Limited, *New Delhi*
Prentice-Hall of Japan, Inc., *Tokyo*
Simon & Schuster Asia Pte. Ltd., *Singapore*
Editora Prentice-Hall do Brasil, Ltda., *Rio de Janeiro*

CONTENTS

TO THE STUDENT – PLEASE READ

This Study Guide should be used with the textbook **Biology: Life on Earth**, fifth edition, by Gerald and Teresa Audesirk. The Study Guide is set up to help you learn the most from your general biology course (and to earn the highest possible grade). However, working through the Study Guide is only a portion of the process. You should take advantage of the three interrelated elements of the course: the lectures, the textbook, and the Study Guide.

1.   **Lecture**: Attend all lectures, be attentive in class, and take good and complete notes. Sitting toward the front of the class will help increase your ability to pay attention. You may want to tape the lectures as well. (Always get your instructor's permission to tape lectures.) On the evening after each lecture, recopy the notes into a more readable and organized form, relying on the textbook and Study Guide for help in understanding confusing and difficult topics and for filling in "gaps" in the notes. (A taped lecture helps here.) You'll be pleasantly surprised by how much you will learn and remember by simply rewriting and adding to class notes.

2.   **Textbook**: Read the textbook chapters before they are covered in lecture. You will find the lecture more comprehensible since the terminology will be somewhat familiar. If you understand the lecture better, it will be less frustrating and more fun to go to class and your confidence level about biology will rise.

3.   **Study Guide**: After reading the textbook and attending lecture, use this Study Guide to help you prepare for the inevitable exams that go with college life. This Study Guide has a detailed overview of each chapter in the textbook, along with a set of review exercises with answers.

FORMAT OF THE STUDY GUIDE

This Study Guide is divided into 41 chapters, corresponding to the organization of the textbook. Each chapter has a similar format, consisting of:

1.   A detailed **overview** of the material in the textbook, organized to allow you to gain additional understanding of the chapter and easily find answers to the review questions. All important concepts from the textbook are included in the overview, along with some helpful hints for remembering the material.

2.   **Key terms and definitions** for that chapter. It has been suggested that the many new terms encountered in an introductory biology course equals that of a first year course in a foreign language. Thus, the key terms and definitions will prove helpful in understanding the content of each chapter. Refer to them as often as needed.

3.   A set of review exercises in a variety of formats, including: **thinking through the concepts**, which may include multiple choice, true/false, and fill-in questions, as well as occasional crossword puzzles, concept maps, tables, or diagrams; and **clues to applying the concepts**, which are essay questions based on ideas covered in the chapter. These questions are designed to increase your ability to think critically about the concepts presented in the chapter. After reading the textbook, attending the lecture, rewriting your lecture notes, reviewing the Study Guide outline and exercises, and studying, you should be able to answer these review questions.

4.   **Answers to exercises**. Refer to these after answering the review exercises. Looking at the answers before trying to answer them yourself can mislead you into thinking you know the answers when you may not. Use the answers provided in this guide to check your answers for accuracy, not to generate your answers.

HINTS FOR STUDYING

You should first learn the definitions of the key terms (and any others that you are unfamiliar with). Then learn how those terms apply or are related to the topic covered. Try to keep in mind the big picture while you are learning the details of a process. While this can be difficult, it will increase your understanding.

Drawing diagrams of your own to outline a process will help increase your understanding, especially if you are a more visual learner. Some students use or devise their own mnemonic devices to help them remember terms or concepts in order. (For example: King Philip Came Over For Good Spaghetti is a great way to remember the taxonomic categories Kingdom, Phylum, Class, Order, Family, Genus, Species.) It doesn't matter how ridiculous it sounds, just so it works for you.

HINTS FOR DOING WELL IN CLASS AND ON LECTURE EXAMS

It is common for colleges and universities have large sections of general biology, and professors may determine grades by giving computer-graded "objective" examinations consisting of multiple choice, matching, and true/false questions. If this describes your course, (and even if it doesn't) the following hints may help improve your grades, even if you think you don't "test well" on **objective exams**.

1.   **Learn the meaning of biological terms**. A listing of key terms with definitions appears after each chapter in the Study Guide. Learn the meaning of these words – they are often asked on objective exams.

2.   **Answer questions you know first**. If you don't know the answer to a question, skip over it and come back to it later. Sometimes you will get a clue to the correct answer from other questions on the exam.

3.   **Read all the choices** for multiple choice questions. Then, try to eliminate as many obviously incorrect choices as possible, and pick the choice that appears most correct from those remaining. If your professor says there is no penalty for guessing, answer every question.

If you are sure that two or more choices are correct and another choice is "all of the above," that is usually the correct answer. (But be careful.)

If part of a statement is false, the entire statement is false.

On **essay exams**, it is always a good idea to write down an outline. This will get your thoughts down on the paper and you can reform them into a comprehensive paragraph. Your answer will be more precise and you will be less likely to leave out important points. Begin with a clear, concise introductory statement summarizing your answer. The rest of the answer should be long enough to answer the question fully.

ACKNOWLEDGMENTS

Our thanks go to the following individuals who were instrumental in bringing this project to reality. Sheri Snavely, Teresa Ryu, and Mary Hornby, from Prentice Hall, offered us the chance to review early drafts of the newly revised Audesirk and Audesirk text and then encouraged us to revise the Study Guide. Most importantly, Jane Loftus spent many hours reviewing the content and accuracy of this supplementary text. It is a much better companion to the Audesirks' text due to her efforts. Of course, any remaining errors are solely our responsibility.

Susan Wadkowski would also like to thank Paige Akins, who suggested I take on this project, and Renee Minot. Due to Renee's skills at formatting and her software expertise, this truly became a partnership effort.

Chapter 1:   An Introduction To Life On Earth

OVERVIEW

This chapter focuses on the characteristics of life and gives a brief overview of the vast diversity of living organisms. The authors introduce basic scientific principles as well as the scientific method. Finally, a brief introduction to the mechanism and evidence for evolution is presented.

1) What Are the Characteristics of Living Things?

Life has **emergent properties** (intangible attributes arising from the complex ordered interactions among the individual characteristics of living things). Life characteristics include (1) complexity based on highly organized **organic** (carbon-based) **molecules** (made up of **elements** with **subatomic particles**); (2) acquisition, conversion, and use of materials (**nutrients**) and **energy** (the ability to do work) from the environment (ultimately by **photosynthesis**, the making of sugar using solar energy) through **metabolism** (the chemical reactions needed to sustain life); (3) **homeostasis** (active maintenance of their complex structure and internal environment); (4) growth involving conversion of environmentally-derived materials into specific molecules in the organism's body; (5) response to stimuli from the environment; (6) reproduction, using the information in molecules of **DNA (deoxyribonucleic acid)** called **genes**; and (7) the capacity for **evolution** making use of **mutations** (random changes in DNA structure) and, usually, **natural selection** (enhanced survival and reproduction of individuals with favorable inherited characteristics).

   The hierarchy of life is (from least complex to most complex): **subatomic particles**, which are grouped into **atoms**, which are grouped into **molecules**, which are grouped into **organelles**, which are grouped into **cells** (the smallest unit of life, each with **plasma membrane**, **cytoplasm**, and **nucleus**), which are grouped into **tissues**, which are grouped into **organs**, which are grouped into **organ systems**, which are grouped into **organisms**.

2) How Do Scientists Categorize the Diversity of Life?

Based on the traits they exhibit (especially cell type, number of cells in each organism, and the mode of nutrition and energy acquisition), organisms are classified into three major categories called **domains**: Bacteria, Archaea, and Eukarya. Bacteria and Archaea have single, simple (**prokaryotic**) cells lacking nuclei. The Eukarya organisms each have one or more complex (**eukaryotic**) cells with nuclei and are divided into four **kingdoms**: Protista, Fungi, Plantae, and Animalia.

   Bacteria, Archaea, and members of the kingdom Protista are mostly **unicellular**, while members of the kingdoms Fungi, Plantae, and Animalia are primarily **multicellular**, their lives depending on the intimate cooperation among cells. Members of different kingdoms have different ways of acquiring energy. Photosynthetic organisms (plants, some bacteria, and some protists) capture solar energy and use it to make sugars and fats; they are called **autotrophs** ("self-feeders"). Organisms that get energy from the bodies of other organisms are called **heterotrophs** ("other feeders"); these include many archaea, some bacteria, all fungi and animals. Bacteria and fungi absorb predigested food molecules, while animals eat chunks of food and break them down in their digestive tracts (ingestion).

3) What Does the Science of Biology Encompass?

A basic principle of modern biology is that living things obey the same laws of physics and chemistry that govern nonliving matter. There are three "scientific principles" (unproven assumptions) essential to biology: (1) **natural causality** (all earthly events can be traced to preceding natural causes); supernatural intervention has no place in science; (2) **uniformity in time and space** (natural laws do not change with time or distance, they apply everywhere and for all time); for example, scientists assume that gravity always has worked as it does today and works the same way everywhere in the universe; and **(3) common perception** (the assumption that all humans individually perceive natural and aesthetic events through their senses in fundamentally the same way). However, our interpretation of such events (like *appreciation* of rock music or the *morality* of abortion) may differ.

The **scientific method** is how scientists study the workings of life. It consists of four interrelated operations:
(1) **observation**, the beginnings of scientific inquiry (for example, maggots appear on fresh meat left uncovered);
(2) **hypothesis**, a tentative testable explanation of an observed event based on an educated guess about its cause (for example, maggots appear on fresh meat left uncovered because flies land on the meat and lay eggs);
(3) **experiment**, a study done under rigidly controlled conditions based on a prediction stemming from the hypothesis (for example, if the fresh meat is covered with a fine gauze to keep the flies away, no maggots should appear in the meat). Simple experiments test the assertion that a single factor, or **variable**, is the cause of a single observation. Scientists design **controls** into their experiments, in which all variables remain constant. Controls are compared to experiments in which only the variable being tested is changed;
(4) **conclusion**, a judgment about the validity of the hypothesis, based on the results of the experiment (for example, maggots did not appear on the meat covered with gauze but did appear on fresh meat left uncovered in the same place at the same time. Thus, the hypothesis is supported by the results of the experiment).

Science is a human endeavor. Besides the scientific method, real scientific advances often involve accidents and acumen, lucky guesses, controversies between scientists and the unusual insights of brilliant scientists. An example of "real science" was the accidental discovery of penicillin by Alexander Fleming.

When a hypothesis is supported by the results of many different kinds of experiments, scientists are confident enough about its validity to call it a **scientific theory**. Scientific conclusions must always remain tentative and be subject to revision if new observations or experiments demand it.

4) Evolution: The Unifying Concept of Biology

Evolution is the theory that present-day organisms descended, with modification, from preexisting forms; it is the unifying concept in biology. As proposed by the English naturalists Charles Darwin and Alfred Russel Wallace in the mid-1800s, evolution occurs due to three natural processes: (1) genetic variation that exists among members of a population; (2) inheritance of variations from parents to offspring (we now know that inherited variations ultimately arise from gene mutation, changes in DNA structure); and (3) natural selection, the survival and enhanced reproduction of organisms with favorable variations (**adaptations**) in structure, physiology, or behavior that best meet the challenges of the environment. Ultimately, natural selection has unpredictable results because environments tend to change dramatically (e.g., ice ages). What helps organisms survive today may be a liability tomorrow.

KEY TERMS AND CONCEPTS

Fill-In: Fill in the crossword puzzle based on the following clues:

Across

2. A relatively stable combination of atoms.
5. The membrane-bound organelle of eukaryotic cells that contains the cell's genetic material.
8. A self-feeder, usually meaning a photosynthetic organism.
9. Any change in the overall genetic composition of a population of organisms from one generation to the next.
11. The smallest unit of life.
13. An organism that eats other organisms.
15. Two or more populations of different species living and interacting in the same area.
16. A structure, usually composed of several tissue types, that acts as a functional unit.

Down

1. A physical unit of inheritance.
2. What a gene does when it undergoes a genetic change.
3. A study done under rigidly controlled conditions based on a prediction stemming from the hypothesis.
4. A new major category of organisms, based on comparisons of cellular, molecular and behavioral similarities and differences.
6. A judgment about the validity of the hypothesis, based on the results of the experiment.
7. In science, an explanation for natural events based on a large number of observations.
9. The ability to do work.
10. The outer membrane of a cell, enclosing its contents.
12. The smallest particle of an element that retains all the properties of the element.
14. A type of molecule that encodes the genetic information of all living cells.

Key Terms and Definitions

autotroph: a "self-feeder," usually meaning a photosynthetic organism.

biodiversity: all living things within a given geographical area and the interrelationships among them.

biosphere: that part of Earth inhabited by living organisms; includes both the living and nonliving components.

cell: the smallest unit of life, consisting, at a minimum, of an outer membrane enclosing a watery medium containing organic molecules, including genetic material composed of DNA.

community: two or more populations of different species living and interacting in the same area.

conclusion: a judgment about the validity of the hypothesis, based on the results of the experiment.

control: that portion of an experiment in which all possible variables are held constant; in contrast to the "experimental" portion, in which a particular variable is altered.

creationism: the hypothesis that all species of organisms on Earth were created in essentially their present form by a supernatural being, and that significant modification of those species, specifically their transformation into new species, cannot occur through natural processes.

domain: a major category of organisms, based on comparisons of cellular, molecular and behavioral similarities and differences; there are three: Bacteria, Archaea, and Eukarya.

deoxyribonucleic acid (DNA): a type of molecule that encodes the genetic information of all living cells.

ecosystem: a community together with its nonliving surroundings.

element: a substance that cannot be broken down to a simpler substance by ordinary chemical means.

emergent property: an intangible attribute that arises as the result of complex ordered interactions among individual parts.

energy: the ability to do work.

eukaryotic: referring to the type of cells that have their genetic material enclosed within a membrane-bound nucleus, contain other membrane-bound organelles, and are usually larger than prokaryotic cells.

evolution: any change in the overall genetic composition of a population of organisms from one generation to the next.

experiment: a study done under rigidly controlled conditions based on a prediction stemming from the hypothesis.

gene: the physical unit of inheritance; part of a DNA molecule.

heterotroph: "other feeder," meaning an organism that eats other organisms.

homeostasis: maintenance of a stable internal environment in the face of changes in the external environment.

hypothesis: in science, a supposition based on previous observations that is offered as an explanation for an event, and is used as the basis for further observations or experiments.

kingdom: the most inclusive category in the classification of organisms. The five kingdoms are Monera, Protista, Fungi, Plantae, and Animalia.

metabolism: the sum of all chemical reactions occurring within a single cell or within all the cells of a multicellular organism.

molecule: a relatively stable combination of atoms.

multicellular: many-celled organisms in which the cells interact and form mutually dependent associations.

mutation: a change in a gene; usually refers to a genetic change that is significant enough to change the appearance or function of the organism.

natural causality: the scientific principle that natural events occur as a result of preceding natural causes.

natural selection: the differential survival and reproduction of organisms due to environmental forces, resulting in the preservation of favorable adaptations; usually refers specifically to differential survival and reproduction based on genetic differences among individuals.

nucleus: the membrane-bound organelle of eukaryotic cells that contains the cell's genetic material.

nutrients: the atoms and molecules that living organisms need to acquire in their diets.

observation: the first step in the scientific method.

organ: a structure within an organism (e.g., intestine), usually composed of several tissue types, that is organized into a functional unit.

organelle: a structure inside a cell that performs a specific function.

organic molecule: molecule found in and usually made by a cell, often containing carbon atoms.

organism: an individual living thing.

organ system: two or more organs working together in the execution of a specific bodily function (e.g., digestive tract).

photosynthesis: the series of chemical reactions in which the energy of light is used to synthesize high-energy molecules.

plasma membrane: the outer membrane of a cell, enclosing its contents and separating them from its environment.

population: a group of organisms of the same species living in the same area and interbreeding.

prokaryotic: referring to cells of the domains Bacteria and Arcaea that do not have their genetic material enclosed

within a membrane-bound nucleus and also lack other membrane-bound organelles.

scientific method: how scientists study the workings of life. It consists of four interrelated operations: observation, hypothesis, experimentation, and conclusion.

scientific theory: in science, an explanation for natural events that is based on a large number of observations and is in accord with scientific principles, especially causality.

species: all of the organisms potentially capable of interbreeding under natural conditions or, if asexually reproducing, are more closely related to one another than to other organisms within the genus.

spontaneous generation: the proposal that living organisms can arise from nonliving matter.

subatomic particle: the particles that make up atoms: electrons, protons, and neutrons.

tissue: a group of (usually similar) cells that together carry out a specific function.

unicellular: an organism consisting of a single cell.

variable: a condition, particularly in a scientific experiment, that is subject to change.

THINKING THROUGH THE CONCEPTS

True or False: Determine if the statement given is true or false. If it is false, change the underlined word so that the statement reads true.

17. _____ Biology is basically different from other sciences.
18. _____ The basic assumptions of science can be proven.
19. _____ The conclusions of science are permanent.
20. _____ Science accepts only natural explanations for natural processes.
21. _____ Creationism is not a science.
22. _____ Organisms that produce their own food are heterotrophic.
23. _____ Bacteria are eukaryotic organisms.
24. _____ Prokaryotic forms do not possess distinct nuclei.
25. _____ Fungi are autotrophic organisms.
26. _____ Redi's experiments supported the theory that life can occur by spontaneous generation.

Matching: The five kingdoms.

27.____   unicellular algae   Choices:
28.____   multicellular and autotrophic   a.   Monera
29.____   unicellular and prokaryotic   b.   Protista
30.____   multicellular, heterotrophic, ingestive   c.   Fungi
31.____   multicellular algae   d.   Plantae
32.____   unicellular and eukaryotic   e.   Animalia
33.____   multicellular, heterotrophic, absorptive
34.____   sponges
35.____   mosses
36.____   bacteria

Matching: Which scientific principles do each of the following violate?

37.____   in biblical times, humans lived to be   Choices:
           900 years old   a.   natural causality
38.____   God created all life on earth in six days   b.   uniformity in time and space
39.____   a six-foot man may look ten feet tall to   c.   common perception
           you
40.____   gravity did not affect the dinosaurs as
           much as it does us
41.____   miracles
42.____   an anorexic woman sees herself as fat

Multiple Choice: Pick the most correct choice for each question.

43.   Which of the following are not characteristics used to categorize organisms into kingdoms?
    a.   types of cells present
    b.   numbers of cells present
    c.   presence or absence of cell walls
    d.   how the organisms acquire energy
    e.   how the organisms move
    f.   choices a and b
    g.   choices c and e
    h.   none of the above

44.   What is the ultimate source of genetic variation?
    a.   mutations in DNA
    b.   adaptations to a changing environment
    c.   natural selection
    d.   spontaneous generation
    e.   homeostasis

45.   The basic difference between a prokaryotic cell and a eukaryotic cell is that the prokaryotic cell
    a.   possesses membrane-bound organelles
    b.   lacks DNA
    c.   lacks a nuclear membrane
    d.   is considerably larger
    e.   is in multicellular organisms

CLUES TO APPLYING THE CONCEPTS

This practice question is intended to sharpen your ability to apply critical thinking and analysis to biological concepts covered in this chapter.

46. Last summer, Zachary grew pepper plants in his garden and decided to use a new fertilizer called UltraGrow. He claims that his plants produced a larger quantity of peppers that were larger in size than those he harvested the year before, and he credited the use of UltraGrow for the improvement. Scientifically speaking, has he proven the effectiveness of the fertilizer? How could he more validly test his hypothesis that UltraGrow works?

ANSWERS TO EXERCISES

1. gene	11. cell	22. false, autotrophic	34. e
2. (Across) molecule	12. atom	23. false, prokaryotic	35. d
2. (Down) mutate	13. heterotroph	24. true	36. a
3. experiment	14. DNA	25. false, heterotrophic	37. b
4. domain	15. community	26. false, disproved	38. a
5. nucleus	16. organ	27. b	39. c
6. conclusion	17. false, essentially	28. d	40. b
7. theory	similar to	29. a	41. a
8. autotroph	18. false, cannot	30. e	42. c
9. (Across) evolution	19. false, temporary	31. d	43. c and e
9. (Down) energy	20. true	32. b	44. a
10. plasma	21. true	33. c	45. c

46. Scientifically speaking, Zachary did not prove his hypothesis that the fertilizer works. He did not perform a controlled experiment, meaning that there could be a number of reasons why he got more and larger peppers last year than the year before: differences in the amounts of sunlight , rainfall, and/or richness of the soil, the types of seeds used, differences in plant pest activity each year, and perhaps others. What Zachary must do is to plant two groups of pepper plants side by side at the same time. To test the effect of UltraGrow, only one factor (the amount of UltraGrow) must differ between the experimental (fed UltraGrow) and the control (not fed UltraGrow) plants. If there is a difference between the fruits produced from the experimental and the control plants, then Zachary may validly conclude that the fertilizer was responsible for the difference.

Chapter 2: Atoms, Molecules, and Life

OVERVIEW

This chapter focuses on matter and energy. The authors describe the structure of atoms and molecules and the three major types of chemical bonding. They also discuss important inorganic molecules, especially water.

1) What Are Atoms?

Atoms are the fundamental structural units of matter. Atoms contain a central dense **nucleus** in which are found positively charged **protons** and uncharged **neutrons**. The **atomic number** of an atom is the number of protons present in its nucleus and this is a constant feature of all atoms of a particular type (e.g., every hydrogen atom has one proton). An **element** is a substance containing the same kind of atoms and cannot be broken down or converted into another substance under ordinary conditions. Similar atoms that differ in the number of neutrons they possess are called **isotopes**.

Atoms also have negatively charged **electrons** that spin about the nucleus in paths called energy levels or shells. Up to two electrons are found in the energy level nearest the nucleus, while the next energy level may contain up to eight electrons. Electrically neutral atoms have equal numbers of protons and electrons (e.g., a carbon atom has six protons and six electrons). Whereas the nuclei of atoms are stable and resistant to change, the electron shells are dynamic, and atoms interact with each other by gaining, losing, or sharing electrons.

2) How Do Atoms Interact to Form Molecules?

Atoms react with other atoms when there are vacancies in their outermost electron shells. Stable atoms have completely filled or empty outer electron shells (like helium) while reactive atoms have partially filled outer shells (like hydrogen). An atom with a partially full outer electron shell is reactive and can become more stable through filling the outer shell by losing, gaining, or sharing electrons with other atoms. These electron interactions with other atoms create attractive forces called **chemical bonds**.

Charged atoms called **ions** interact to form **ionic bonds**. Atoms with almost full or almost empty outer electron shells will interact by gaining or losing electrons, respectively, forming charged ions that will attract each other (e.g., sodium [Na^+] and chloride [Cl^-] ions), forming molecules by the formation of ionic bonds ($NaCl$ is table salt). Ionic molecules tend to form crystals, although ionic bonds are relatively weak and easily broken.

Uncharged atoms can become stable by sharing electrons, forming **covalent bonds**. An atom of hydrogen has one electron in a shell that can hold two. Two hydrogen atoms each share an electron with the other, forming a molecule of hydrogen gas (H_2) held together by one covalent bond. Oxygen atoms need two electrons to fill their outer shells; thus, two oxygen atoms form a molecule of O_2 gas by forming two covalent bonds. Covalent bonds are relatively strong. Most biological molecules are held together by covalent bonding. The atoms C,H,N,O,P,S (pronounced "chanops") are often found in cellular molecules.

Polar covalent bonds form when atoms share electrons unequally. A molecule having only one type of atom, like H_2 gas, is held together by a **nonpolar covalent bond** since each hydrogen nucleus exerts equal attraction on the shared electrons. In molecules made of different atoms, like water (H_2O), **polar covalent bonds** form since the oxygen nucleus has more protons than the hydrogen nuclei and exert a greater attraction on the electrons which spend more time orbiting the stronger nucleus. Thus, water is electrically neutral overall, but the oxygen end is more negatively charged than the hydrogen end; water is a **polar** molecule. Consequently, the negative end of one water attracts the positive end of another, forming a **hydrogen bond**. Many other molecules in cells form hydrogen bonds.

3) Why Is Water So Important to Life?

Water interacts with many other molecules. Since water is a polar molecule, it can dissolve many other substances (it is a good **solvent**). Water will surround positive and negative ions, dissolving crystals of polar molecules. Water is attracted to and dissolves molecules containing polar covalent bonds (called **hydrophilic** molecules) such as sugars and amino acids. Uncharged and nonpolar molecules, like fats and oils, are **hydrophobic** and do not dissolve in water. Water can break apart into H^+ and OH^- ions. A volume of pure water contains equal amounts of these ions and is said to have a value of 7 on the **pH** scale. Solutions with $H^+ > OH^-$ is **acidic** (pH less than 7) and a solution with $OH^- > H^+$ is **basic** (pH greater than 7). Each pH scale unit represents a 10-fold increase or decrease in the concentration of H^+ ions. **Buffers** like bicarbonate help organisms maintain a constant pH in their cells by accepting or releasing H^+ ions in response to small changes in pH.

Water moderates the effects of temperature changes due to three properties. It has high **specific heat** (it takes a lot of energy to raise the temperature of water) due to the presence of hydrogen bonds between water molecules. Water has high **heat of vaporization** (it takes a lot of heat to evaporate a molecule of water, leaving the remaining water cooler). Water also has a high **heat of fusion** (much energy is removed from water as it forms ice, thus heating up its surroundings). Since ice floats on cold liquid water, lakes remain liquid on the bottom in winter, allowing aquatic life to continue.

Due to hydrogen bonding, water molecules stick together (**cohesion**), producing the **surface tension** at the surface of lakes and pools that allows light insects to walk on water. The cohesion of water allows trees to pull water from the roots up into the leaves.

KEY TERMS AND CONCEPTS

Fill-In: From the following list of terms, fill in the blanks in the following statements.

acidic	covalent	hydrophilic	neutrons	released
atomic number	electrons	hydrophobic	nonpolar	single covalent
atoms	energy levels	ion	pH	specific heat
basic	fusion	ionic	polar	triple covalent
buffers	hydrogen	ions	protons	vaporization
chemical bonds				

(1)_____ contain a central dense nucleus in which are found positively charged (2)_____ and electrically neutral (3)_____.

Atoms also have negatively charged (4)_____, which spin about the nucleus in paths called (5)_____.

The (6)_____ of an atom is the number of protons present in its nucleus and this is constant for all atoms of a particular type.

Atoms enter into chemical reactions when there are vacancies in their outermost (7)_____.

(8)_____ are attractive forces between atoms due to interactions of their (9)_____.

(10)_____ result when atoms lose or gain electrons. The attraction between a Na^+ (11)_____ and a Cl^- (11)_____ is called (12)_____ bonding.

Atoms that interact by sharing electrons form molecules by (13)_____ bonding. If two atoms share one electron each, they form a (14)_____ bond; if two atoms share three electrons each, they form a (15)_____ bond.

If two similar atoms share electrons, they form a (16)_____ covalent bond, but if dissimilar atoms share electrons, as in water, they may form (17)_____ covalent bonds.

When two water molecules electrically attract each other, the attraction is called a (18)_____ bond. Charged molecules that attract water are called (19)_____ molecules, while uncharged molecules that do not attract water are called (20)_____ molecules.

A solution with equal amounts of H^+ and OH^- (21)_____ has a (22)_____ of 7.

(23)_____ solutions have a pH that is less than 7, while (24)_____ solutions have a pH greater than 7.

(25)_____ are important because they help maintain the pH of a cell at approximately 7. Water moderates temperature changes because it has a high (26)_____ (a lot of energy is needed to increase the temperature of water), it has a high heat of (27)_____ (a lot of energy is needed to evaporate water), and it has high heat of (28)_____ (a lot of energy is (29)_____ when water becomes ice).

Key Terms and Definitions

acid: a substance that releases hydrogen ions (H+) into solution; a solution with a pH of less than 7.

acidic: the property of having a higher concentration of H^+ ions than of OH- ions.

atom: the smallest particle of an element that retains the properties of the element.

atomic nucleus: the central region of an atom, composed of protons and neutrons.

atomic number: the number of protons in the nuclei of all atoms of a particular element.

base: a substance that is capable of combining with and neutralizing H ions, producing a solution with a pH greater than 7.

basic: the property of having a higher concentration of OH- ions than of H^+ ions.

buffer: a compound that minimizes changes in pH by reversibly taking up or releasing H+ ions.

calorie: the amount of energy required to raise the temperature of 1 gram of water $1°C$.

chemical bond: the force of attraction between neighboring atoms that holds them together in a molecule.

cohesion: the tendency of the molecules of a substance to hold together.

compound: a substance composed of two or more elements that can be broken into its constituent element by chemical means.

covalent bond: a chemical bond between atoms in which electrons are shared.

double covalent bond: a covalent bond that occurs when two atoms share two pairs of electrons.

electron: a subatomic particle, found in energy levels outside the nucleus of an atom, bearing a unit of negative charge and very little mass.

electron shell: all the electrons in a given energy level at a given distance from the nucleus of an atom.

element: a substance that cannot be broken down to a simpler substance by ordinary chemical means.

energy level: the specific amount of energy characteristic of a given electron shell in an atom.

heat of fusion: the energy that must be removed from a compound to transform it from a liquid into a solid at its freezing temperature.

heat of vaporization: the energy that must be supplied to a compound to transform it from a liquid into a gas at its boiling temperature.

hydrogen bond: the weak attraction between a hydrogen atom bearing a partial positive charge (due to polar covalent bonding with another atom) and another atom, usually oxygen or nitrogen, bearing a partial negative charge. Hydrogen bonds may form between atoms of a single molecule or of different molecules.

hydrophilic: pertaining to a substance that dissolves readily in water, or to parts of a large molecule that form hydrogen bonds with water.

hydrophobic: pertaining to a substance that does not dissolve in water.

hydrophobic interaction: the tendency for hydrophobic molecules to cluster together when immersed in water.

ion: an atom or molecule that has either an excess of electrons (and hence is negatively charged) or has lost electrons (and is positively charged).

ionic bond: a chemical bond formed by the electric attraction between positively and negatively charged ions.

isotope: one of several forms of a single element, the nuclei of which contain the same number of protons but different numbers of neutrons.

molecule: a particle composed of one or more atoms held together by chemical bonds. A molecule is the smallest particle of a compound that displays all the properties of that compound.

neutron: a subatomic particle found in the nuclei of atoms, bearing no charge and having mass approximately equal to that of a proton.

nonpolar covalent bond: a covalent bond with an equal sharing of electrons.

pH scale: a scale with values from 0 to 14, used for measuring the relative acidity of a solution. At pH 7 a solution is neutral, pH 0 to 7 is acidic, and pH 7 to 14 is basic. Each unit on the pH scale represents a tenfold change in the concentration of hydrogen ions.

polar covalent bond: a covalent bond with an unequal sharing of electrons, so that one atom is relatively negative while the other is relatively positive.

proton: a subatomic particle found in the nuclei of atoms, bearing a unit of positive charge and a relatively large mass roughly equal to the mass of the neutron.

radioactive: pertaining to an atom with an unstable nucleus that spontaneously disintegrates with the emission of radiation.

single covalent bond: a covalent bond that occurs when two atoms share a single pair of electrons.

solvent: a liquid that is capable of dissolving (uniformly dispersing) other substances in itself.

specific heat: the amount of energy required to raise the temperature of 1 gram of a substance $1°C$.

surface tension: the property of a liquid to resist penetration by objects at its interface with the air, due to cohesion between molecules of the liquid.

triple covalent bond: a covalent bond that occurs when two atoms share three pairs of electrons.

THINKING THROUGH THE CONCEPTS

Label the types of chemical bonding in the following diagram of water molecules:

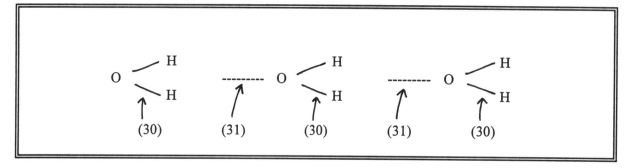

Refer to the figure to the right when answering the following questions.

32. Which is more chemically stable, a sodium atom or a sodium ion?

33. Which has the higher atomic number, a chlorine atom or a chloride ion?

34. The type of chemical bond depicted in part (b) is called a(n)

    _____ bond.

35. The structure shown in part (c) is called a

    _____.

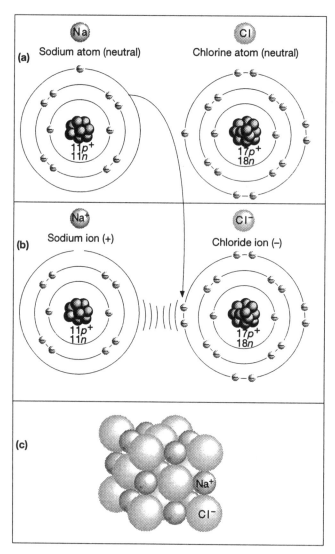

Matching: Chemical bonding.

36. ____ Two atoms share a pair of electrons.
37. ____ One atom donates an electron and another atom accepts it.
38. ____ It holds a crystal of salt together.
39. ____ The attraction between the polar regions of molecules.
40. ____ It results in atoms with unbalanced electrical charges.
41. ____ It holds a molecule of water together.
42. ____ The attraction between two water molecules.
43. ____ The electrical attraction between oppositely charged atoms.
44. ____ It holds most biological molecules together.
45. ____ It can result in polar molecules.

Choices:
    a.. ionic bonding
    b. covalent bonding
    c. hydrogen bonding

Matching: Properties of water

46. ____ why leaves "float" on water
47. ____ why a lot of energy is needed to convert liquid water to water vapor
48. ____ why water is such a good solvent
49. ____ why a lot of energy is needed to raise the temperature of water
50. ____ why water can break down salt crystals
51. ____ why fish can live in lakes in the wintertime
52. ____ why a lot of heat must be lost before water turns to ice
53. ____ why evaporation cools us in the summertime
54. ____ why belly-flopping dives hurt.

Choices:
    a. a polar molecule
    b. high specific heat
    c. high heat of vaporization
    d. high heat of fusion
    e. solid less dense than liquid
    f. cohesion

Multiple Choice: Pick the most correct choice for each question.

55 The nucleus of an atom <u>never</u> contains
    a. protons
    b. neutrons
    c. electrons

56 What determines the atomic number of an atom?
    a. The number of electrons in its outermost energy level
    b. The total number of energy levels of electrons
    c. The arrangement of neutrons in the atomic nucleus
    d. The number of protons in the atomic nucleus

57. Which of the following is an example of hydrogen bonding? The bond between
    a. O and H in a single molecule
    b. O of one water molecule and the H of a second water molecule
    c. O of one water molecule and the O of a second water molecule
    d. H of one water molecule and the H of a second water molecule

58. For an atom to achieve maximum stability and become chemically unreactive, what must occur?
    a. Its outermost energy level must be filled with electrons.
    b. The number of electrons must equal the number of protons.
    c. Sharing of electrons between atoms must occur.
    d. Ionization of atoms is required.
    e. Hydrogen bonds must form.

59. How is the formation of ions explained?
    a. Different atoms share electrons.
    b. Different atoms gain and lose electrons.
    c. Different atoms gain and lose protons.
    d. Different atoms share protons.
    e. Different atoms share neutrons.

60. If a substance measures 7.0 on the pH scale, that substance
    a. has equal concentrations of H^+ and OH^- ions
    b. may be very acidic
    c. has greater concentration of H^+ and less of OH^- ions
    d. probably lacks OH^- ions
    e. may be very basic

61. A glass of lemon juice had a pH of 2 while a glass of grapefruit juice has a pH of 3. How much higher is the H^+ ion concentration in lemon juice than in grapefruit juice?
    a. ten times as much
    b. two-thirds as much (2/3)
    c. one and a half times as much (3/2)
    d. one-tenth as much

62. As ice melts, it
    a. releases heat into its surroundings
    b. absorbs heat from its surroundings
    c. increases its property of cohesion
    d. increases its heat of vaporization
    e. immediately vaporizes

True or False: Determine if the statement given is true or false. If it is false, change the underlined word so that the statement reads true.

63. _____   The smallest unit of matter in an element is the underlined molecule.
64. _____   A positive unit in an atom is the proton.
65. _____   An electron is heavier than a proton.
66. _____   The number of protons in an atom determines its atomic weight.
67. _____   Helium is more likely to explode than hydrogen.
68. _____   Electrons close to a nucleus have less energy than those farther away from the nucleus.
69. _____   In salt, sodium and chlorine atoms attract each other by forming covalent bonds.
70. _____   The sodium atoms in salt tend to take on electrons.
71. _____   In a polar water molecule, the hydrogen region has a positive electrical charge.
72. _____   When atoms share electrons, they form ionic bonds.
73. _____   Electrons in a shell closer to the nucleus have more energy than those in a shell farther away.
74. _____   Atomic reactivity depends on the number of electrons in the innermost electron

Fill in the Table. Fill in the blanks in the following table.

Atom/Ion	Atomic Number	Number of Protons	Number of Electrons	Number of Electrons in the Outermost Energy Level
Hydrogen (H)	1	1	1	1
75. Oxygen (O)	8		8	
76. Carbon (C)		6		
77. Chloride ion (Cl-)	17		18	

CLUES TO APPLYING THE CONCEPTS

These practice questions are intended to sharpen your ability to apply critical thinking and analysis to biological concepts covered in this chapter.

78. Why are winter temperatures near a large lake often a few degrees warmer than in a land-locked city in the same state, while during the summer the reverse is true?

79. Why does water bead up when it is spilled onto the surface of a newly waxed automobile or counter top, whereas alcohol spreads out without beading up?

ANSWERS TO EXERCISES

1. atoms
2. protons
3. neutrons
4. electrons
5. energy levels
6. atomic number
7. energy levels
8. chemical bonds
9. electrons
10. ions
11. ion
12. ionic
13. covalent
14. single covalent
15. triple covalent
16. non-polar
17. polar
18. hydrogen
19. hydrophilic
20. hydrophobic
21. ions
22. pH
23. acidic
24. basic
25. buffers

26. specific heat
27. vaporization
28. fusion
29. released
30. covalent bonds
31. hydrogen bonds
32. sodium ion
33. neither
34. ionic
35. crystal
36. b
37. a
38. a
39. c
40. a
41. b
42. c
43. a
44. b
45. b
46. f
47. c
48. a
49. b
50. a

51. e
52. d
53. c
54. f
55. c
56. d
57. b
58. a
59. b
60. a
61. a
62. b
63. false, atom
64. true
65. false, lighter
66. false, number
67. false hydrogen, helium
68. true
69. false, ionic
70. false, give up
71. true
72. false, covalent
73. true
74. false, outermost

75. Oxygen has 8 protons, and 6 electrons in the outer shell.

76. Carbon has atomic number of 6, 6 electrons, and 4 electrons in the outer shell.

77. Chloride ion has 17 protons, and 8 electrons in the outer shell.

78. During the winter, the water on the surface of the lake freezes into ice, giving off heat into the surrounding atmosphere (heat of fusion). This raises the temperature around the lake in winter. During the summer, the water on the surface of the lake evaporates into the atmosphere, absorbing heat from the surrounding atmosphere (heat of vaporization). This lowers the temperature around the lake in summer.

79. Water molecules are polar and alcohol molecules are nonpolar. A waxed surface also is nonpolar. Thus, water molecules are attracted to each other by cohesion. The positive H ends of some water molecules are attracted to negative O ends of other water molecules, forming spheres of water with high surface tension (the water molecules on the surface of the sphere are not attracted to the waxy surface, only to other water molecules). Alcohol molecules spread out on the waxy surface because they are not attracted to each other, so that gravity forces them to spread out on the surface.

Chapter 3: Biological Molecules

OVERVIEW

This chapter focuses on the basic molecules that make up living things. The authors describe the structure, synthesis, and function of four types of large organic molecules: carbohydrates, lipids, proteins, and nucleic acids.

1) Why Is Carbon So Important in Biological Molecules?

Organic molecules contain carbon and functional groups, and are made using a "modular approach." Organic molecules include a skeleton of carbon atoms, often with hydrogen atoms attached. Since each carbon can form four covalent bonds, molecules with many carbons can form complex shapes including chains, rings, and branches. Organic molecules also contain **functional groups** of atoms including hydroxyl (-OH), carboxyl (-COOH), amino ($-NH_2$), phosphate ($-H_2PO_4$), and methyl ($-CH_3$) groups. These groups help determine the chemical properties of organic molecules. Cells use a "modular approach" to make large organic molecules by joining together many smaller **subunits** (many **monomers** like **sugar** are joined to make a **polymer** like **starch**, for instance).

2) How Are Organic Molecules Synthesized?

Biological molecules are built up or broken down by adding or removing water molecules. Subunits are linked together to form large molecules by a chemical reaction called **dehydration synthesis**. One subunit loses a hydrogen (-H) and another loses a hydroxyl (-OH) group; the two subunits form a covalent bond linking them together and the (-H) and (-OH) join to form water (H_2O). The reverse reaction is called **hydrolysis**: Water splits into (-H) and (-OH) and each covalently bonds to one or another subunit of a polymer, resulting in the breakdown of the polymer by removal of individual monomers.

3) What Are Carbohydrates?

Carbohydrates have carbon, hydrogen, and oxygen in an approximate 1:2:1 ratio. Carbohydrates are single **sugars** (like **glucose**) called **monosaccharides**, double sugars called **disaccharides**, or longer chains of sugars (like **starch** and **cellulose**) called **polysaccharides**. Carbohydrates (sugars and starch) are used to provide energy to cells or to provide structural support. Different monosaccharides have slightly different structures even though they contain the same types of atoms: Glucose, fructose, and galactose each are $C_6H_{12}O_6$.

Disaccharides such as **sucrose** (table sugar), **lactose** (milk sugar), and **maltose** (made from starch) are used for short-term energy storage in plants. Some polysaccharides are used for long-term energy storage in plants (starch) and animals (**glycogen**), while others provide structural support for plants (cellulose) and certain insects and fungi (**chitin**).

dehydration synthesis

4) What Are Lipids?

Lipid molecules are insoluble in water and contain mainly carbons and hydrogens. Some lipids (**fats** and **oils**) store energy, some (**waxes**) form waterproof coatings on plants and animals, some (**phospholipids**) are found in cell membranes, and some (**steroids**) act as hormones (made in one area part of an organism and used in another).

Oils, fats, and waxes contain only carbon, hydrogen, and oxygen. They contain **fatty acid** subunits and do not form ringed structures. Fats and oils form from **glycerol** and three fatty acid molecules through dehydration synthesis and are called **triglycerides**. Fats and oils store a much higher concentration of chemical energy than do carbohydrates and proteins. Fats are solid and oils are liquid at room temperature, because fats contain fatty acids **saturated** with H (making them more compact) and oils contain fatty acids **unsaturated** with H (making them more kinky due to the presence of double covalent bonds, and thus less compact). Waxes are chemically similar to fats but contain alcohol subunits.

Phospholipids have water-soluble "heads" and water insoluble "tails" and are found in high concentration in cell membranes. Phospholipids are similar to oils but with one fatty acid "tail" replaced by a phosphate "head" group attached to a polar charged functional group containing nitrogen. Thus, phospholipids have dissimilar ends: the "tail" is nonpolar while the "head" is polar. Steroids (like cholesterol, male and female sex hormones, and bile) have four rings of carbons with various functional groups attached.

5) What Are Proteins?

Proteins are polymers of **amino acids**; some function as **enzymes**. Depending on the sequences of amino acids in proteins, they may function as enzymes (controlling all chemical reaction within cells) or as structural components (elastin or keratin), energy storage (albumin), transport (hemoglobin), cell movement (muscle proteins), hormones (insulin, growth hormone), antibodies to fight infection, or poisons (snake venom).

All amino acids have a similar structure: a central carbon bonded to an amino group ($-NH_2$), a carboxyl group ($-COOH$), a hydrogen, and a variable (or R) group which differs among the 20 types of amino acids and gives each its distinctive properties. Amino acids are joined to form protein chains by dehydration synthesis. The $-NH_2$ of one amino acid is joined to the $-COOH$ of the next by a covalent bonds called a **peptide bond**, resulting in a **peptide** molecule with two amino acids. As more amino acids are added one by one, a **polypeptide** develops until the protein is complete.

A protein may have up to four levels of three-dimensional structure since they are highly organized molecules. The **primary structure** is the sequence of amino acids in a protein and is coded by the genes. The **secondary structure** is a coiled or **helix** structure caused by hydrogen bonding between the C=O and N-H regions of different amino acids in the sequence. The helical coil is distorted into the **tertiary structure** when the R groups of different amino acids interact, particularly when covalent bonds form between the R groups of different cysteine amino acids in a protein. Interactions between the R groups of different polypeptides can form huge proteins (like hemoglobin) with two or more polypeptide subunits and such complex proteins have **quaternary structure**. Within a protein, the exact type, position, and number of amino acids bearing specific R groups determines the three-dimensional structure of the protein which in turn determines its biological function.

6) What Are Nucleic Acids?

The genetic material is composed of nucleic acids which are polymers of subunits called **nucleotides**. Each nucleotide has a five-carbon sugar (ribose in **RNA** or deoxyribose in **DNA**), a phosphate group, and a variable nitrogen-containing base. The four types of **ribonucleic acid** (RNA) nucleotides contain ribose and either adenine (A), cytosine (C), guanine (G), or uracil (U) bases, whereas the 4 types of **deoxyribonucleic acid** (DNA) nucleotides contain deoxyribose and either A, C, G, or thymine (T) bases. Nucleotides are covalently bonded together into long chains to form DNA and RNA molecules. DNA is found in the chromosomes of all living things with the sequence of bases providing the genetic information needed for cells to make specific proteins. RNA molecules are copied from DNA in the nucleus, move into the cytoplasm, and direct the construction of proteins there.

Other nucleotides (like cyclic adenosine monophosphate) act as intracellular messengers. Some nucleotides (like **adenosine triphosphate** or **ATP**) have extra phosphate groups and carry energy from one place to another within cells. Some nucleotides (called **coenzymes**), usually in conjunction with vitamins, assist enzymes in their functions.

KEY TERMS AND CONCEPTS

Fill in the crossword puzzle with key terms, based on the following clues.

Across:

1. The most common monosaccharide, with the formula $C_6H_{12}O_6$.
3. A triglyceride lipid that is solid at room temperature.
5. A water-insoluble organic molecule such as a wax.
9. Contains a phosphate group, a five-carbon sugar, and a nitrogen-containing base.
11. A chain made of several amino acids joined covalently.
12A. A molecule with a central carbon joined to -H, -NH$_2$, -COOH, and a variable R group.
14. A molecule with primary, secondary, and tertiary structure.
15. A lipid coating containing alcohol that plants use to repel water.

Down:

2. A polysaccharide used by fungi and some animals for structural support.
4. A disaccharide found in mammalian milk.
6. An organic molecule composed of many nucleotides.
7. The shape of a secondary structure of a protein.
8. A simple carbohydrate, such as a monosaccharide.
10. A protein catalyst that speeds up the rate of specific biological reactions.
12D. A nucleotide with three phosphate groups for energy transfer.
13. Abbreviation for the genetic material.

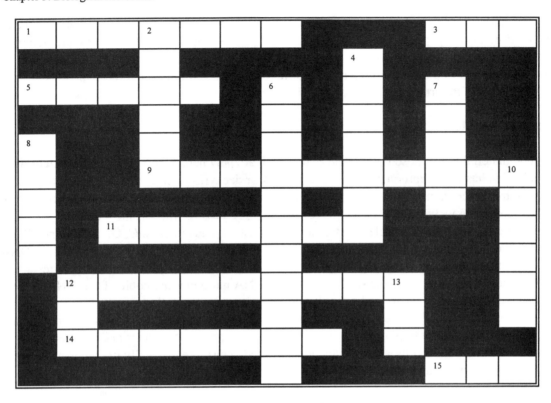

Key Terms and Definitions

adenosine triphosphate (ATP): a nucleotide with three phosphate groups that serves as an energy-carrier molecule in cells.

amino acid: the individual subunit that make up proteins, composed of a central carbon atom to which is bonded an amino group (--NH2), a carboxylic acid group (--COOH), a hydrogen atom, and a variable group of atoms denoted by R.

carbohydrate: a compound composed of carbon, hydrogen, and oxygen, with the approximate chemical formula $(CH_2O)n$, that includes sugars and starches.

cellulose: an insoluble carbohydrate composed of glucose subunits; it forms plant cell walls.

chitin: a compound found in the cell walls of fungi and the exoskeletons of insects and some other arthropods, composed of chains of nitrogen-containing, modified glucose molecules.

coenzyme: an organic molecule that assists enzymes in their actions.

cyclic nucleotide: a nucleotide in which the phosphate group is bonded to the sugar at two points, forming a ring.

Cyclic nucleotides serve as intracellular messengers.

dehydration synthesis: a chemical reaction in which two molecules are joined by a covalent bond with the simultaneous removal of a hydrogen from one molecule and a hydroxyl group from the other, forming water.

deoxyribonucleic acid (DNA): a molecule composed of deoxyribose nucleotides; the genetic information of all living cells.

disaccharide: a carbohydrate formed by the covalent bonding of two monosaccharides.

disulfide bridge: the covalent bond formed between the sulfur atoms of two cysteines in a protein; it often causes the protein to fold, bringing otherwise distant parts of the protein close together.

enzyme: a protein catalyst that speeds up the rate of specific biological reactions.

fat: a lipid composed of three saturated fatty acids covalently bonded to glycerol; fats are solid at room temperature.

fatty acid: an organic molecule composed of a long chain of carbon atoms with a carboxylic acid (--COOH) group at one end; fatty acids may be saturated (all single bonds between the carbon atoms) or unsaturated (one or more double bonds between the carbon atoms).

functional group: one of several groups of atoms commonly found in organic molecules, including hydrogen, hydroxyl, amino, carboxyl, and phosphate groups.

glucose: the most common monosaccharide, with the molecular formula $C_6H_{12}O_6$; most polysaccharides, including cellulose, starch, and glycogen, are made of glucose subunits covalently bonded together.

glycerol: a three-carbon alcohol to which fatty acids are covalently bonded to make fats and oils.

glycogen: a polysaccharide composed of branched chains of glucose subunits, used as a carbohydrate storage molecule in animals.

helix: a spiral structure similar to a corkscrew or a spiral staircase; a type of secondary structure of a protein.

hydrolysis: the chemical reaction that breaks a covalent bond through the addition of hydrogen to the atom forming one side of the original bond, and a hydroxyl group to the atom on the other side.

lactose: a disaccharide composed of glucose and galactose found in mammalian milk.

lipid: one of a number of water-insoluble organic molecules, containing large regions composed solely of carbon and hydrogen; lipids include oils, fats, waxes, phospholipids, and steroids.

maltose: a disaccharide composed of two glucose molecules.

monomer: a small organic molecule, several of which may be bonded together to form a chain called a polymer.

monosaccharide: the basic molecular unit of all carbohydrates, usually composed of a chain of carbon atoms to which are bonded hydrogen and hydroxyl groups.

nucleic acid: an organic molecule composed of nucleotide subunits. The two common types of nucleic acids are ribonucleic acids (RNA) and deoxyribonucleic acids (DNA).

nucleotide: an organic molecule made of a phosphate group, a five-carbon monosaccharide (ribose or deoxyribose), and a nitrogen-containing base.

oil: a lipid composed of three fatty acids, some of which are unsaturated, covalently bonded to a molecule of glycerol; oils are liquid at room temperature.

organic molecule: a molecule that contains both carbon and hydrogen.

peptide: a chain composed of two or more amino acids linked together by peptide bonds.

peptide bond: the covalent bond between the amino group nitrogen of one amino acid and the carboxyl group carbon of a second amino acid, which joins the two amino acids together in a peptide or protein.

phospholipid: a lipid consisting of glycerol to which two fatty acids and one phosphate group are bonded. The phosphate group bears another group of atoms, often containing nitrogen, and often bearing an electric charge.

polymer: a molecule composed of three or more (to thousands) smaller subunits called monomers. The monomers that make up a single polymer may be identical (e.g., the glucose monomers of starch) or different (e.g., the amino acids of a protein).

polysaccharide: a large carbohydrate molecule composed of branched or unbranched chains of repeating monosaccharide subunits, usually glucose or modified glucose molecules; polysaccharides include starches, cellulose, and glycogen.

primary structure: the amino acid sequence of a protein.

protein: an organic molecule composed of one or more chains of amino acids.

quaternary structure: the complex three-dimensional structure of a protein that is composed of more than one peptide chain.

ribonucleic acid (RNA): a molecule composed of ribose nucleotides; transfers hereditary instructions from the nucleus to the cytoplasm; also the genetic material of some viruses.

saturated: referring to a fatty acid with as many hydrogen atoms as possible bonded to the carbon backbone; a fatty acid with no double bonds in its carbon backbone.

secondary structure: a repeated regular structure assumed by protein chains, held together by hydrogen bonds; usually either a helix or a pleated sheet.

starch: a polysaccharide composed of branched or unbranched chains or glucose molecules, used by plants as a carbohydrate-storage molecule.

steroid: a lipid composed of four fused rings of carbon atoms to which functional groups are attached.

subunit: a small organic molecule, several of which may be bonded together to form a larger molecule. See also monomer.

sucrose: a disaccharide composed of glucose and fructose.

sugar: a simple carbohydrate molecule, either a mono- or a disaccharide.

tertiary structure: the complex three-dimensional structure of a single peptide chain. The tertiary structure is held in place by disulfide bonds between cysteine amino

acids, by attraction and repulsion among amino acid side groups, and by interaction between the cellular environment (water or lipids) and the amino acid side groups.

triglyceride: a lipid composed of three fatty acid molecules bonded to a single glycerol molecule.

unsaturated: referring to a fatty acid with fewer than the maximum number of hydrogen atoms bonded to its carbon backbone; a fatty acid with one or more double bonds in its carbon backbone.

wax: a lipid composed of fatty acids covalently bonded to long-chain alcohols.

THINKING THROUGH THE CONCEPTS

True or False: Determine if the statement given is true or false. If it is false, change the underlined word so that the statement reads true.

16. _____ Sucrose is a monosaccharide.
17. _____ In carbohydrates, the amount of C equals the amount of H.
18. _____ When two monosaccharides become a disaccharide, a water molecule is added.
19. _____ Two glucose molecules may join together to form a polysaccharide.
20. _____ Animals store their food in the form of glycogen.
21. _____ Carbohydrates have more energy per gram than fats.
22. _____ Fats are made of fatty acids and cholesterol.
23. _____ The water-attracting portion of a phospholipid is located in the middle of the cell membrane.
24. _____ The acid portion of an amino acid is the NH_2 group.
25. _____ The sequence of amino acids is the secondary structure of a protein.

Matching: Subunits making up large organic molecules

26. ____ nitrogen bases
27. ____ amino acids
28. ____ glycerol
29. ____ phosphoric acid
30. ____ five-carbon sugars
31. ____ monosaccharides
32. ____ fatty acids

Choices:

a.   carbohydrates
b.   nucleic acids
c.   proteins
d.   fats

Matching: Carbohydrates

33. ____ One sugar molecule per carbohydrate.
34. ____ Two sugar molecules per
carbohydrate.
35. ____ Many sugar molecules per
carbohydrate.
36. ____ Glucose, fructose, and galactose.
37. ____ Starch, glycogen, and cellulose.
38. ____ Sucrose, lactose, and maltose.
39. ____ Ribose and deoxyribose.

Choices:
    a.   Disaccharide
    b.   Monosaccharide
    c.   Polysaccharide

Matching: Specific carbohydrates

40. ____ Energy storage polysaccharide in
plants.
41. ____ Plant cell wall component.
42. ____ Milk sugar.
43. ____ Found in insect skeletons.
44. ____ Most common sugar.
45. ____ Table sugar.
46. ____ Found in some nucleic acids.
47. ____ Energy storage polysaccharide in
animals.

Choices:
    a.   Glucose
    b.   Sucrose
    c.   Lactose
    d.   Deoxyribose
    e.   Starch
    f.   Glycogen
    g.   Cellulose
    h.   Chitin

Matching: Lipids

48. ____ Structure contains alcohol.
49. ____ Cholesterol.
50. ____ Liquid at room temperature.
51. ____ Contains all saturated fatty acids.
52. ____ Contains unsaturated fatty acids.
53. ____ Energy storage in plants.
54. ____ Have hydrophilic and hydrophobic
ends.
55. ____ Have four fused rings of carbon atoms.
56. ____ A major component of cell
membranes.

Choices:
    a.   Fats
    b.   Oils
    c.   Waxes
    d.   Phospholipids
    e.   Steroids

Matching: Protein structure

Choices:
a. Primary structure
b. Secondary structure
c. Tertiary structure
d. Quaternary structure

57. ____ Interaction of R groups of different amino acids, causing contortions in shape.
58. ____ Sequence of amino acids in a polypeptide chain.
59. ____ Joining of several polypeptide chains.
60. ____ Helical shape of a peptide chain due to hydrogen bonding.

Multiple Choice: Pick the most correct choice for each question.

61. What type of chemical reaction results in the breakdown of organic polymers into their respective subunits?
    a. dehydration synthesis
    b. oxidation
    c. hydrolysis

62. Which of the following reactions requires the removal of water to form a covalent bond?
    a. glycogen ➡ glucose subunits
    b. dipeptide ➡ two amino acids
    c. cellulose ➡ glucose
    d. glucose + galactose ➡ lactose
    e. triglyceride ➡ 3 fatty acids + glycerol

63. What maintains the secondary structure of a protein?
    a. peptide bonds

b. disulfide bonds
c. hydrogen bonds
d. ionic bonds
e. covalent bonds

64. What determines the specific function of a protein?
    a. the exact sequence of its amino acids
    b. the number of disulfide bonds
    c. having a hydrophilic head and a hydrophobic tail region
    d. having fatty acids as monomers
    e. the length of the molecule

65. Complex 3-dimensional tertiary structures of globular proteins are characterized by:
    a. an absence of hydrophilic amino acids
    b. a helical shape
    c. a lack of cysteines in the amino acid sequence
    d. the presence of disulfide bridges
    e. a pleated sheet shape

CLUES TO APPLYING THE CONCEPTS

These practice questions are intended to sharpen your ability to apply critical thinking and analysis to biological concepts covered in this chapter.

66. A bear wandered into town and died, apparently of natural causes, and the forensics lab wants to determine the types of food the bear ate just before it died. When the contents of its stomach are analyzed, the following types of molecules are found: glucose, adenine, galactose, long chains of carbon with -COOH at one end, ribose, fragments with both phosphate and nitrogen components, deoxyribose, cytosine, and chitin. Determine which of the following types of large molecules the bear consumed: lipids, proteins, carbohydrates, nucleic acids. Also, indicate whether specific dietary products such as beef or milk were present, and what types of plants or animals might have contributed to the stomach contents. Explain your choices.

ANSWERS TO EXERCISES

1. glucose
2. chitin
3. fat
4. lactose
5. lipid
6. nucleic acid
7. helix
8. sugar
9. nucleotide
10. enzyme
11. peptide
12A. amino acid
12D. ATP
13. DNA
14. protein
15. wax
16. false, disaccharide
17. false, oxygen
18. false, taken away
19. false, disaccharide
20. true
21. false, less

22. false, glycerol
23. false, on the edges
24. false, COOH
25. false, primary
26. b
27. c
28. d
29. b
30. b
31. a
32. d
33. b
34. a
35. c
36. b
37. c
38. a
39. b
40. e
41. g
42. c
43. h

44. a
45. b
46. d
47. f
48. c
49. e
50. b
51. a
52. b
53. b
54. d
55. e
56. d
57. c
58. a
59. d
60. b
61. c
62. d
63. c
64. a
65. d

66. The bear had eaten lipids (long chains of carbon with -COOH at one end are fatty acids), carbohydrates (glucose and galactose are monosaccharides; chitin is a polysaccharide), and nucleic acids (ribose and deoxyribose are sugars in nucleic acids, adenine and cytosine are nitrogen bases in nucleic acids, and nucleic acid fragments have both phosphate and nitrogen components). No proteins were consumed since no amino acids are present. Milk might have been consumed since galactose sugar is present, and some insects (indicated by the presence of chitin) were eaten.

Chapter 4:   Energy Flow in the Life of a Cell

OVERVIEW

This chapter focuses on energy flow through the universe and particularly through living cells and organisms. The authors discuss the basic laws of thermodynamics and outline the basic types of metabolic reactions. They also explain how cells use enzymes to control chemical reactions.

1) What Is Energy and What Are the Basic Properties of Energy?

Energy is the capacity to do work like making molecules, moving them around, and generating light and heat. **Kinetic energy** is the energy of movement including light, heat, and electricity. **Potential energy** is stored energy, including chemical energy stored in the bonds of molecules.

The **laws of thermodynamics** define the basic properties and behavior of energy. The **first law** (called conservation of energy) is that energy cannot be created or destroyed, although it can be changed from one form to another (chemical energy in gasoline ➜ heat and movement of cars, for instance). The **second law of thermodynamics** states that when energy is converted from one form to another, the amount of useful energy decreases, since heat usually is given off. Spontaneous energy conversions in nature produce an increase in randomness and disorder, called entropy. Disorder spreads through the universe, and life alone battles against it by using energy from the sun to maintain orderliness within cells.

2) How Does Energy Flow in Chemical Reactions?

A chemical reaction converts **reactant** substances into **product** molecules. If the reactant energy is greater than the product energy, the reaction is **exergonic** because energy is released during the chemical reaction. For example, when sugar is heated (**activation energy** is added) with oxygen until it burns, chemical energy within the sugar molecules is then released as heat and light (fire), and the molecules produced (carbon dioxide and water) have less energy.

If the product energy is greater than the reactant energy, the reactant is **endergonic** because energy is added to the reaction as it occurs. For example green plants use solar energy to make high-energy sugar and oxygen from low-energy water and carbon dioxide. Cells make complex biological molecules like proteins using endergonic reactions.

In a coupled reaction, an exergonic reaction provides the energy for an endergonic reaction. If the coupled reactions occur in different places within cells, the energy usually is transferred from place to place by energy-carrier molecules like ATP.

3) How Do Cells Control Their Metabolic Reactions?

Cell **metabolism** refers to the chemical reactions within cells; these often occur in sequences called **metabolic pathways**. Cells regulate chemical reactions by using proteins called **enzymes,** which act as catalysts. Most reactions can be accelerated by raising the temperature, thus supplying more **activation energy**.

4) What Are Enzymes?

Enzymes act as **catalysts** since they lower the activation energy needed to begin exergonic chemical reactions. Due to its three-dimensional shape, a particular enzyme is very specific, catalyzing at most only a few types of reactions. The **active site** region of an enzyme is where the reactant molecules fit into the enzyme, similar to a key (enzyme) fitting into a lock (reactant). The enzyme is not changed by the reaction it catalyzes. The activity of enzymes is influenced by their environment, such as pH, temperature, salt concentration, or the availability of **coenzyme** molecules (often vitamins) necessary to aid enzymes in interacting with reactants. The activity of some enzymes in cells is reduced by the molecules they produce (a process called **feedback inhibition**), so that a cell doesn't produce too much product.

5) How Is Cellular Energy Carried Between Coupled Reactions?

The energy from glucose (exergonic reactions) is transferred to reusable **energy-carrier molecules** for transfer to the muscle protein that uses energy to contract (endergonic reactions). **Adenosine triphosphate (ATP)** is the principal energy-carrier molecule. Energy from exergonic reactions is used to make ATP from **adenosine diphosphate (ADP)** and inorganic phosphate (P_i). ATP carries the energy to various cellular sites where energy-requiring reactions occur. The ATP then is broken down into ADP and P_i, releasing energy to drive the endergonic reactions. Heat is released during these energy transfers, resulting in a loss of usable energy (an increase in entropy). Energy may be transported within a cell by other carrier molecules as well as by ATP. Some energy may be captured by electrons, which are carried by electron-carrier molecules to other parts of the cell to be released to drive endergonic reactions.

KEY TERMS AND CONCEPTS

Fill-In: From the following list of terms, fill in the blanks below.

active site
adenosine triphosphate
(ATP)
catalyst
coenzyme
coupled reaction
endergonic

energy
entropy
enzyme
exergonic
feedback inhibition
first law of thermodynamics
kinetic energy

metabolic pathway
metabolism
potential energy
product
reactant
second law of thermodynamics

(1) _____ This region of an enzyme binds substrates.
(2) _____ Any change in an isolated system causes the amount of useful energy to decrease, and the amount of randomness and disorder (entropy) to increase.
(3) _____ This is the energy of movement.
(4) _____ A molecule composed of ribose sugar, adenine, and three phosphate groups.
(5) _____ This is a protein that speeds up the rate of specific biological reactions.
(6) _____ This is a molecule resulting from a chemical reaction.
(7) _____ The total of all chemical reactions occurring within a cell or organism.
(8) _____ The product of a reaction inhibits an enzyme involved in making the product.
(9) _____ A pair of reactions, one exergonic and one endergonic, are linked together so

that the energy produced by one provides the energy needed for the other.

(10)_____ This is a chemical reaction requiring an input of energy to proceed.

(11)_____ This is the major energy carrier in cells.

(12)_____ This is a substance that speeds up a chemical reaction without itself being permanently changed in the process.

(13)_____ This is a molecule bound to an enzyme and required for its proper functioning.

(14)_____ This is a measure of the amount of randomness and disorder in a system.

(15)_____ This is a chemical reaction that liberates energy and increases entropy.

(16)_____ Within any isolated system, energy can be neither created nor destroyed.

(17)_____ This is stored energy, such as chemical energy in the bonds of molecules.

(18)_____ This is a molecule used up in a chemical reaction to form a product.

(19)_____ This is a sequence of chemical reactions within a cell.

(20)_____ Energy can be converted from one form to another.

(21)_____ The capacity to do work.

(22)_____ This includes light, heat, and mechanical movement.

Key Terms and Definitions

activation energy: in a chemical reaction, the energy needed to force the electron clouds of reactants together, prior to the formation of products.

active site: the region of an enzyme molecule that binds substrates and performs the catalytic function of the enzyme.

adenosine diphosphate (ADP): a molecule composed of ribose sugar, adenine, and two phosphate groups. When a third phosphate is added, the high-energy molecule ATP is formed.

adenosine triphosphate (ATP): a molecule composed of ribose sugar, adenine, and three phosphate groups. The last two phosphate groups are attached by energy-carrier high-energy bonds. ATP is the major energy carrier in cells.

catalyst: a substance that speeds up a chemical reaction without itself being permanently changed in the process; catalysts lower the activation energy of a reaction.

coenzyme: an organic molecule, often derived from a water-soluble vitamin, that is bound to certain enzymes and is required for their proper functioning.

coupled reaction: a pair of reactions, one exergonic and one endergonic, that are linked together so that the energy produced by the exergonic reaction provides the energy needed to drive the endergonic reaction.

electron carrier: a molecule that can reversibly gain and lose electrons. Electron carriers generally accept high-energy electrons produced during an exergonic reaction and donate the electrons to acceptor molecules that use the energy to drive endergonic reactions.

endergonic: pertaining to a chemical reaction that requires an input of energy to proceed; an "uphill" reaction.

energy: the capacity to do work.

energy-carrier molecules: a molecule that stores energy in high-energy chemical bonds and releases the energy again to drive coupled endothermic reactions. ATP is the most common energy carrier in cells.

entropy: a measure of the amount of randomness and disorder in a system.

enzyme: a protein catalyst that speeds up the rate of specific biological reactions.

exergonic: pertaining to a chemical reaction that liberates energy (either heat energy or in the form of increased entropy); a "downhill" reaction.

feedback inhibition: in enzyme-mediated chemical reactions, the condition in which the product of a reaction inhibits one or more of the enzymes involved in synthesizing the product.

first law of thermodynamics: a principle of physics that within any isolated system, energy can be neither created nor destroyed, but can be converted from one form to another.

kinetic energy: the energy of movement; includes light, heat, mechanical movement, and electricity.

Laws of thermodynamics: principles of physics that explain the relationships among the various forms of energy.

metabolic pathway: a sequence of chemical reactions within a cell, in which the products of one reaction are the reactants for the next.

metabolism: the sum of all chemical reactions occurring within a single cell or within all the cells of a multicellular organism.

potential energy: stored energy; includes chemical energy and the energy of position within a gravitational field.

product: an atom or molecule resulting from a chemical reaction.

reactant: an atom or molecule that is used up in a chemical reaction to form a product.

second law of thermodynamics: a principle of physics that any change in an isolated system causes the quantity of concentrated, useful energy to decrease, and the amount of randomness and disorder (entropy) to increase.

substrate: the atoms or molecules that are the reactants for an enzyme-catalyzed chemical reaction.

THINKING THROUGH THE CONCEPTS

True or False: Determine if the statement given is true or false. If it is false, change the underlined word so that the statement reads true.

23. _____ Within a closed system, the amount of energy is variable over time.
24. _____ A fire creates energy.
25. _____ The second law of thermodynamics is concerned with entropy.
26. _____ Photosynthesis and similar reactions decrease entropy.
27. _____ Eventually, all molecules in the universe will become randomly dispersed.
28. _____ Reactions that release energy are endothermic.
29. _____ Enzymes increase the activation energy needed for chemical reactions to occur.
30. _____ ATP contains a six-carbon sugar.
31. _____ Conversion of ATP to ADP releases energy.
32. _____ In coupled reactions, the "downhill" reaction liberates less energy than the "uphill" reaction.

Matching: Chemical reactions.

33.____ Once started, these will continue by themselves.
34.____ These need "activation energy" to get started.
35.____ Reactants have more energy than products.
36.____ Photosynthesis is classed as this.
37.____ Products have more energy than reactants.
38.____ Energy is released from the reaction.
39.____ This usually involves energy-carrier molecules.
40.____ Burning wood in a fireplace is classed as this.

Choices:
a. exergonic reactions
b. endergonic reactions
c. both of these
d. coupled reactions

Matching: Catalysts and enzymes.

41._____ All are protein molecules.
42._____ These can speed up chemical reactions.
43._____ These are not changed in the reactions they affect.
44._____ Generally, these are very specific as to the reaction they affect.
45 _____. These cannot cause energetically unfavorable reactions to occur.
46._____ Their activity can be regulated.

Choices:
  A.   Catalysts
  B.   Enzymes
  C.   Both
  D.   Neither

Multiple Choice: Pick the most correct choice for each question.

47.   In exergonic chemical reactions
   a.   reactants have more energy than products
   b.   reactants have less energy than products
   c.   reactants and products have equal amounts of energy
   d.   energy is stored in the reactions
   e.   enzymes are not necessary

48.   Which of the following statements about catalysts is <u>false</u>?
   a.   Biological catalysts usually are enzymes.
   b.   Catalysts increase energy of activation requirements.
   c.   Catalysts often increase the rate of reaction.
   d.   Catalysts are not permanently altered during the reaction.
   e.   Enzymes affect the amount of activation energy required in reactions.

49.   Cells regulate enzyme activity in all the following ways <u>except</u>
   a.   the amount of enzyme manufactured may be regulated
   b.   enzymes may be synthesized in an inactive form
   c.   feedback inhibition may occur
   d.   energy carrier molecules may be used to regulate enzyme activity

50.   When a muscle cell requires energy for contraction, what happens to ATP?
   a.   ATP makes more ATP.
   b.   ATP enters a metabolic pathway.
   c.   ATP is hydrolyzed.
   d.   ATP is phosphorylated.
   e.   ATP is synthesized.

51.   Which is the most common short-term energy-storage molecule?
   a.   glycogen
   b.   fat
   c.   sucrose
   d.   adenosine triphosphate

52.   The statement that "energy is neither created nor destroyed" is part of
   a.   entropy
   b.   first law of thermodynamics
   c.   second law of thermodynamics
   d.   allosteric inhibition

53.   The second law of thermodynamics states that
   a.   light can be converted into heat
   b.   within an isolated system, the total amount of energy remains constant
   c.   energy always flows from an area of higher concentration to an area of lower concentration
   d.   useful energy increases within an isolated system
   e.   useful energy decreases within an isolated system

Refer to the figure below to answer the following questions.

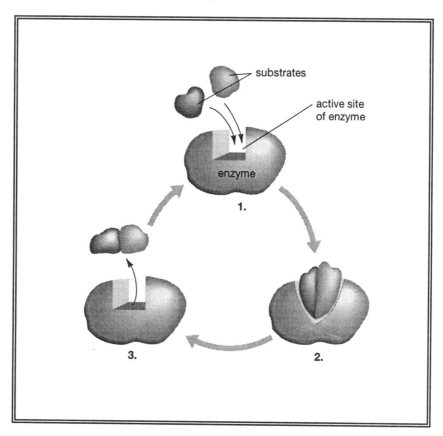

54. Which of the following statements are true?
    a. Substrates enter an enzyme's active site in a random orientation.
    b. While in the active site, substrates change their shape.
    c. The active site changes its shape once the substrates are present.
    d. Products fit the active site just as well as substrates do.
    e. Enzymes are permanently changed during chemical reactions.

55. If the figure depicts an endergonic reaction, which of the following statements are true?
    a. The substrates have more energy than the products.
    b. The products have more energy than the substrates.
    c. ATP could be produced from energy released by this reaction
    d. ATP could be used up to provide energy for this reaction
    e. The enzyme will be destroyed by the reaction.

CLUES TO APPLYING THE CONCEPTS

These practice questions are intended to sharpen your ability to apply critical thinking and analysis to biological concepts covered in this chapter.

56. When paper burns, it gives off both heat and light. Thus, the reaction is exergonic. Why, then, doesn't the paper that this book contains spontaneously burst into flames? If you touched the book with a burning match, you could set it on fire. What role does the energy supplied by the match play in this process?

ANSWERS TO EXERCISES

1.	active site	14.	entropy	27.	true	42.	c
2.	second law of	15.	exergonic	28.	false, exothermic	43.	c
	thermodynamics	16.	first law of	29.	false, decrease	44.	b
3.	kinetic energy		thermodynamics	30.	false, five-carbon	45.	c
4.	adenosine	17.	potential energy	31.	true	46.	b
	triphosphate (ATP)	18.	reactant	32.	false, more	47.	a
5.	enzyme	19.	metabolic pathway	33.	a	48.	b
6.	product	20.	first law of	34.	c	49.	d
7.	metabolism		thermodynamics	35.	a	50.	c
8.	feedback inhibition	21.	energy	36.	d	51.	d
9.	coupled reaction	22.	kinetic energy	37.	b	52.	b
10.	endergonic	23.	false, constant	38.	a	53.	e
11.	ATP	24.	false, changes	39.	d	54.	b and c
12.	catalyst	25.	true	40.	a	55.	b and d
13.	coenzyme	26.	true	41.	b		

56. The paper in this book doesn't spontaneously burst into flames because a sufficient amount of activation energy is not present to begin the intense exergonic reaction called burning. Holding a lit match to the paper would supply sufficient activation energy to begin the burning process. Often, as paper ages, it turns brown and becomes brittle. This occurs because the paper is reacting with oxygen in much the same way as in burning, except that the process is very slow due to insufficient activation energy.

Chapter 5:   Cell Membrane Structure and Function

OVERVIEW

This chapter focuses on the characteristics and functions of cell walls and cell membranes. The authors discuss the transport of molecules across cell membranes, especially the processes of diffusion and osmosis. Cell connections and communication between cells also are discussed.

1) How Is the Structure of a Membrane Related to Its Function?

A **plasma membrane** surrounds a cell, protecting and isolating it while allowing it extensive communication with its surroundings. The basic structure of membranes is a "**fluid mosaic**" of proteins (which regulate exchange of substances and communication) floating within a double layer of phospholipids and cholesterol (which isolates the cell from its watery environment).

Within membranes, phospholipids arrange themselves into a double layer (**phospholipid bilayer**) with the hydrophilic head forming the outer borders and the hydrophobic tails facing each other inside. Polar water-soluble molecules (salts, amino acids, sugars) cannot pass through the phospholipid bilayer. A variety of proteins are embedded within or attached to the phospholipid bilayer to regulate the molecule movement through the membrane and communicate with the environment. Some have carbohydrates attached, forming **glycoproteins**, which aid in cell communication.

Transport proteins regulate movement of water-soluble molecules through the plasma membrane. **Channel proteins** form pores to allow small molecules and ions (Ca^{++}, K^+, Na^+) to pass through. **Carrier proteins** bind molecules and, by changing shape, pass them across the membrane. **Receptor proteins** trigger cell responses and/or communication between cells when certain molecules (hormones or nutrients) bind to them. **Recognition proteins** often are glycoproteins on the outer membrane surface of certain cells (immune system cells, for instance) and serve as identification tags and attachment sites for other cells and molecules. Plasma membranes show **differential permeability**, allowing some molecules to pass through, but not others.

2) How Are Substances Transported Across Cell Membranes?

Molecules in fluids move in response to **concentration gradients**, from regions of greater concentration to regions of lower concentration. Movement of molecules across membranes occurs by both **passive transport** (down the concentration gradient [high ➡ low] by **diffusion** and requiring no cell energy) and **active transport** (against the concentration gradient [low➡ high] and requiring cell energy). The greater the concentration gradient, the faster diffusion occurs, but diffusion cannot move molecules rapidly over long distances. In **simple diffusion**, water, dissolved gases, or lipid-soluble molecules pass freely through the phospholipid bilayer. In **facilitated diffusion**, molecules cross the membrane in a way that doesn't use energy, assisted by transport or receptor proteins embedded in the membrane.

Diffusion of water across differentially permeable membranes is called **osmosis**. Extracellular fluids in animals usually are equal in water concentration (**isotonic**) to cellular fluids, so water diffuses equally into and out of cells. If a solution has a lower water/higher dissolved molecules concentration (**hypertonic**) than a cell (higher water/lower dissolved molecules concentration, or **hypotonic**), water will leave the cell faster than it enters and the cytoplasm will shrink. If, however, a solution is hypotonic to cells that are hypertonic, water will enter the cells faster than it leaves and the cytoplasm will expand. In such cases, certain animal cells have contractile vacuoles that pump the excess water out, while plant cells have central vacuoles which expand and allow those cells to become rigid.

Active transport uses energy to move substances against their concentration gradients into or out of cells. Digestive cells concentrate nutrients and brain cells get rid of excess ions by active transport. Active transport proteins, often called "pumps," span plasma membranes and use energy (usually from breaking down ATP molecules) to transport molecules across the membrane against the concentration gradient.

Many cells acquire particles too large to pass through membranes by **endocytosis** (using energy to surround the substance with plasma membrane and pinching it off internally to form a **vesicle**). In **pinocytosis**, a small area of membrane pinches inward to surround extracellular fluid and buds off into the cytoplasm to form a tiny vacuole. In **receptor-mediated endocytosis**, depressed areas of membrane called coated pits contain many copies of a receptor protein. These attach to specific extracellular molecules and the coated pit deepens into a U-shaped area that pinches off into the cytoplasm forming a coated vesicle. In **phagocytosis**, cells (such as *Amoeba* and white blood cells) can ingest entire microorganisms or large molecules by extending sections of plasma membrane to form **pseudopods** that surround the object and enclose it within a food vacuole in the cytoplasm for digestion. Through **exocytosis**, cells eliminate unwanted materials (like digestive waste) or secrete molecules (like hormones) into the extracellular fluid. A membrane-bound vesicle moves within the cytoplasm to the cell surface, where its membrane fuses with the plasma membrane, excreting the contents.

3) How Are Cells Specialized?

Desmosomes attach cells together, particularly within animal tissues, by gluing together the plasma membranes of adjacent cells with proteins or carbohydrates. Protein filaments run from the desmosomes into the interiors of each cell, adding additional strength to the attachment. **Tight junctions** waterproof the membranes of cells forming tubes and sacs that must remain watertight (like the urinary bladder). The membranes of adjacent cells nearly fuse along a series of ridges, forming leakproof gaskets. **Gap junctions** are clusters of protein channels directly connecting the cytoplasm of adjacent cells to help communication (via the flow of hormones, nutrients, ions, or electrical signals) among heart muscle cells, gland cells, or brain cells, among others. **Plasmodesmata** are cytoplasmic strands, surrounded by plasma membrane, that pass through openings in the walls of adjacent plant cells, allowing water nutrients, and hormones to pass freely from one cell to another.

Cell walls cover the outer surfaces of many cells. In protists, cell walls are made of cellulose, protein, or glassy silica. In plants, they are made of cellulose and other polysaccharides, in fungi they are made of chitin, and in bacteria they are made of a chitin-like material. In plants, cells secrete a **primary cell wall** of cellulose through their plasma membranes, and later secrete a thicker **secondary cell wall** of cellulose and other polysaccharides beneath the primary wall. The primary cell walls of adjacent cells are joined by a middle lamella layer made of a polysaccharide called pectin. Cells walls are strong yet porous.

KEY TERMS AND CONCEPTS

Fill-In: From the following list of key terms, fill in all the boxes in the following concept map.

active transport
carrier proteins
channel proteins
cholesterol
desmosomes

differential
   permeability
endocytosis
gap junctions
passive transport
phagocytosis

phospholipid
pinocytosis
plasma membrane
plasmodesmata
proteins
receptor proteins

receptor-mediated
   endocytosis
recognition proteins
tight junctions
transport proteins

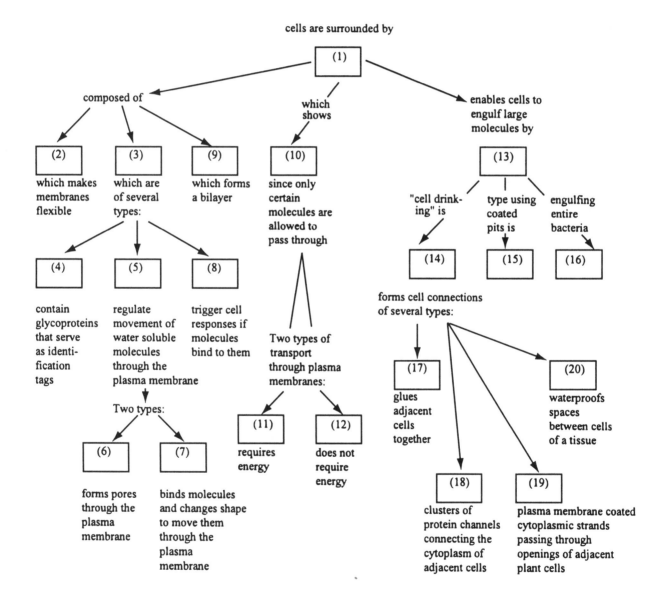

Key Terms and Definitions

active transport: the movement of materials across a membrane through cellular energy, usually against a concentration gradient.

carrier protein: a membrane protein that facilitates diffusion of specific substances across the membrane. The molecule to be transported binds to the outer surface of the carrier protein; the protein then changes shape, allowing the molecule to move across the membrane through the protein.

cell wall: a layer of material, usually made up of cellulose or cellulose-like materials, found outside the plasma membranes of plants, fungi, bacteria, and plantlike protists.

channel protein: a membrane protein that forms a channel or pore completely through the membrane and that is usually permeable to one or a few water-soluble molecules, especially ions.

concentration: the number of particles in a dissolved substance per unit volume of fluid.

concentration gradient: the difference in concentration of a substance between one region and another.

cytoplasm: the material contained within the plasma membrane of a cell, exclusive of the nucleus.

desmosome: a strong cell-to-cell junction that attaches cells to one another.

differential permeability: membrane property by which some substances permeate more readily than other substances.

diffusion: net movement of particles from a region of high concentration of that particle to a region of low concentration, driven by the concentration gradient. May occur entirely within a fluid or across a barrier such as a membrane.

endocytosis: the movement of material into a cell by a process in which the plasma membrane engulfs extracellular material, forming membrane-bound sacs that enter the cytoplasm.

exocytosis: the movement of material out of a cell by a process in which intracellular material is enclosed within a membrane-bound sac that moves to the plasma membrane and fuses with it, releasing the material outside the cell.

facilitated diffusion: diffusion of molecules across a membrane, assisted by protein pores or carriers embedded in the membrane.

fluid: a liquid or gas.

fluid mosaic model: a model of membrane structure. According to this model, membranes are composed of a double layer of phospholipids inside which is embedded a variety of proteins. The phospholipid bilayer is a somewhat fluid matrix that allows movement of proteins within it.

gap junction: a type of cell-to-cell junction in animals in which channels connect the cytoplasm of adjacent cells.

glycoprotein: a protein to which a carbohydrate is attached.

gradient: a difference in concentration, pressure, or electrical charge between two regions of space.

hypertonic: referring to a solution that has a higher concentration of dissolved particles (and therefore a lower free water concentration) than the cytoplasm of a cell.

hypotonic: referring to a solution that has a lower concentration of dissolved particles (and therefore a higher free water concentration) than the cytoplasm of a cell.

isotonic: referring to a solution that has the same concentration of dissolved particles (and therefore the same free water concentration) as the cytoplasm of a cell.

middle lamella: a thin layer of pectin and other carbohydrates that separates and sticks together the primary cell walls of adjacent plant cells.

osmosis: the diffusion of water across a differentially permeable membrane, usually down a concentration gradient of free water molecules. Water moves to the solution that has a lower free water concentration.

passive transport: movement of materials across a membrane down a gradient of concentration, pressure, or electrical charge without using cellular energy.

phagocytosis: a type of endocytosis in which extensions of a plasma membrane engulf extracellular particles and transport them into the interior of the cell.

phospholipid bilayer: a double layer of phospholipids that forms the basis of all cellular membranes; the phospholipid heads face the water of extracellular fluid or the cytoplasm, and the tails are buried in the middle of the bilayer.

pinocytosis: nonselective movement of extracellular fluid into a cell, enclosed within a vesicle formed from the plasma membrane.

plasma membrane: the outer membrane of a cell, composed of a bilayer of phospholipids in which proteins are embedded.

plasmodesma (pl. plasmodesmata): a cell-to-cell junction in plants that connects the cytoplasm of adjacent cells.

primary cell wall: cellulose and other carbohydrates secreted by a young plant cell between the middle lamella and the plasma membrane.

pseudopod: a temporary extension of the plasma membrane used for locomotion or phagocytosis in certain cells such as the protist *Amoeba* or white blood cells of vertebrates.

receptor-mediated endocytosis: selective uptake of molecules from the extracellular fluid by binding to a receptor located at a coated pit on the plasma membrane and pinching off the coated pit into a vesicle that moves into the cytoplasm.

receptor protein: a protein or glycoprotein, located on a membrane (or in the cytoplasm), that recognizes and binds to specific molecules. Binding by receptors often triggers a response by a cell, such as endocytosis, protein synthesis, or exocytosis.

recognition protein: a protein or glycoprotein protruding from the outside surface of a plasma membrane that identifies a cell as belonging to a particular species, to a specific individual of that species, and often to a specific organ within the individual.

secondary cell wall: a thick layer of cellulose and other polysaccharides secreted by certain plant cells between the primary cell wall and the plasma membrane.

simple diffusion: diffusion of water, dissolved gases, or lipid-soluble molecules through the phospholipid bilayer of a membrane.

tight junction: a type of cell-to-cell junction in animals that prevents the movement of materials through the spaces between cells.

transport protein: a protein in a plasma membrane that binds specific molecules and facilitates their transport across the membrane.

vesicle: a small membrane-bound sac within the cytoplasm.

THINKING THROUGH THE CONCEPTS

True or False: Determine if the statement given is true or false. If it is false, change the underlined word so that the statement reads true.

21. _____ As a cell increases in size, its surface area increases more rapidly than its internal volume.
22. _____ Red blood cells will burst when placed in fresh water.
23. _____ The water-loving portion of a compound is hydrophobic.
24. _____ The rate of diffusion is increased by decreasing the temperature.
25. _____ In diffusion, molecules move toward regions of higher concentration.
26. _____ More water will enter a cell if it is placed in a hypotonic solution.
27. _____ Solutions with higher salt concentrations than a cell are hypertonic when compared to the cell.
28. _____ Freshwater organisms deal with the tendency of their cells to gain water.
29. _____ Endocytosis is the movement of substances into cells.
30. _____ The movement of a solid substance into a cell is pinocytosis.

Matching: Cell walls and cell membranes

31.____ contains cellulose in plants
32.____ isolates the cytoplasm from the external environment
33.____ regulates flow of materials into and out of cells
34.____ contains chitin in fungi
35.____ communicates with other cells
36.____ stiff, porous, and non-living
37.____ "fluid-mosaic model"
38.____ has a lipid bilayer

Choices:

a.   cell wall
b.   cell membrane

Matching: Diffusion and osmosis

39.____ effect of movement of all molecules down the concentration gradient
40.____ effect of water moving down its concentration gradient across a differentially permeable membrane
41.____ movement of O_2 into a cell and CO_2 out of a cell
42.____ a cell expands when placed in pure water
43.____ can cause cells to shrink

Choices:

a.   osmosis
b.   diffusion

Matching: Effect of osmosis on cells

44.____ Animal cells will expand.
45.____ Animal cells will shrivel up.
46.____ Red blood cells will burst.
47.____ Celery will wilt.
48.____ Lettuce leaves will become turgid (rigid, crisp) in fluid.
49.____ Red blood cells will neither shrivel up nor swell up.

Choices:

a.   cells placed in hypotonic solution
b.   cells placed in hypertonic solution
c.   cells placed in isotonic solution

Matching: Cell connections and communications

50.____ clusters of protein channels for communication between cells
51.____ waterproof gaskets between cells
52.____ attachments between cells that are stretched, compressed, or bent as organisms move
53.____ large, membrane-bound tubes for water passage between cells

Choices:

a.   tight junctions
b.   plasmodesmata
c.   gap junctions
d.   desmosomes

Multiple Choice: Pick the most correct choice for each question.

54. The hydrophobic tails of a phospholipid bilayer are oriented toward the
    a. interior of the plasma membrane
    b. extracellular fluid surrounding the cell
    c. cytoplasm of the cell
    d. nucleus of the cell

55. Molecules that permeate a plasma membrane by facilitated diffusion
    a. require the use of energy
    b. require the aid of transport proteins
    c. move from areas of low concentration to areas of high concentration
    d. do so much more quickly than those crossing by simple diffusion
    e. are water molecules

56. A molecule that can diffuse freely through a phospholipid bilayer is probably
    a. water-soluble
    c. positively charged
    c. nonpolar
    d. negatively charged
    e. a membrane-spanning protein

57. The preferential movement of water molecules across a differentially permeable membrane is termed
    a. facilitated diffusion
    b. osmosis
    c. active transport
    d. exocytosis
    e. a concentration gradient

58. If red blood cells are placed in a hypotonic solution, what happens?
    A. The cells swell and burst.
    b. The cells shrivel up and shrink.
    c. The cells remain unchanged in volume.
    d. The cells take up salt molecules from the hypotonic solution.
    e. The cells release salt molecules into the hypotonic solution.

59. Solutions that cause water to preferentially enter cells by osmosis are called
    a. hypertonic
    b. isotonic
    c. hypotonic
    d. endosmotic
    e. exosmotic

CLUES TO APPLYING THE CONCEPTS

These practice questions are intended to sharpen your ability to apply critical thinking and analysis to biological concepts covered in this chapter.

60. Suppose you are taking a cruise from San Francisco to Hawaii. About halfway there, the ship begins to sink and all passengers and crew board lifeboats and are floating around in the ocean waiting to be rescued. After several days, you are so thirsty that you bend over the side of your life boat and drink some of the seawater. Did you do a wise thing? Explain what you think will happen to your body within a few hours of drinking the ocean water, and explain the biological basis for your reactions.

ANSWERS TO EXERCISES

1. plasma membrane
2. cholesterol
3. proteins
4. recognition proteins
5. transport proteins
6. channel proteins
7. carrier proteins
8. receptor proteins
9. phospholipid
10. differential permeability
11. active transport
12. passive transport
13. endocytosis
14. pinocytosis
15. receptor-mediated endocytosis
16. phagocytosis
17. desmosomes
18. gap junctions
19. plasmodesmata

20. tight junctions
21. false, less rapidly
22. true
23. false, hydrophilic
24. false, increasing
25. false, lower
26. true
27. true
28. true
29. true
30. false, phagocytosis
31. a
32. b
33. b
34. a
35. b
36. a
37. b
38. b
39. b

40. a
41. b
42. a
43. a
44. a
45. b
46. a
47. b
48. a
49. c
50. c
51. a
52. d
53. b
54. a
55. b
56. c
57. b
58. a
59. c

60. Although you were thirsty and your cells craved water, drinking the salty seawater was unwise because seawater is hypertonic (has a higher concentration of salts) to hypotonic cellular cytoplasm. So, in your stomach, cells will begin to lose water due to osmosis, since water flows through a selectively permeable membrane from regions of greater water concentration (the hypotonic cytoplasm) toward regions of lesser water concentration (the hypertonic seawater). Soon, your stomach cells will have lost so much water that they will begin to die, causing you to go into convulsions and, perhaps, die as well.

Chapter 6: Cell Structure

OVERVIEW

This chapter describes and compares prokaryotic and eukaryotic cells. The authors cover the organization of eukaryotic cells and emphasize the various organelles found in these cells.

1) What Are the Basic Features of Cells?

In the 1850s, Virchow proclaimed "All cells come from cells." Modern cell theory principles state: (1) every organism is made of at least one cell; (2) cells are the functional units of life; and (3) all cells arise from preexisting cells. All cells obtain energy and nutrients from their environment, make molecules necessary for growth and repair, get rid of wastes, interact with other cells, and reproduce.

The **plasma membrane** (phospholipid bilayer with embedded proteins) encloses the cell and mediates interactions between the cell and its environment. It (1) isolates cytoplasm from external environment, (2) regulates flow of materials between cytoplasm and its environment, and (3) allows interactions with other cells. **DNA** determines cell structure and function and allows the cell to reproduce. In **eukaryotic cells** (plants, animals, fungi, and protists), DNA is found within the **nucleus** (a membrane-bound structure). In **prokaryotic cells** (bacteria), DNA is in a non-membrane enclosed space, the **nucleoid**. All cells contain **cytoplasm** (all material inside plasma membrane and outside the nucleus/nucleoid). Cytoplasm includes water, salts, and organic molecules, and contains, in eukaryotic cells, a variety of **organelles** (membrane-bound structures performing distinct cell functions).

Cell function limits cell size. Most cells are small (1 to 100 micrometers in diameter) because they need to exchange nutrients and wastes through their plasma membranes mainly by diffusion, a slow process that takes place over long distances. Larger cells have greater needs for exchange of molecules with the environment, but they have smaller surface area/internal volume ratio than do smaller cells. Thus, cells tend to remain small.

2) What Are the Features of Prokaryotic Cells?

Prokaryotic cells are small (less than 5 micrometers long) with simple internal features (no nucleus or membrane-bound organelles). They contain **ribosomes** (made of RNA and proteins and on which protein synthesis occurs) and usually a stiff cell wall. Photosynthetic bacteria have inner membranes containing light-capturing proteins and enzymes.

3) What Are the Features of Eukaryotic Cells?

Eukaryotic cells are larger (greater than 10 micrometers in diameter) and contain organelles and a cytoskeleton (network of protein fibers for cellular shape and organization). The **nucleus** is the cellular control center, containing genetic material (DNA). Nuclear components are: (1) the **nuclear envelope** (two membranes riddled with pores to control flow of informational molecules), which separates nuclear material from the cytoplasm; (2) **chromatin** (DNA and associated proteins organized into **chromosomes**); and (3) one or more **nucleoli**, the sites of ribosome assembly.

Eukaryotic cells contain a complex system of internal membranes. The **plasma membrane** isolates a cell and allows selective interactions between a cell and its environment. The **endoplasmic reticulum** (ER) forms interconnected membrane-bound tubes and channels within the cytoplasm and is continuous with the nuclear membrane. Numerous ribosomes are embedded in **rough ER** (site of protein synthesis), while **smooth ER** (major site of lipid synthesis) lacks ribosomes. Proteins made by ribosomes in rough ER move through ER channels and accumulate in regions that bud off to form **vesicles** (membrane-bound cytoplasmic sacs).

The **Golgi complex** (membranous sacs derived from smooth ER) has three functions: (1) it separates out lipids and proteins obtained from ER according to their destinations; (2) it chemically alters some molecules; and (3) it packages molecules into vesicles for transport. **Lysosomes** are cellular digestive centers containing digestive enzymes to break down proteins, fats, and carbohydrates taken into cells as food. Lysosomes fuse with food vacuoles and lysosomal enzymes then digest the food into small molecules. Lysosomes also digest defective organelles. Membranes flow through a cell in an orderly way, for example from ER to the Golgi complex to a vesicle which fuses with plasma membrane.

Vacuoles (fluid filled membrane-bound sacs) serve many functions, including water regulation, support, and storage. **Contractile vacuoles**, found in freshwater microorganisms, use energy to pump out water constantly entering due to osmosis. **Central vacuoles**, found in plant cells, may: (1) collect cellular wastes; (2) store poisons that deter feeding animals; (3) store sugars and amino acids for cellular use; and (4) collect pigments that give flowers their colors. Central vacuoles contents become hypertonic to cytoplasm and take in water through osmosis. The pressure of the expanding central vacuole, called **turgor pressure**, stiffens the cell, providing support for nonwoody plant parts.. Houseplants stiffen when watered and wilt when sufficient water is lacking.

Mitochondria extract energy from food molecules and **chloroplasts** capture solar energy. They: (1) are oblong, about 1-5 micrometers long; (2) are surrounded by a double membrane; (3) have DNA; and (4) make ATP. Mitochondria ("powerhouses of the cell" found in all eukaryotic cells) make ATP using energy stored in food molecules. **Anaerobic** (without oxygen) metabolism of sugar in the cytoplasm produces little ATP energy. Mitochondria use **aerobic** (with oxygen) metabolism to generate about 18 or 19 times as much ATP. The inner mitochondrial membrane loops back and forth to form deep folds (**cristae**) so that there are two regions: the **intermembrane compartment** between outer and inner membrane and the inner **matrix** region.

Chloroplasts (specialized **plastids**) are the sites of photosynthesis. Their inner membranes enclose semifluid stroma. Stroma contains **thylakoids** (interconnected stacks of hollow membranous discs containing **chlorophyll** and other pigments; a stack of thylakoids is a **granum**. Chlorophyll captures sunlight energy and transfers it to other molecules that make ATP and other energy-carrier molecules. In the stroma, these energy molecules are used to combine carbon dioxide and water into sugars. Plastids store various types of molecules, including pigments and starch.

The **cytoskeleton** provides shape, support, and movement. It includes several types of protein fibers: thin **microfilaments**, medium-sized **intermediate filaments**, and thick **microtubules**. Cytoskeleton functions: (1) cell shape (especially in animal cells); (2) cell movement (through assembly, disassembly, and sliding of microfilaments); (3) organelle movement (especially vesicles, by microfilaments and microtubules); and (4) cell division (microtubules move chromosomes into daughter nuclei, and division of the cytoplasm in animal cells results from contraction of a ring of microfilaments).

Cilia (short, with many per cell) and **flagella** (long, with few per cell) move cells or move fluid past cells. These are slender extensions of plasma membrane containing a ring of nine fused pairs of microtubules with an unfused pair in the center of the ring ("9+2" arrangement). They are powered by ATP made by many mitochondria at their bases. Some prokaryotic cells have "flagella" but these do not contain microtubules.

KEY TERMS AND CONCEPTS

Fill-In: Fill in the names of the structures indicated in this diagram of an animal cell, identifying the following:

cytoplasm

chromatin

Golgi complex

lysosome

mitochondrion

nucleolus

nucleus

plasma membrane

rough endoplasmic reticulum

smooth endoplasmic reticulum

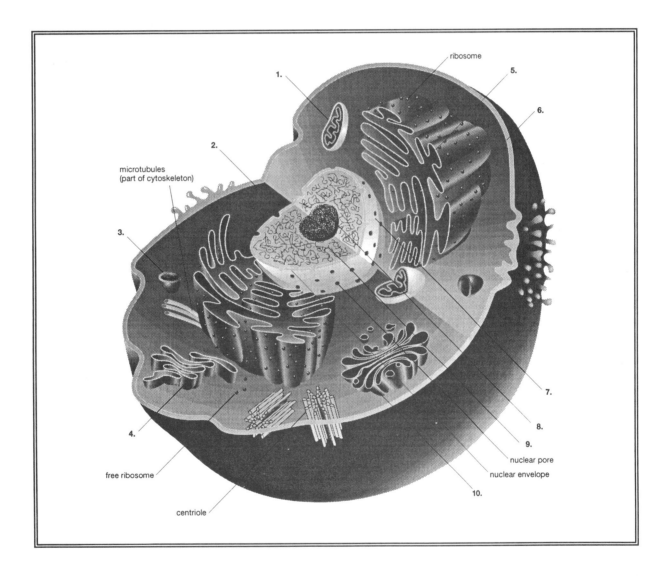

Key Terms and Definitions

aerobic: using oxygen.

anaerobic: not using oxygen.

capsule: a polysaccharide or protein coating that surrounds the cell walls of some bacteria.

central vacuole: fluid filled membrane-bound sacs found in plant cells with various functions: (1) collect cellular wastes; (2) store poisons that deter feeding animal; (3) store sugars and amino acids for cellular use; and (4) collect pigments that give flowers their colors.

centriole: in animal cells, a microtubule-containing structure found at the microtubule organizing center and the base of each cilium and flagellum. Gives rise to the microtubules of cilia and flagella, and may be involved in spindle formation during cell division.

chlorophyll: a pigment found in chloroplasts that captures light energy during photosynthesis.

chloroplast: the organelle of plants and plantlike protists that is the site of photosynthesis; surrounded by a double membrane and containing an extensive internal membrane system bearing chlorophyll.

chromatin: the complex of DNA and proteins that makes up the chromosomes of eukaryotic cells.

chromosome: threadlike bodies that contain the genetic material, DNA; eukaryotic chromosomes also contain proteins bound to the DNA.

cilium (pl. cilia): short, hairlike projection from the surface of some eukaryotic cells, containing microtubules in a 9 + 2 arrangement. Movement of cilia may propel cells through fluid medium or move fluids over a stationary surface layer of cells.

contractile vacuole: fluid filled membrane-bound sacs found in cytoplasm of freshwater microorganisms; use energy to pump out water constantly entering due to osmosis.

crista (pl. cristae): a fold in the inner membrane of a mitochondrion.

cytoplasm: the material contained within the plasma membrane of a cell, exclusive of the nucleus.

cytoskeleton: a network of protein fibers in the cytoplasm that gives shape to a cell, holds and moves organelles, and is often involved in cell movement.

deoxyribonucleic acid (DNA): the molecules that are the genetic material of all living cells.

endoplasmic reticulum (ER): a system of membranous channels within eukaryotic cells; the site of most protein and lipid biosynthesis.

eukaryotic: referring to cells of organisms of the kingdoms Protista, Fungi, Plantae, and Animalia. Eukaryotic cells have their genetic material enclosed within a membrane-bound nucleus and contain other membrane-bound organelles.

flagellum (pl. flagella): a long, hairlike extension of the cell membrane. In eukaryotic cells, contains microtubules arranged in a 9 + 2 pattern. Movement of flagella propel cells through fluid media.

food vacuole: a membranous sac containing food obtained by phagocytosis.

Golgi complex: a stack of membranous sacs found in most eukaryotic cells, which is the site of processing and separation of membrane components and secretory materials.

granum (pl. grana): in chloroplasts, a stack of thylakoids.

intermediate filament: part of the cytoskeleton of eukaryotic cells, probably functioning mainly for support.

intermembrane compartment: the fluid-filled space between the inner and outer membranes of a mitochondrion.

lysosome: a membrane-bound organelle containing intracellular digestive enzymes.

matrix: fluid contained within the inner membrane of a mitochondrion.

microfilament: part of the cytoskeleton of eukaryotic cells, composed of the proteins actin and (sometimes) myosin; functions in the movement of cell organelles and in locomotion by pseudopods.

microtubule: a hollow, cylindrical strand found in eukaryotic cells, composed of the protein tubulin; part of the cytoskeleton used in movement of cell organelles, cell growth, and construction of cilia and flagella.

mitochondrion: an organelle, bounded by two membranes, that is the site of the reactions of aerobic metabolism.

nuclear envelope: the double membrane system surrounding the nucleus of eukaryotic cells. The outer membrane is often continuous with the endoplasmic reticulum.

nucleoid: the location of the genetic material in prokaryotic cells; not membrane-enclosed.

nucleolus: the region of the eukaryotic nucleus engaged in ribosome synthesis, consisting of the genes encoding ribosomal RNA, newly synthesized ribosomal RNA, and ribosomal proteins.

nucleus: the membrane-bound organelle of eukaryotic cells that contains the cell's genetic material.

organelle: a structure found in the cytoplasm of eukaryotic cells that performs a specific function; sometimes used to refer specifically to membrane-bound structures such as the nucleus or endoplasmic reticulum.

pilus: a hairlike projection made of protein and found on the surface of certain bacteria; often used to attach the bacterium to another cell.

plasma membrane: the outer membrane of a cell, composed of a phospholipid bilayer in which proteins are embedded.

plastid: in plant cells, an organelle bounded by two membranes that may be involved in photosynthesis (chloroplasts), pigment storage, or food storage.

prokaryotic: referring to cells of the kingdom Monera. Prokaryotic cells do not have their genetic material enclosed within a membrane-bound nucleus and also lack other membrane-bound organelles.

ribonucleic acid (RNA): a molecule composed of ribose nucleotides, each of which consists of a phosphate group, the sugar ribose, and one of the bases adenine, cytosine, guanine, or uracil; transfers heritary instructions from the nucleus to the cytoplasm; also the genetic material of some viruses.

ribosome: a particle composed of RNA and protein that is the site of protein synthesis in both eukaryotic and prokaryotic cells.

rough endoplasmic reticulum: endoplasmic reticulum lined on the outside with ribosomes.

slime layer: polysaccharide or protein coatings that some disease-causing bacteria secrete outside their cell wall.

smooth endoplasmic reticulum: endoplasmic reticulum without ribosomes.

stroma: the semi-fluid material of chloroplasts in which the grana are embedded.

thylakoid: a membranous sac within a chloroplast; the thylakoid membranes contain chlorophyll.

turgor pressure: the pressure of the central vacuole, which expands by osmosis, to stiffen plant cells, providing support for nonwoody plant parts.

vacuole: a vesicle, often large, consisting of a single membrane enclosing a fluid-filled space.

vesicle: a small membrane-bound sac within the cytoplasm.

THINKING THROUGH THE CONCEPTS

True or False: Determine if the statement given is true or false. If it is false, change the underlined word so that the statement reads true.

11._____ More primitive types of cells are called <u>eukaryotic</u> cells.

12._____ The presence of large numbers of ribosomes is characteristic of <u>rough</u> endoplasmic reticulum.

13._____ The majority of the hereditary material is found in the <u>cytoplasm</u>.

14._____ <u>Mitochondria</u> are associated with release of energy from sugar.

15._____ <u>Mitochondria</u> are associated with the storage of energy in sugar.

16._____ Lipid-producing enzymes are more common in <u>rough</u> endoplasmic reticulum.

17._____ <u>Animals</u> store food in the form of starch.

18._____ <u>Animal</u> cells are more likely to have vacuoles.

19._____ Higher <u>plants</u> lack ciliated or flagellated cells.

20._____ Cilia are <u>longer</u> than flagella.

Matching: Cell types.

21.____	lack a membrane-bound nucleus		Choices:	
22.____	have many chromosomes with DNA and protein		a.	eukaryotic cells
23.____	lack most cytoplasmic organelles		b.	prokaryotic cells
24.____	larger and more complex		c.	both
25.____	have nucleoid regions in the cytoplasm			
26.____	have DNA			
27.____	have flagella with 9+2 structure			

Matching: Mitochondria and chloroplasts.

28.____	make ATP using energy		Choices:	
29.____	capture sunlight energy to make sugar		a.	mitochondria
30.____	convert sugar energy into ATP energy		b.	chloroplasts
31.____	have DNA		c.	both
32.____	have thylakoid membranes and semi-fluid stroma			
33.____	extract energy from food molecules.			
34.____	found in plants			
35.____	have cristae membranes and semi-fluid matrix			
36.____	have chlorophyll			
37.____	function in photosynthesis			
38.____	function in cellular respiration			

Matching: Cilia and flagella.

39.____	microtubular extensions through the cell membrane		Choices:	
40.____	shorter and more numerous per cell		a.	cilia
41.____	have a "9 + 2" arrangement of microtubular pairs		b.	flagella
42.____	bend perpendicular to the cell membrane		c.	both
43.____	bend parallel to the cell membrane			
44.____	used for food gathering and for movement			

Fill-In: Fill in the names of the structures indicated in the following figure.

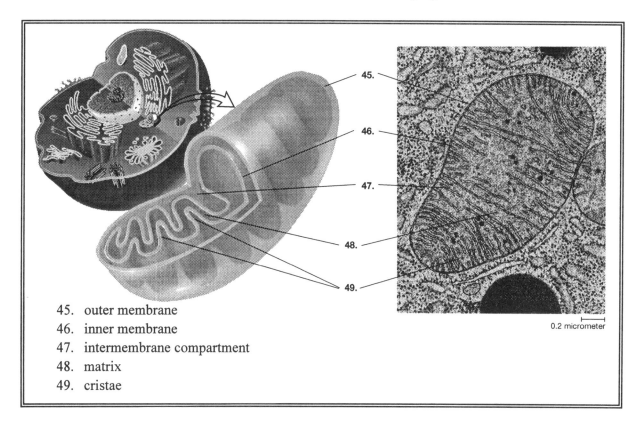

45. outer membrane
46. inner membrane
47. intermembrane compartment
48. matrix
49. cristae

Matching: Organelles that manufacture or digest proteins and lipids.

50.____ digest food particles.
51.____ interconnected membrane tubes and channels in the cytoplasm.
52.____ stacks of membranes in the cytoplasm.
53.____ made of RNA and proteins.
54.____ function in the nucleus.
55.____ membrane-bound vesicles.
56.____ "workbenches" for protein synthesis.
57.____ sort out various lipids and proteins.
58.____ may be rough or smooth in appearance.
59.____ sites of lipid synthesis.
60.____ large and small subunits are assembled in the nucleolus.
61.____ packages proteins and lipids into vesicles for transport out of the cell.
62.____ digests defective organelles.

Choices:

a.   ribosomes
b.   endoplasmic reticulum
c.   Golgi complexes
d.   lysosomes
e.   none of these

Multiple Choice: Pick the most correct choice for each question.

63. All cells possess all of the following <u>except</u>:
    a.   cytoplasm
    b.   genetic material
    c.   nuclear membrane
    d.   chromosome
    e.   plasma (cell) membrane

64. Which of the following is <u>not</u> a similarity of mitochondria and chloroplasts?
    a.   both make ATP
    b.   both capture solar energy and convert it into chemical energy
    c.   both possess their own DNA
    d.   both have a double membrane
    e.   both probably evolved from bacteria long ago

65. Both prokaryotic and eukaryotic cells possess:
    a.   mitochondria
    b.   chloroplasts
    c.   cytoskeleton
    d.   ribosomes
    e.   lysosomes

66. Which organelle extracts energy from food molecules and uses it to make ATP?
    a.   mitochondrion
    b.   chloroplast
    c.   ribosome
    d.   centriole
    e.   nucleus

CLUES TO APPLYING THE CONCEPTS

These practice questions are intended to sharpen your ability to apply critical thinking and analysis to biological concepts covered in this chapter.

67. Why do cells seldom grow large enough to be seen without the aid of a microscope?

68. Suppose you discovered a chemical that had all of the following effects on a cell: cell growth slowed down, the movement of cilia slowed down, cell divisions occurred less frequently, proteins were made less often, and endocytosis occurred less often. Which of these organelles do you think the chemical affected most severely: lysosomes, Golgi complex, or mitochondria? Briefly explain your answer.

ANSWERS TO EXERCISES

1. mitochondrion
2. cytoplasm
3. lysosome
4. smooth endoplasmic reticulum
5. rough endoplasmic reticulum
6. plasma membrane
7. nucleus
8. nucleolus
9. chromatin
10. Golgi complex
11. false, prokaryotic
12. true
13. false, nucleus
14. true
15. false, chloroplasts
16. false, smooth

17. false, plants
18. false, plant
19. true
20. false, shorter
21. b
22. a
23. b
24. a
25. b
26. c
27. a
28. c
29. b
30. a
31. c
32. b
33. a

34. c
35. a
36. b
37. b
38. a
39. c
40. a
41. c
42. b
43. a
44. c
45. outer membrane
46. inner membrane
47. intermembrane compartment
48. matrix
49. cristae

50. d
51. b
52. c
53. a
54. e
55. d
56. a
57. c
58. b
59. b
60. a
61. c
62. d
63. c
64. b
65. d
66. a

67. Cells seldom grow this large because of two factors. First, a cell this size would have too much volume for the amount of surface area present, so that not enough plasma membrane area would be present for the movement of molecules into and out of such a giant cell. Second, the time it would take for molecules deep inside of such a cell to move to the surface or vice versa would be extremely long because diffusion over relatively long distances is too slow to support life processes.

68. Mitochondria, because all the processes affected require energy and mitochondria are the chief organelles where the energy in sugar is converted into cellular energy in the form of ATP molecules.

Chapter 7: Capturing Solar Energy: Photosynthesis

OVERVIEW

This chapter details the process of photosynthesis. The authors outline the light-dependent and light-independent steps of photosynthesis, and present the two major ways that plants trap carbon dioxide.

1) What Is Photosynthesis?

Starting with carbon dioxide and water, **photosynthesis** converts sunlight energy into chemical energy stored in the bonds of glucose and oxygen ($6\ CO_2 + 6\ H_2O$ + light energy $\rightarrow C_6H_{12}O_6 + 6\ O_2$.) It occurs in plants, algae, and some bacteria. In plants, leaves and chloroplasts are adaptations for photosynthesis. Leaves obtain CO_2 from the air through adjustable pores in the epidermis called stomata. **Mesophyll** cells within leaves contain most of the chloroplasts. Vascular bundles (veins) carry water and minerals to the mesophyll cells and carry sugars to other plant parts.

2) Light-Dependent Reactions: How Is Light Energy Converted to Chemical Energy?

Photosynthesis consists of **light-dependent reactions** (chlorophyll in thylakoid membranes capture sunlight energy to split water and make O_2 and some ATP and NADPH energy-carriers) and **light-independent reactions** (stroma enzymes use chemical energy in ATP and NADPH to make glucose from CO_2).

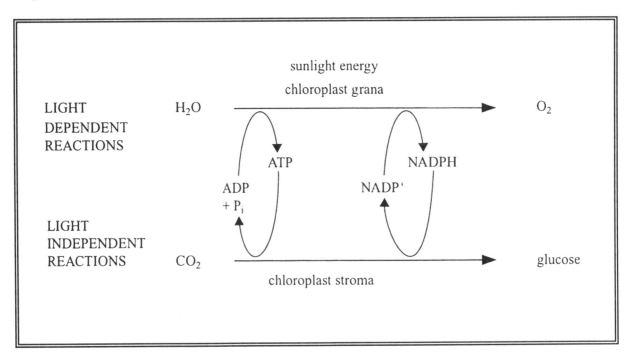

3) Light-Dependent Reactions: How is Light Energy Converted to Chemical Energy?

Light is first captured by pigments in chloroplasts. Wavelengths of light are composed of **photons** (individual packets of energy). Chloroplasts contain several pigments that absorb different wavelengths of light: **chlorophyll** absorbs violet, blue, and red light and reflects green (this is why leaves are green) and **accessory pigments** (**carotenoids** absorb blue and green and reflect yellow, orange, and red, and **phycocyanins** absorb green and yellow and reflect blue and purple) absorb light energy and transfer it to chlorophyll.

Light-dependent reactions occur in clusters of molecules called **photosystems** (proteins including chlorophyll, accessory pigments, and electron-carrying molecules in thylakoids). Each photosystem has two major parts: (1) **light-harvesting complex** (300 pigment molecules that absorb light and pass the energy to a specific chlorophyll called the **reaction center**); and (2) **electron transport system** (**ETS**, series of electron carrier molecules embedded in the thylakoid membrane). When reaction center chlorophyll receives energy, one of its electrons enters the ETS and moves from one carrier to the next, releasing energy that allows ADP to form ATP (through a hydrogen ion gradient called **chemiosmosis** in **photosystem II**) and $NADP^+$ to form NADPH (in **photosystem I**). Electrons from the ETS of system II replenish those lost by the reaction center chlorophyll of system I.

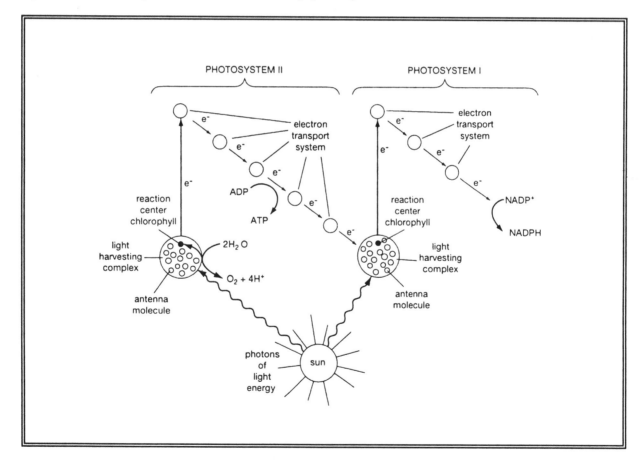

Splitting water maintains the flow of electrons through the photosystems. Electrons flow one way: from splitting water through reaction center of system II through ETS of system II to reaction center of system I through ETS of system I to NADPH.

4) Light Independent Reactions: How is Chemical Energy Stored in Glucose Molecules?

As long as sufficient ATP and NADPH are available, the light-independent reactions do not require sunlight. The C_3 (three-carbon or Calvin-Benson) cycle captures carbon dioxide. It requires (1) CO_2 from the air; (2) CO_2-capturing sugar, ribulose bisphosphate (RuBP); (3) stroma enzymes ; and (4) energy from ATP and NADPH (from light-dependent reactions). The three steps of the C_3 cycle are carbon fixation (CO_2 attaches to RuBP), making intermediate molecules using energy from ATP and NADPH, and regeneration RuBP Carbon fixed during the C_3 cycle is used to make glucose.

5) Water, CO_2, and the C_4 Pathway.

For land plants, having leaves porous to CO_2 also allows water to evaporate. Large waterproof leaves with adjustable pores (stomata) help compensate for this. When stomata close to reduce water loss, however, less CO_2 enters and less O_2 leaves. At these times, oxygen combines with RuBP (**photorespiration**) but no useful cellular energy results and no glucose is made. During hot dry weather, plants may die due to lack of sufficient glucose. C_4 plants (those with chloroplasts in mesophyll and **bundle-sheath cells** around the leaf veins) reduce photorespiration, using a two-stage carbon fixation process called the **C_4 pathway**. In C_4 plants, CO_2 reacts with PEP (phosphoenolpyruvate) instead of RuBP in the mesophyll, making oxaloacetic acid (4-carbon molecules) that travels to bundle sheath cells where it breaks down to release CO_2 there, where the regular C_3 cycle occurs. The remnant molecule returns to the mesophyll where ATP energy is used to regenerate PEP. C_4 uses more energy than C_3. Thus, C_4 plants (crabgrass, for instance) thrive in deserts and in midsummer (much light energy, scarce water) but C_3 plants (Kentucky bluegrass, for instance) thrive in cool, wet, cloudy climates.

KEY TERMS AND CONCEPTS

Fill-In: From the following list of key terms, fill in the blanks in the sentences below.

ATP	electron transport system	phenylenolpyruvate (PEP)	stroma
bundle-sheath cells	glucose	photorespiration	sunlight
chemical	light-dependent	photosynthesis	thylakoids
chlorophyll	light-harvesting complex	photosystems	
chemiosmosis	light-independent	photosystem I	
C_3	oxygen	photosystem II	
C_4	oxaloacetic acid	reaction center	

Starting with carbon dioxide and water, (1)_____ converts (2)_____ energy into (3)_____ energy stored in the bonds of glucose and oxygen.

Photosynthesis consists of: (a) (4)_____ reactions (5)_____ in thylakoid membranes capture sunlight energy to split water and make (6)_____ gas and some (7)_____ and NADPH energy-carriers); and (b) (8)_____ reactions ((9)_____ enzymes use chemical energy in ATP and NADPH to make (10)_____ from CO_2).

Light-dependent reactions occur in clusters of molecules called (11)_____ consisting of proteins including chlorophyll, accessory pigments, and electron-carrying molecules in the (12)_____.

Each photosystem has two major parts: (a) (13)_____ (300 pigment molecules that absorb light and pass the energy to a specific chlorophyll called the (14)_____; and (b) (15)_____ (series of electron carrier molecules embedded in the thylakoid membrane).

In photosystem (16)_____ ADP becomes ATP through a hydrogen ion gradient called (17)_____. In photosystem (18)_____ NADP$^+$becomes NADPH. Electrons from the electron transport center photosystem (19)_____ and replenish those lost by the reaction center chlorophyll of photosystem (20)_____.

Plants with chloroplasts in both mesophyll and (21)_____ around the leaf veins reduce - (22)_____ using a two-stage carbon fixation process called the (23)_____ pathway. In these plants, CO_2 reacts with (24)_____ instead of RuBP in the mesophyll, making (25)_____ (4-carbon molecules) that travels to the (26)_____ cells where it breaks down to release CO_2 there, where the regular (27)_____ cycle occurs. The remnant molecule returns to the mesophyll where (28)_____ energy is used to regenerate (29)_____.

Key Terms and Definitions

accessory pigments: found in chloroplasts; carotenoids (absorb blue and green and reflect yellow, orange, and red light) and phycocyanins (absorb green and yellow and reflect blue and purple light) that absorb light energy and transfer it to chlorophyll.

bundle-sheath cells: cells that surround the veins of plants. In C_4 plants, but not C_3 plants, these cells contain chloroplasts.

C_3 cycle: the cyclic series of reactions whereby carbon dioxide is fixed into carbohydrates during the light-independent reactions of photosynthesis. Also called Calvin-Benson cycle.

C_4 pathway: the series of reactions in certain plants that fixes carbon dioxide into organic acids for later use in the C_3 cycle of photosynthesis.

Calvin-Benson cycle: see C_3 cycle.

carbon fixation: the initial steps in the C_3 cycle, in which carbon dioxide reacts with ribulose bisphosphate to form a stable organic molecule.

carotenoid: red, orange, or yellow pigments found in chloroplasts that serve as accessory light-gathering molecules in thylakoid photosystems.

chemiosmosis: a process of ATP generation in chloroplasts and mitochondria. The movement of electrons down an

electron transport system is used to pump hydrogen ions across a membrane, thereby building up a concentration gradient of hydrogen ions across the membrane. The hydrogen ions diffuse back across the membrane through the pores of ATP-synthesizing enzymes. The energy of their movement down their concentration gradient drives ATP synthesis.

chlorophyll: the primary light-trapping molecule in photosynthesis.

electron transport system: molecules found in the thylakoid membranes of chloroplasts and the inner membrane of mitochondria, which extract energy from electrons and generate ATP.

granum (pl. grana): in chloroplasts, a stack of thylakoids.

light-dependent reactions: the first stage of photosynthesis, in which the energy of light is captured as ATP and NADPH; occurs in thylakoids of chloroplasts.

light-harvesting complex: in photosystems, the assembly of pigment molecules (chlorophyll and often carotenoids or phycocyanins) that absorb light energy and transfer the energy to electrons.

light-independent reactions: the second stage of photosynthesis, in which the energy obtained by the light-dependent reactions is used to fix carbon dioxide into carbohydrates; occurs in the stroma of chloroplasts.

mesophyll: cells within leaves that contain most of the chloroplasts.

photon: the smallest unit of light.

photorespiration: a series of reactions in plants in which O_2 replaces CO_2 during the C_3 cycle, preventing carbon fixation. This wasteful process dominates when C_3 plants are forced to close their stomata to prevent water loss.

photosynthesis: the complete series of chemical reactions in which the energy of light is used to synthesize high-energy organic molecules, usually carbohydrates, from low-energy inorganic molecules, usually carbon dioxide and water.

photosystem: in thylakoid membranes, a light-harvesting complex and its associated electron transport system.

phycocyanin: a bluish pigment found in the membranes of chloroplasts and used as an accessory light-gathering

molecule in thylakoid photosystems.

reaction center: in the light-harvesting complex of a photosystem, the chlorophyll molecule to which light energy is transferred by the antenna pigments. The captured energy ejects an electron from the reaction center chlorophyll, and the electron is transferred to the electron transport system.

stoma (pl. stomata): adjustable openings in plant leaves. Most gas exchange between leaves and the air occurs through the stomata.

stroma: the semi-fluid medium of chloroplasts, in which the grana are embedded.

thylakoid: a disk-shaped, membranous sac found in chloroplasts, the membranes of which contain the photosystems and ATP-synthesizing enzymes used in the light-dependent reactions of photosynthesis.

THINKING THROUGH THE CONCEPTS

True or False: Determine if the statement given is true or false. If it is false, change the underlined word so that the statement reads true.

30. _____ Aerobic respiration probably evolved before photosynthesis.
31. _____ Chloroplasts are associated with stroma.
32. _____ Glucose is synthesized in the grana.
33. _____ Blue is a more energetic wavelength of light than red.
34. _____ Ultraviolet light is visible to insects.
35. _____ Light energy is first captured in photosystem I.
36. _____ As electrons are transferred from one carrier to another, the electrons gain energy.
37. _____ Photosynthesis uses O_2 and produces CO_2.
38. _____ In the light dependent reactions, chlorophyll captures sunlight energy and uses it to make glucose.
39. _____ The electron transport system is part of the light-dependent process of photosynthesis.

Matching: Photosynthesis.

40.____ CO$_2$ is captured and converted into sugars.
41.____ light energy is converted into chemical energy of ATP and NADPH.
42.____ occurs in chloroplast grana.
43.____ uses chemical energy to make glucose.
44.____ uses chlorophyll, carotenoids, and phycocyanins to trap light energy.
45.____ Calvin-Benson, or C$_3$ cycle.
46.____ energy obtained from NADPH and ATP.
47.____ produces oxygen gas.
48.____ thylakoid membranes.
49.____ photosystems I and II.
50.____ carbon fixation occurs.
51.____ involves electron transport.
52.____ occurs in chloroplast stroma.
53.____ water is split into oxygen and hydrogen.

Choices:

a. light dependent reactions
b. light independent reactions

Multiple Choice: Pick the most correct choice for each question.

54. Molecules of chlorophyll are located in the membranes of sacs called
   a.   cristae
   b.   thylakoids
   c.   stroma
   d.   grana
   e.   chloroplasts

55. A pigment that absorbs red and blue light and reflects green light is
   a.   phycocyanin
   b.   carotenoid
   c.   chlorophyll
   d.   melanin
   e.   colored orange

56. Light dependent photosynthetic reactions produce
   a.   ATP, NADPH, oxygen gas
   b.   ATP, NADPH, carbon dioxide gas
   c.   glucose, ATP, oxygen gas
   d.   glucose, ATP, carbon dioxide gas
   e.   ADP, NADP, glucose

57. Where does the oxygen gas produced during photosynthesis come from?
   a.   carbon dioxide
   b.   water
   c.   ATP
   d.   glucose
   e.   the atmosphere

58. Carbon fixation requires which of the following?
   a.   sunlight
   b.   products of energy-capturing reactions
   c.   high levels of oxygen gas and low levels of carbon dioxide gas
   d.   water, ADP, and NADP

59. The immediate source of hydrogen atoms for the production of sugar during photosynthesis comes from
   a.   ATP
   b.   water
   c.   NADPH
   d.   glucose
   e.   chlorophyll

CLUES TO APPLYING THE CONCEPTS

These practice questions are intended to sharpen your ability to apply critical thinking and analysis to biological concepts covered in this chapter.

60. Suppose you wanted to devise an experiment to determine whether the oxygen gas generated by photosynthesis came from oxygen molecules released from water or from carbon dioxide, each of which are broken down during photosynthesis. Briefly describe an idea you might have for such a study. Hint: perhaps you might want to use a radioactive isotope of oxygen in your study.

61. Suppose you wanted to determine which color of light had the greatest effect on the amount of photosynthesis occurring in plants. How would you set up such an experiment? What variables would you use and what aspects would serve as controls?

ANSWERS TO EXERCISES

1.	photosynthesis	21.	bundle-sheath cells	41.	a
2.	sunlight	22.	photorespiration	42.	a
3.	chemical	23.	C_4	43.	b
4.	light-dependent	24.	phenylenolpyruvate (PEP)	44.	a
5.	chlorophyll	25.	oxaloacetic acid	45.	b
6.	oxygen	26.	bundle-sheath cells	46.	b
7.	ATP	27.	C_3	47.	a
8.	light-independent	28.	ATP	48.	a
9.	stroma	29.	phenylenolpyruvate (PEP)	49.	a
10.	glucose	30.	false, after	50.	b
11.	photosystems	31.	false, grana	51.	a
12.	thylakoids	32.	false, stroma	52.	b
13.	light-harvesting complex	33.	true	53.	a
14.	reaction center	34.	true	54.	b
15.	electron transport system	35.	false, photosystem II	55.	c
16.	photosystem II	36.	false, lose	56.	a
17.	chemiosmosis	37.	false, produces, uses	57.	b
18.	photosystem I	38.	false, oxygen gas	58.	b
19.	photosystem II	39.	true	59.	c
20.	photosystem I	40.	b		

60. You could set up one study where the plants got normal CO_2 and water containing radioactive oxygen (H_2O^*), and another study where the plants got normal (H_2O) and carbon dioxide containing radioactive oxygen (CO_2^*). In each study, you would collect the oxygen gas and the sugars produced by photosynthesis and determine where the radioactive oxygen atoms show up.

61. After keeping a group of plants in the dark for several days, expose them, in separate experiments, to light from bulbs of different colors but equal intensities for equal amounts of time, keeping the plants equally watered and at constant temperatures, and compare the amounts of oxygen gas each plant produces.

Chapter 8:   Harvesting Energy: Glycolysis and Cellular Respiration

OVERVIEW

This chapter covers the processes of glycolysis and cellular respiration. The authors explain glycolysis and fermentation and discuss the role of mitochondria in converting the chemical energy of organic molecules, especially glucose, into the usable energy of ATP during aerobic respiration.

1) How Is Glucose Metabolized?

The chemical equations for glucose formation by photosynthesis and the complete metabolism of glucose are nearly symmetrical:

Photosynthesis:
$$6\ CO_2 + 6\ H_2O + \text{solar energy} \rightarrow C_6H_{12}O_6 + 6\ O_2 \uparrow$$

Complete glucose metabolism:
$$C_6H_{12}O_6 + 6\ O_2 \rightarrow 6\ CO_2 \uparrow + 6\ H_2O + \text{some chemical energy and much heat energy}$$

2) How Is the Energy of Glucose Harvested During Glycolysis?

In all living cells, the first step of glucose metabolism (**glycolysis**) proceeds the same either in the presence (aerobic) or absence (anaerobic) of oxygen. Glycolysis splits the six-carbon glucose into two three-carbon molecules of pyruvate and some released energy is used to make two ATP molecules. Under anaerobic conditions, **fermentation** occurs: the pyruvate is converted into lactate or ethanol in the cytoplasm. Under aerobic conditions, **cellular respiration** occurs: lactate enters the mitochondria and is broken down into CO_2 and H_2O, generating 34 to 36 ATP molecules.

Glycolysis consists of two major steps: glucose activation (glucose is converted into fructose bisphosphate from energy provided by 2 ATP) and energy harvest (4 ATP molecules made, and NAD^+ is converted into NADH electron carriers using energy released when fructose bisphosphate is converted into pyruvate). Under anaerobic conditions in animal muscle, fermentation of pyruvate to produce lactate occurs, gaining electrons and hydrogen ions when NADH is converted into NAD^+. Lactate is toxic when concentrated, causing discomfort and fatigue. When oxygen is present, lactate is converted into pyruvate which enters cellular respiration. Anaerobic conditions in many microorganisms produce alcoholic fermentation: pyruvate is converted into ethanol + CO_2, gaining electrons and hydrogen ions when NADH is converted into NAD^+. Alcoholic fermentation in yeast is useful in the brewing (ethanol) and baking (CO_2 makes bread rise) industries.

3) How Does Cellular Respiration Generate Still More Energy from Glucose?

During aerobic cellular respiration in the mitochondria of eukaryotic cells, pyruvate is converted into

$CO_2 + H_2O$, plus many ATP molecules. The final reactions require oxygen because it is the final acceptor of electrons. Most ATP made during cellular respiration is generated by reactions catalyzed by enzymes in the mitochondrial **matrix**, electron transfer proteins in the inner membrane, and movement of hydrogen ions through ATP-synthesizing proteins in the inner membrane. Steps involved are as follows:

(1)   The pyruvate enters mitochondria by diffusion through pores;

(2)   Each pyruvate + coenzyme-A + NAD^+ → CO_2 + NADH + acetyl-coenzyme-A which enters the **Kreb's cycle** and is converted into two CO_2 and one ATP, and donates energetic electrons to several electron-carrying molecules (three NADH and one $FADH_2$);

(3)   The electron carriers donate their energetic electrons to the electron transport system (ETS) of the inner mitochondrial membrane where the energy is used to transport hydrogen ions from the matrix to the **intermembrane compartment** where electrons + hydrogen ions + oxygen → H_2O;

(4)   In **chemiosmosis**, the hydrogen ion gradient (high in the intermembrane compartment, low in the matrix) created by the ETS discharges through pores in the ATP-synthesizing enzymes located in the inner membrane, and the energy is used to produce 32 to 34 molecules of ATP;

(5)   The ATP leaves the mitochondria by diffusion and enters the cytoplasm.

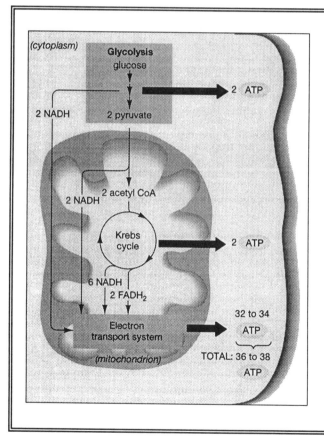

A SUMMARY OF THE ENERGY HARVEST FROM THE COMPLETE METABOLISM OF ONE GLUCOSE MOLECULE

Glycolysis and the citric acid cycle each produce two ATP molecules. The reactions within the mitochondrial matrix produce eight NADH molecules and two $FADH_2$ molecules. By donating its electrons to the electron transport system, each NADH molecule yields three ATP molecules, for a total of 24 ATPs. Each $FADH_2$ molecule yields two ATP molecules, for a total of 4 ATPs. The electrons from the two NADH molecules produced in the cytoplasm during glycolysis must be transported into the mitochondrion to reach the electron transport system. In heart and liver cells, this transport is "free"; in most cells, transport costs one ATP per NADH. The two "glycolytic NADH" molecules therefore yield either 4 or 6 ATP molecules, depending on the cell. Therefore, the energy harvest from electron transport is 32 to 34 ATPs. Including 2 ATPs from glycolysis and 2 ATPs from the citric acid cycle, the total energy yield from glucose metabolism is 36 to 38 ATPs.

KEY TERMS AND CONCEPTS

Fill-In: Write the answers in the numbered blanks in the table below.

Name of metabolic process	Is oxygen necessary?	Part of a cell where it occurs	Net Number of ATP molecules produced	Types of molecules produced
Glycolysis	(1)	(2)	(3)	(4)
Alcoholic fermentation	(5)	(6)	(7)	(8)
Lactate fermentation	(9)	(10)	(11)	(12)
Krebs cycle	(13)	(14)	(15)	(16)
Electron transport system and chemiosmosis	(17)	(18)	(19)	(20)

Key Terms and Definitions

cellular respiration: the oxygen-requiring reactions occurring in mitochondria that break down the end products of glycolysis into carbon dioxide and water, while capturing large amounts of energy as ATP.

chemiosmosis: a process of ATP generation in chloroplasts and mitochondria. The movement of electrons down an electron transport system is used to pump hydrogen ions across a membrane, thereby building up a concentration gradient of hydrogen ions across the membrane. The hydrogen ions diffuse back across the membrane through the pores of ATP-synthesizing enzymes. The energy of their movement down their concentration gradient drives ATP synthesis.

electron transport system: a series of molecules found in the inner membrane of mitochondria and the thylakoid membranes of chloroplasts that extract energy from electrons and generate ATP or other energetic molecules.

fermentation: anaerobic reactions that convert the pyruvic acid produced by glycolysis into lactic acid or alcohol and CO_2.

glycolysis: reactions carried out in the cytosol that break down glucose into two molecules of pyruvic acid, producing two ATP molecules. Glycolysis does not require oxygen, but can proceed when oxygen is present.

intermembrane compartment: the space contained between the inner and outer membranes of a mitochondrion.

Krebs cycle: (in honor of Hans Krebs, who discovered many of its biochemical details); a cyclic series of reactions in which the acetyl groups from the pyruvic acids produced by glycolysis are broken down to CO_2, accompanied by the formation of ATP and electron carriers. It occurs in the matrix of mitochondria.

matrix: the fluid contained within the inner membrane of mitochondria.

pyruvate: the three-carbon molecule formed from the splitting of a glucose molecule during glycolysis.

THINKING THROUGH THE CONCEPTS

True of False: Determine if the statement given is true or false. If it is false, change the underlined word so that the statement reads true.

21._____   Aerobic forms of life evolved before anaerobic forms.
22._____   Aerobic respiration uses O_2 and produces CO_2.
23._____   Glycolysis requires oxygen in order to function.
24._____   Glycolysis occurs in the mitochondria of a cell.
25._____   Pyruvic acid is produced by glycolysis.
26._____   The chemical energy in sugar is used to make O_2.
27._____   When NADH becomes NAP^+ the hydrogens are used to make sugar.
28._____   Lactic acid fermentation occurs when oxygen is abundant in muscle cells.
29._____   When each pyruvic acid is completely broken down, six CO_2 molecules are released.
30._____   Each glucose molecule releases enough energy to make 100 molecules of ATP.

Matching: Glucose metabolism.

31._____ most of the ATP is made
32._____ occurs only under anaerobic conditions
33._____ occurs only under aerobic conditions
34._____ occurs under either anaerobic or aerobic conditions
35._____ glucose is split into 2 pyruvate molecules
36._____ occurs in mitochondria
37._____ lactate
38._____ occurs in the cytoplasm
39._____ produces CO_2 and ATP
40._____ requires some ATP energy to get started
41._____ produces ethanol
42._____ acetyl-CoA
43._____ Krebs cycle
44._____ fructose diphosphate

Choices:

a. glycolysis
b. fermentation
c. both a. and b.
d. cellular respiration
e. none of these

Short answer.

45. Explain, using chemical equations, how photosynthesis and aerobic cellular respiration are "complementary" processes.

Multiple choice: Pick the most correct choice or choices for each question:

46. During glycolysis, what provides the initial energy to break down glucose?
    a. ATP
    b. pyruvate
    c. NADH
    d. cytoplasmic enzymes
    e. mitochondria

47. At the end of glycolysis, where are the original carbons of the glucose molecule located?
    a. in six molecules of carbon dioxide
    b. in two molecules of NADH
    c. in two molecules of pyruvate
    d. in two molecules of citric acid

48. The anaerobic breakdown of glucose is called
    a. artificial respiration
    b. glycolysis
    c. photosynthesis
    d. fermentation
    e. Krebs cycle

49. What happens when pyruvate is converted into lactate?
    a. the lactate enters the Krebs cycle
    b. the mitochondria are activated
    c. NAD^+ is regenerated for use in glycolysis
    d. oxidation of pyruvate occurs
    e. oxygen gas is liberated

50. Oxygen is necessary for cellular respiration because oxygen
    a. combines with hydrogen ions to form water
    b. combines with carbon to form carbon dioxide
    c. combines with carbon dioxide and water to form glucose
    d. breaks down glucose into carbon dioxide and water
    e. allows glucose to be converted into pyruvic acid

51. When oxygen is present
    a. most cells utilize aerobic cellular respiration
    b. most animal cells carry out lactate fermentation
    c. most bacteria and yeast carry out alcoholic fermentation
    d. Glucose is broken down to produce 2 ATP molecules
    e. mitochondria are less likely to function normally

CLUES TO APPLYING THE CONCEPTS

These practice questions are intended to sharpen your ability to apply critical thinking and analysis to biological concepts covered in this chapter.

52. Why can drowning, suffocation, or carbon monoxide poisoning lead to death?   The obvious initial response is that they prevent oxygen from reaching our cells, but go beyond that to explain why this can cause death.

53. Some animals that live in deserts survive without actually drinking water. They do eat food containing a little water, but most of the water they need is made within their cells and called "metabolic water." From what you have read in this chapter, explain one way by which metabolic water is produced.

ANSWERS TO EXERCISES

1. no
2. cytoplasm
3. two
4. pyruvate, ATP, NADH
5. no
6. cytoplasm
7. zero
8. CO_2, ethanol, NAD^+
9. no
10. cytoplasm
11. zero
12. lactate, NAD^+
13. yes
14. mitochondrial matrix
15. one ATP per pyruvate
16. CO_2, ATP, NADH,

     $FADH_2$
17. yes
18. inner mitochondrial membrane and intermembrane compartment
19. 32 to 34
20. H_2O, ATP, NAD^+, FAD
21. false, after
22. true
23. false, does not require
24. false, cytoplasm
25. true
26. false, ATP
27. false, water
28. false, absent

29. false, three
30. false, 36
31. d
32. b
33. d
34. a
35. a
36. d
37. b
38. c
39. b and d
40. a
41. b
42. d
43. d
44. a

45.

Photosynthesis

$$6\ CO_2 + 6\ H_2O \qquad + \text{solar energy} \rightarrow \quad C_6H_{12}O_6 + 6\ O_2 \uparrow$$

used in photosynthesis   made in photosynthesis
made in cell respiration   used in cell respiration

Aerobic Cell Respiration

$$C_6H_{12}O_6 + 6\ O_2 \qquad\qquad \rightarrow \qquad\quad 6\ CO_2 \uparrow + 6\ H_2O \qquad + \text{energy}$$

made in photosynthesis   used in photosynthesis
used in cell respiration   made in cell respiration

46. a
47. c

48. b,d
49. c

50. a
51. a

52. Oxygen is the final electron acceptor in aerobic cellular respiration, receiving electrons and hydrogen ions from the electron transport chain and forming water. Without sufficient oxygen, the electrons and hydrogen ions clog up the electron transport chain, not allowing NADH and $FADH_2$ molecules to give off their electrons and hydrogen ions, and as cells run out of these molecules, mitochondria shut down. This forces cells to rely on glycolysis and fermentation to produce very little ATP from glucose metabolism, and without sufficient energy from ATP, cells cannot continue functioning and die.

53. When oxygen combines with electrons and hydrogen ions at the end of the electron transport chain, "metabolic water" is produced. Some animals can use this cell-made water to help meet the water demands of their cells, and thus survive.

Chapter 9: DNA: The Molecule of Heredity

OVERVIEW

This chapter focuses on the molecular basis of inheritance, namely DNA. The authors describe the structure of DNA and explain how DNA replicates.

1) What Is the Composition of Chromosomes?

Eukaryotic **chromosomes** consist of long **DNA (deoxyribonucleic acid)** molecules complexed with proteins, and become visible within the nucleus during cell division. Nucleic acids consist of four types of smaller molecules called **nucleotides**, each with a pentose sugar (S) called deoxyribose in DNA, a phosphate group (P), and a nitrogen-containing **base** (B).

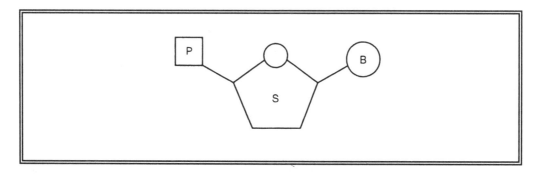

DNA nucleotides have four types of bases: **purine** bases (two rings of atoms) are adenine (A) and guanine (G); and **pyrimidine** bases (one ring of atoms) are thymine (T) and cytosine (C). The sequence of nucleotide bases in DNA can encode the vast amounts of information needed to make about 50,000 different proteins in human cells. A stretch of DNA 10 nucleotides long can have more than a million possible sequences of the four bases ($4^{10} = 1,048,576$), and typical human DNA molecules are billions of nucleotides long.

2) What Is the Structure of DNA?

Chemical analysis showed that, although the amounts of each of the four bases vary considerably from species to species, the DNA of any given species contains equal amounts of A and T and equal amounts of G and C. X-ray analysis showed that DNA is a twisted into a helix (corkscrew shaped) and has repeating subunits. In 1953, Watson and Crick figured out that a DNA molecule has two strands twisted into a **double helix**, each made of a series of nucleotides: The phosphate group of one nucleotide bonds to the sugar of the next, forming a "backbone" of alternating P and S with the bases sticking out.

```
—P-S —P-S —P-S —P-S —P-S —P-S—
    |      |      |      |      |      |
    B      B      B      B      B      B
```

In a DNA double helix, the backbones are on the outside (like uprights of a ladder) and the bases are paired up inside (like rungs of a ladder). Each rung is a **complimentary base pair** (either A-T or G-C) held together by hydrogen bonds.

3) How does DNA replication ensure genetic constancy?

Each chromosome has a long DNA molecule together with proteins that help organize and fold up the DNA. DNA duplication (**replication**) produces two identical double helices of DNA. DNA replication involves separating the two DNA strands and making new strands from nucleotides with bases complimentary to the parental strands. (New DNA = one parental strand + one new complimentary strand). DNA replication involves three steps, each controlled by enzymes: (1) the two original (parental) DNA strands unwind and separate from each other, using DNA helicase enzyme; (2) using **DNA polymerase** enzyme, each parental strand is used as a template for the new DNA strand, made by connecting nucleotides in an order determined by the nucleotide sequence of the parental strand so that A-T and G-C base pairs occur (DNA ligase enzyme connects the nucleotides in the new strand); and (3) parental and daughter complimentary strands wind together into double helices. This is **semiconservative replication** (one parental strand is conserved in each new double helix of DNA).

Proofreading produces almost error-free replication of DNA. DNA polymerases occasionally attaches bases incorrectly during DNA replication (1 in 10,000 times), but most of these are corrected by several "DNA repair enzymes" so that completely replicated mammalian DNA has only about one error per billion base-pairs. But environmental forces (such as X-rays, ultraviolet radiation from the sun, and certain chemicals) also damages DNA.

KEY TERMS AND CONCEPTS

From the information in this chapter and previous chapters, fill in the crossword puzzle based on the following clues.

Across
2. base that pairs with thymine in DNA
4. letter symbols for purine bases in DNA
6. complimentary base sequence for TTAG
10. enzyme that connects the nucleotides in a newly made DNA strand
11. type of nitrogenous base with two rings of atoms
12. DNA replication is ____-conservative
14. letter symbols for pyrimidine bases in DNA

15. DNA has a double-_____ 3-D structure
16. (Across) complimentary base sequence for CAATG
17. type of nucleic acid with ribose sugar
18. what genes are made of
20. (Across) number of sizes of nitrogenous bases in nucleic acids
22. new strands of DNA are made using this enzyme
24. the smallest unit of life
25. base that pairs with guanine in DNA

Down

1. co-discoverer of the double helix DNA structure
3. has a phosphate group, a sugar, and a nitrogenous base
5. the opposite complementary base pair to 20 down
7. type of nitrogenous base with one ring of atoms
8. an example is adenine
9. the enzyme that unwinds DNA during replication
13. cell structure in nucleus containing genetic material
16. (Down) base that pairs with cytosine in DNA
19. number of different bases in a DNA molecule
20. (Down) a complimentary base pair in DNA
21. co-discoverer of the double helix DNA structure
23. number of types of sugar molecules in DNA

Key Terms and Definitions

base: in molecular genetics, one of the nitrogen-containing, single- or double-ringed structures that distinguish one nucleotide from another. In DNA, the bases are adenine, guanine, cytosine, and thymine.

chromosome: in eukaryotes, a linear strand composed of DNA and protein, found in the nucleus of a cell, that contains the genes; in prokaryotes, a circular strand composed solely of DNA.

complementary base pair: in nucleic acids, bases that pair by hydrogen bonding; in DNA, adenine is complementary to thymine, and guanine is complementary to cytosine.

deoxyribonucleic acid (DNA): a molecule composed of deoxyribose nucleotides; the genetic information of all living cells.

DNA polymerase: an enzyme that covalently bonds DNA nucleotides together into a continuous strand, using a preexisting DNA strand as a template. DNA polymerase catalyzes the replication of the DNA of chromosomes during interphase prior to mitosis and meiosis.

double helix: a corkscrew-shaped object, as if a wire were wrapped around a cylinder.

gene: the unit of heredity encoding the information needed to specify the amino acid sequenceof proteins and hence particular traits; a segment of DNA located at a particular

place on a chromosome.

nucleotide: an individual subunit of which nucleic acids are composed. A single nucleotide consists of a phosphate group covalently bonded to a sugar (deoxyribose in DNA), which is in turn covalently bonded to a nitrogenous base (adenine, guanine, cytosine, or thymine). Nucleotides are linked together by covalent bonds between the phosphate of one and the sugar of the next, to form a strand of nucleic acid.

purine: a type of nitrogen-containing base found in nucleic acids that consists of two fused rings; includes adenine and guanine in both DNA and RNA.

pyrimidine: a type of nitrogen-containing base found in nucleic acids that consists of a single ring; includes cytosine (both DNA and RNA), thymine (DNA only) and uracil (RNA only).

replication: the copying of the double-helical DNA during chromosomal duplication.

semiconservative replication: the process of replication of the DNA double helix; the two DNA strands separate, and each is used as a template for the synthesis of a complementary DNA strand. Each daughter double helix therefore consists of one parental strand and one new strand.

THINKING THROUGH THE CONCEPTS

True or False: Determine if the statement given is true or false. If it is false, change the underlined word so that the statement reads true.

26._____   A molecule of DNA is <u>single</u> stranded.
27._____   DNA contains four types of <u>sugars</u>.
28._____   Sugars found in DNA have <u>five</u> carbons each.
29._____   DNA is found in cellular <u>chromosomes</u>.
30._____   DNA contains sugars, bases, and <u>sulfur</u> groups.
31._____   Purine bases are <u>larger</u> molecules than pyrimidine bases.
32._____   The concentration of DNA is <u>constant</u> for different body cells of the same species.
33._____   Adenine pairs with <u>guanine</u> in DNA.
34._____   The duplication of DNA is called <u>fully</u> conservative replication.
35._____   The building blocks of nucleic acids are <u>amino acids</u>.

Matching: Purine and pyrimidine bases. (Some questions may have more than one answer.)

36.____ Purine bases
37.____ Pyrimidine bases
38.____ Base pairs in DNA
39.____ Single ring bases
40.____ Double ring bases

Choices:

    a. adenine (A)   d. thymine (T)
    b. cytosine (C)   e. A-T and G-C
    c. guanine (G)   f. A-G and T-C

Matching: DNA structure

Choices:

    a. nucleotide
    b. deoxyribose
    c. phosphate group
    d. purine base
    e. pyrimidine base

Multiple Choice: Pick the most correct choice for each question.

46. If amounts of bases in a DNA molecule are measured, we find
    a. A = C and G = T
    b. A = G and C = T
    c. T = A and C = G
    d. that no two bases would be equal in amount
    e. that all bases are equal in amount

47. When comparing DNA and RNA, we find
    a. no sugar is present in either molecule
    b. hydrogen bonding is important only in DNA
    c. only DNA has a backbone of sugars and phosphates
    d. adenine pairs with different bases in DNA and RNA
    e. thymine pairs with different bases in DNA and RNA

48. The DNA of a certain organism has guanine as 30% of its bases. What percentage of its bases would be adenine?
    a. 0%
    b. 10%
    c. 20%
    d. 30%
    e. 40%

49. The correct structure of a nucleotide is
    a. phosphate-ribose-adenine
    b. phospholipid-sugar-base
    c. phosphate-sugar-phosphate-sugar
    d. adenine-thymine and guanine-cytosine
    e. base-phosphate-sugar

50. The two polynucleotide chains in a DNA molecule are attracted to each other by
    a. covalent bonds between carbon atoms
    b. hydrogen bonds between bases
    c. peptide bonds between amino acids
    d. ionic bonds between "R" groups in amino acids
    e. covalent bonds between phosphates and sugars

51. Using an analogy of DNA as a twisted ladder, the rungs (steps) of the ladder are
    a. phosphate groups
    b. sugar groups
    c. paired nitrogenous bases
    d. oxygen-carbon double bond

52. All the cells of a specific organism contain equal amounts of
    a. adenine and guanine
    b. guanine and cytosine
    c. adenine and cytosine
    d. thymine and cytosine

53. The purines bases in DNA are
    a. adenine and guanine
    b. adenine and thymine
    c. cytosine and adenine
    d. guanine and cytosine
    e. cytosine and thymine

54. The sequence of subunits in the DNA backbone is
    a. --base--phosphate--sugar--base --phosphate--sugar--
    b. --base--phosphate--base-- phosphate--base--phosphate--
    c. --phosphate--sugar--phosphate--sugar- -phosphate--sugar--
    d. --sugar--base--sugar--base-- sugar--base--sugar--base--
    e. --base--sugar--phosphate--base --sugar--phosphate--

55. A pyrimidine base always base-pairs with a
    a. single-ring pyrimidine
    b. double-ring pyrimidine
    c. single-ring purine
    d. double-ring purine

56. Figuratively speaking, a double helix is comparable to
    a. coiled rope
    b. stacked up plates
    c. braided hair
    d. twisted ladder
    e. tangled threads

CLUES TO APPLYING THE CONCEPTS

These practice questions are intended to sharpen your ability to apply critical thinking and analysis to biological concepts covered in this chapter.

57. Instead of DNA being a double-stranded molecule, suppose DNA was a single-stranded molecule, so that each complete molecule of DNA consisted of one chain of nucleotides. If DNA were single-stranded, describe how DNA replication could take place, using the same DNA polymerase enzyme that creates complimentary base pairing, so that exact copies of genes could be made. Would this be more efficient or less efficient than replication involving double-stranded DNA? Briefly explain.

ANSWERS TO EXERCISES

1. Watson
2. adenine
3. nucleotide
4. AG
5. GC
6. AATC
7. pyrimidine
8. base
9. helicase
10. ligase
11. purine
12. semi
13. chromosome
14. TC
15. helix
16. (Across) GTTAC
16. (Down) guanine
17. RNA
18. DNA
19. four

20. (Across) two
20. (Down) TA
21. Crick
22. polymerase
23. one
24. cell
25. cytosine
26. false, double
27. false, bases
28. true
29. true
30. false, phosphorus
31. true
32. true
33. false, thymine
34. false, semi
35. false, nucleotides
36. a and c
37. b and d

38. e
39. b and d
40. a and c
41. d
42. b
43. c
44. a
45. e
46. c
47. d
48. c
49. a
50. b
51. c
52. b
53. a
54. c
55. d
56. d

57. Suppose a hypothetical single-stranded DNA molecule has the base sequence AAAAAAAAAA. During replication, the DNA polymerase would make a complimentary DNA molecule with the base sequence TTTTTTTTTT. Then, the DNA polymerase would have to make a complementary copy of the TTTTTTTTTT DNA molecule, which would be AAAAAAAAAA, the same as the original gene. Then, the TTTTTTTTTT molecule would have to be broken down, since it isn't a normal DNA molecule for that organism. This scheme, where two replications yield one new DNA molecule, is much less efficient than replicating a double-stranded DNA molecule, where one round of semiconservative replication yields two DNA molecules.

Chapter 10: Gene Expression and Regulation

OVERVIEW

This chapter covers how genes are expressed and regulated. The authors introduce the "one-gene one-protein" hypothesis. They present the processes of transcription (information in DNA makes RNA) and translation (information in RNA makes protein) along with the three types of RNA and their functions. Transcriptional regulation of genes in eukaryotic cells is discussed, as is the effects of mutation.

1) How Are Genes and Proteins Related?

A **gene** is a segment of DNA with a nucleotide sequence specifying the amino acid sequence of a protein. There are thousands of genes per chromosome. Using red bread mold (*Neurospora*) with single doses of chromosomes and genes in its cells, Beadle and Tatum used X-rays to cause **mutations** (changes in the base sequence of DNA). Each mutation caused the loss of the ability of *Neurospora* to make one enzyme. Data supported the hypothesis that each gene encodes information (as sequence of bases) needed for making one specific protein (an amino acid sequence): the **one-gene, one-protein hypothesis**. Some genes code for structural proteins or for types of **RNA** (ribonucleic acid), but most code for enzymes.

2) What Is the Role of RNA in Protein Synthesis?

Eukaryotic DNA is in the nucleus while protein synthesis occurs on **ribosomes** in the cytoplasm. RNA molecules carry information from the DNA to the ribosomes. RNA (1) is single-stranded; (2) has ribose sugar in its backbone; (3) has uracil (U) instead of thymine; and (4) exists as **messenger RNA (mRNA), transfer RNA (tRNA)**, and **ribosomal RNA (rRNA)**. Information from DNA is used to make proteins in a two-step process: **transcription** (DNA makes mRNA, which carries genetic information, and tRNA and rRNA which help the mRNA make proteins at the ribosomes) and **translation** (proteins are made at the ribosomes through interaction of mRNA, tRNA, and rRNA).

In the genetic code, base sequences stand for amino acids. The **genetic code** uses three bases to specify each amino acid. Since there are four types of bases in RNA and 20 types of amino acids in proteins, sequences of three consecutive bases in mRNA (called **codons**) code for amino acids, yielding $4^3 = 64$ possible code words for proteins (for instance, UUU in mRNA = phenylalanine in a protein). The mRNA **start codon** is AUG, coding for the first amino acid in a protein. Three codons (UAG, UAA, UGA) are mRNA **stop codons**, signals that the protein's amino acid sequence is completed.

Transcription copies the DNA of only selected genes into mRNA, and copies only one (the **template strand**) of the two strands of DNA into mRNA. In a long DNA molecule, one strand may be the template strand for some genes, and the other strand may be template strand for other genes. Transcription is a three-step process: (1) initiation (using **RNA polymerase** enzyme, which attaches to a gene at the **promotor region**, a short sequence of bases at the beginning of the gene); (2) elongation (RNA polymerase moves along the gene, synthesizing a single strand of RNA that is complementary to the template strand of DNA (base pairing as usual except that RNA has uracil (U) instead of thymine);

and (3) termination (RNA polymerase reaches the termination signal, a sequence of DNA bases that causes the RNA molecule to separate from the DNA and from the RNA polymerase, and the RNA polymerase to detach from the DNA).

The mRNA conveys the code (base sequence) for protein synthesis from the nucleus into the cytoplasm through pores in the nuclear envelope. The mRNA binds to ribosomes where the sequence of mRNA codons are translated into the sequence of amino acids in proteins. The rRNA forms part of the **ribosomes** (composites of rRNA and proteins). Each ribosome has two subunits: the small subunit recognizes and binds mRNA and tRNA; and the large subunit has an enzymatic region for adding amino acids to the growing protein chain, and regions for binding tRNA. The tRNA molecules decode the mRNA codons into protein amino acids, by binding to free amino acids and delivering them to the ribosomes where they are incorporated into proteins according to mRNA instructions. Each tRNA bears three exposed bases (the **anticodon**) that pair in a complementary manner to codon bases that specify where the tRNA's amino acid is to be added to the protein chain. For example, the start codon (AUG) pairs with tRNA with anticodon UAC, bringing in methionine to be the first amino acid in the protein.

Like transcription, translation has three steps: (1) initiation of protein synthesis; (2) elongation of the protein chain; and (3) termination. Initiation begins with the binding of protein "initiator factors" and the initiator tRNA to the small subunit of a ribosome, which then binds to an mRNA molecule and moves to the start codon. The large ribosomal subunit then binds to the small subunit and the initiator tRNA binds to the first binding site of the large subunit. Elongation begins when the second tRNA recognizes and binds to the second codon in the second binding site of the ribosome. A peptide bond forms between the first and second amino acid, the first amino acid detaches from the first tRNA, which then leaves the ribosome; and the ribosome shifts the second tRNA to the first binding site and attracts the third tRNA into the second binding site. The third amino acids forms a peptide bond with the second amino; the second amino acid detaches from the second tRNA, which then leaves the ribosome; and the ribosome shifts the third tRNA to the first binding site and attracts the fourth tRNA into the second binding site. This process continues until, near the end of the mRNA, a stop codon is reached. Then enzymes cut the protein off the last tRNA, releasing it from the ribosome.

3) How Do Mutations in DNA Affect the Function of Genes?

Mutations are changes in the sequence of bases in DNA, often through a mistake in base pairing during DNA replication. In a **point mutation**, a pair of bases becomes incorrectly matched. An **insertion mutation** occurs when one or more new nucleotide pairs are added into a gene. A **deletion mutation** occurs when one or more nucleotide pairs are removed from a gene. Deletions and insertions can have quite harmful effects on a gene because all the codons that follow the deletion or insertion will be misread.

Four types of effects may result from mutations: (1) the protein is unchanged; (2) the new protein is equivalent if the active site is unchanged and the rest of the molecule is changed in an unimportant way; (3) protein function is changed by an altered amino acid sequence; or (4) protein function is destroyed by a misplaced stop codon, resulting in a much shortened protein chain.

Mutations provide the raw material for evolution. Mutation rates vary from 1 in 10,000 to 1 in 1,000,000. If mutations in gametes are not lethal, they may be passed on to future generations. Mutation is the source for genetic variation and thus is essential for evolution.

4) How Are Genes Regulated?

Most cells in the body have identical DNA (60,000 to 100,000 genes) but they don't use all the DNA all the time. Gene expression changes over time, and an organism's environment can determine which genes are translated. Four of the steps at which the rate of gene activity may be regulated are: (1) transcription of individual genes; (2) translation of various mRNAs; (3) modification of inactive proteins into their active forms; and (4) enzyme activity.

Eukaryotic cells may regulate the transcription of (1) individual genes, through the action of regulatory proteins such as steroid hormones; (2) regions of chromosomes with several genes, by condensing those regions into compact DNA that is inaccessible to RNA polymerase; or (3) entire chromosomes with thousands of genes, such as one of the X chromosomes in mammalian females to form a **Barr Body**. Which X chromosome is inactivated in any cell is random, but all its daughter cells will have the same condensed chromosome. For example, separate patches of orange and black fur in female calico cats are due to fur-color genes on the X chromosome.

KEY TERMS AND CONCEPTS

Fill-In: From the following list of key terms, fill in the blanks in the following statements.

anticodon	mutation	start codon
codon	one-gene, one-protein	stop codons
deletion mutation	hypothesis	transcription
genetic code	point mutation	transfer RNA
insertion mutation	ribosome	translation
messenger RNA	RNA polymerase	

The (1)_____ is the amino acid meanings of all the codons, each of which directs the incorporation of an amino acid during protein synthesis.

The enzyme that catalyzes the covalent bonding of free RNA nucleotides into a continuous strand, using the sequence of bases in DNA as a template, is called (2)_____.

The hypothesis that each gene encodes information (as a sequence of bases) needed for making one specific protein (amino acid sequence) is the (3)_____ .

A (4)_____ is a sequence of three nucleotides in transfer RNA that is complementary to the three nucleotides in messenger RNA.

The process in which a sequence of nucleotide bases in mRNA is converted into a sequence of amino acids in a protein is called (5)_____.

The mRNA (6)_____ is AUG, coding for the first amino acid in a protein. Three codons (UAG, UAA, UGA) are mRNA (7)_____, signals that the protein's amino acid sequence is completed.

A molecule of (8)_____ is a strand of nucleotides, complementary to the DNA of a gene, that conveys genetic information to ribosomes to be used to sequence amino acids during protein synthesis.

(9)_____ is the synthesis of an RNA molecule from a DNA template.

A change in the base sequence of DNA is a (10)_____.

A molecule that binds to a specific amino acid and has a set of three nucleotides complementary to the codon for that amino acid is known as (11)_____.

In a (12)_____, a pair of bases becomes incorrectly matched; (13)_____ occur when one or more new nucleotide pairs are added into a gene; a (14)_____ occurs when one or more nucleotide pairs are removed from a gene.

A (15)_____ is a sequence of three nucleotides of mRNA that specifies a particular amino acid to be incorporated into a protein.

An organelle with two subunits, each composed of RNA and protein, that serves as the site of protein synthesis is a (16)_____.

Key Terms and Definitions

anticodon: a sequence of three nucleotides in transfer RNA that is complementary to the three nucleotides of a codon of messenger RNA.

androgen insensitivity: a condition in which a genetic male appears to be a female as a result of a mutation that renders androgen receptors in cells nonfunctional.

Barr body: an inactive X chromosome found in somatic cells of mammals that have at least two X chromosomes (usually females). The Barr body usually appears as a dark spot in the nucleus.

codon: a sequence of three nucleotides of messenger RNA that specifies a particular amino acid to be incorporated into a protein. Certain codons also signal the beginning and end of protein synthesis.

deletion mutation: a mutation in which one or more nucleotides are removed from a gene.

exon: segment of DNA in a eukaryotic gene that codes for amino acids in a protein.

gene: a functional segment of DNA located at a particular place on a chromosome; the sequence of nucleotides in a gene specifies the sequence of amino acids in a particular protein.

genetic code: the collection of codons of mRNA, each of which directs the incorporation of a particular amino acid into a protein during protein synthesis.

insertion mutation: a mutation in which one or more nucleotides are inserted within a gene.

intron: segment of DNA in a eukaryotic gene that does not code for amino acids in a protein.

messenger RNA (mRNA): a strand of RNA, complementary to the DNA of a gene, that conveys the genetic information in DNA to the ribosomes to be used during protein synthesis. Sequences of three nucleotides (codons) in mRNA specify particular amino acids to be incorporated into a protein.

mutation: a change in the base sequence of DNA.

neutral mutation: a mutation (change in DNA sequence) that has little or no phenotypic effect.

one-gene, one-protein hypothesis: the proposition that each gene encodes the information for the synthesis of a specific protein.

point mutation: a mutation in which a single base pair in DNA has been changed.

promoter: a specific sequence of DNA to which RNA polymerase binds, initiating gene transcription.

ribonucleic acid (RNA): a single-stranded nucleic acid molecule composed of nucleotides, each of which consists of a phosphate group, the sugar ribose, and one of the bases adenine, cytosine, guanine, or uracil.

ribosomal RNA (rRNA): a type of RNA that combines with proteins to form ribosomes.

ribosome: an organelle consisting of two subunits, each composed of ribosomal RNA and protein. Ribosomes are the site of protein synthesis, in which the sequence of nucleotides of messenger RNA is translated into the sequence of amino acids in a protein.

RNA polymerase: an enzyme that catalyzes the covalent bonding of free RNA nucleotides into a continuous strand, using RNA nucleotides that are complementary to those of a strand of DNA.

start codon: a codon in messenger RNA that signals the beginning of protein synthesis on a ribosome.

stop codon: a codon in messenger RNA that stops protein synthesis and causes the completed protein chain to be released from the ribosome.

template strand: the strand of the DNA double helix from which RNA is transcribed.

transcription: the synthesis of an RNA molecule from a DNA template.

transfer RNA (tRNA): a type of RNA that (1) binds to a specific amino acid and (2) bears a set of three nucleotides (the anticodon) complementary to the mRNA codon for that amino acid. Transfer RNA carries its amino acid to a ribosome during protein synthesis, recognizes a codon of mRNA, and positions its amino acid for incorporation into the growing protein chain.

translation: the process whereby the sequence of nucleotides of messenger RNA is converted into the sequence of amino acids of a protein.

Werner syndrome: a condition in which individuals age prematurely, normally resulting in death by age 50; caused by a mutation in the gene for DNA replication/repair enzymes.

THINKING THROUGH THE CONCEPTS

True or False: Determine if the statement given is true or false. If it is false, change the underlined word so that the statement reads true.

17._____   Genes are made of RNA in human cells.
18._____   Transfer RNA carries amino acids to the ribosomes.
19._____   Protein synthesis occurs in the ribosome.
20._____   Messenger RNA is double stranded.
21._____   Messenger RNA is manufactured in the cytoplasm.
22._____   The triplets of bases in messenger RNA are called anticodons.
23._____   Proteins are made during transcription.
24._____   Proteins contain many nucleotide subunits.
25._____   Barr bodies are found in normal mammalian females.
26._____   Barr bodies are active X chromosomes found in mammals.

Matching: Types of RNA molecules.

27.____ has anticodons
28.____ deciphers the genetic code
29.____ carries the genetic code to make
        proteins
30.____ picks up and transports amino acids
31.____ part of ribosomes
32.____ has codons
33.____ twisted into a cloverleaf shape
34.____ fits into binding sites in ribosomes

Choices:

a. mRNA
b. tRNA
c. rRNA

Matching: Transcription and translation.

35.____   information from DNA makes RNA
36.____   information from RNA makes protein
37.____   occurs in the nucleus of eukaryotic cells
38.____   occurs in the cytoplasm of eukaryotic cells
39.____   involves RNA polymerase
40.____   involves amino acids
41.____   involves ribosomes
42.____   involves codon-anticodon interactions
43.____   involves copying the genetic code
44.____   involves deciphering the genetic code

Choices:

   a.   transcription
   b.   translation

Multiple Choice: Pick the most correct choice for each question.

45. Inherited disorders induced by X-rays in red bread mold by Beadle and Tatum
    a. are caused by errors in mitosis
    b. are related to enzyme deficiencies
    c. can always be cured by dietary restrictions
    d. are environmental and not genetic in origin
    e. can never be cured by supplying the missing end product

46. Which of these choices is coded for by the shortest piece of DNA?
    a. a tRNA having 75 nucleotides
    b. an mRNA having 50 codons
    c. a protein having 40 amino acids
    d. a protein with 2 polypeptides, each having 35 amino acids
    e. an mRNA having 100 bases

47. If a bacterial protein has 30 amino acids, how many nucleotides are needed to code for it?
    a. 30
    b. 60
    c. 90
    d. 120
    e. 600

48. Blood cells and muscle cells make different enzymes because
    a. blood cells contain only genes for blood cell proteins and muscle cells contain only muscle protein genes
    b. all cells of an organism have all genes
    c. not every gene acts in every type of cell
    d. blood cells have hemoglobin while muscle cells have microtubules
    e. adult red blood cells lack nuclei in mammals

49. Because of random X chromosome inactivation, one of the X chromosomes of a mammalian female
    a. is functionally inactive
    b. is present in each cell in three doses
    c. does not divide during meiosis
    d. disappears from each cell early during development
    e. is genetically identical to the other X chromosome

Fill-In. Based on the figure to the right, answer the following questions.

50. The step indicated by (1) in the figure
    is commonly known as

    _____.

51. The step indicated by (2) in the figure
    is commonly known as

    _____.

52. The step indicated by (3) in the figure
    is commonly known as

    _____.

53. The step indicated by (4) in the figure
    is commonly known as

    _____.

Place text Figure 10-8 here.
Drop off the labels numbered
1, 2, 3, and 4.

Fill in the table.

Type of molecules	Sequences of bases or amino acids (See text Table 10-2 for genetic code table)
54. DNA template strand	___ TAG ___ AGC ___ TCA
55. DNA non-template strand	GAA ___ TTA ___ CCG ___
56. messenger RNA codons	___ ___ ___ ___ ___ ___
57. transfer RNA anticodons	___ ___ ___ ___ ___ ___
58. protein amino acid sequence	___ ___ ___ ___ ___ ___

Fill-In:

Suppose a section of DNA from a normal gene has the following sequences of bases in one of its polynucleotide strands:

Normal base sequence: TACTTTACGTCGTGAAAACGGTAT
If this strand is used to make a mRNA molecule:

- the base sequence in the mRNA is (59)_____
- the normal amino acid sequence in the polypeptide is (use text Table 10-2)

    (60)_____

Now, suppose a point mutation occurs and a single base (**C***) is added (an insertion mutation) to the gene sequence, causing a new altered sequence to occur:

MUTANT BASE SEQUENCE: TAC**C***TTTACGTCGTGAAAACGGTAT

If this mutant strand is used to make a mRNA molecule

- the base sequence in the mRNA is (61)_____
- the abnormal amino acid sequence in the polypeptide is (use text Table 10-2)

    (62)_____

CLUES TO APPLYING THE CONCEPTS

This practice question is intended to sharpen your ability to apply critical thinking and analysis to biological concepts covered in this chapter.

63. Briefly explain why in questions 59-62 the addition of a single base in the DNA caused so many amino acids to change in the polypeptide.

ANSWERS TO EXERCISES

1. genetic code
2. RNA polymerase
3. one-gene one-protein hypothesis
4. anticodon
5. translation
6. start codon
7. stop codons
8. messenger RNA (mRNA)
9. transcription
10. mutation
11. transfer RNA (tRNA)
12. point mutation
13. insertion mutation
14. deletion mutation
15. codon
16. ribosome
17. false, DNA

18. true
19. true
20. false, single
21. false, nucleus
22. false, codons
23. false, translation
24. false, amino acid
25. true
26. false, inactive
27. b
28. b
29. a
30. b
31. c
32. a
33. b
34. b
35. a

36. b
37. a
38. b
39. a
40. b
41. b
42. b
43. a
44. b
45. b
46. a
47. c
48. c
49. a
50. transcription
51. translation
52. modification
53. catalysis

54. CTT-TAG-AAT-AGC-GGC-TCA

55. GAA-ATC- TTA-TCG-CCG-AGT

56. GAA-AUC-UUA-UCG-CCG-AGU

57. CUU-UAG-AAU-AGC-GGC-UCA

58. glutamic acid-isoleucine-leucine-serine-proline-serine-

59. AUG-AAA-UGC-AGC-ACU-UUU- GCC-AUA

60. Methionine-lysine-cysteine-serine-threonine-phenylalanine-alanine-isoleucine

61. AUG-**G**AA-AUG-CAG-CAC-UUU-UGC-CAU-A

62. Methionine-glutamic acid-methionine-glutamine-histidine-phenylalanine-cysteine-histidine-

63. The reason so many amino acids are changed is that the addition of one base causes all the codons to become different because they are read by ribosomes as three consecutive mRNA bases and the addition of one base results in a shift in the "reading frame" the ribosome uses to determine the codons.

Chapter 11: The Continuity of Life: Cellular Reproduction

OVERVIEW

This chapter focuses on mitosis and meiosis. After briefly discussing the cell cycle in prokaryotes, the authors cover mitosis, a basic type of eukaryotic cell division. They describe the structure of eukaryotic chromosomes and review the typical chromosome numbers in body cells and sex cells. The eukaryotic cell cycle is discussed, as well as interphase, mitosis, and cytokinesis. Then, after contrasting asexual and sexual reproduction, the authors focus on meiosis. They describe the stages of meiosis I and meiosis II as they relate to the production of sex cells. Meiosis I and II are compared, as are mitosis and meiosis. Finally, three basic types of life cycles among eukaryotic organisms are outlined and contrasted.

1) What Are the Essential Features of Cell Division?

Eukaryotic cellular DNA is packaged in **chromosomes** (darkly staining nuclear rods of condensed protein and DNA). In nondividing cells, chromosomal material is called **chromatin** (thin, hard to see chromosomes). When a cell divides, it passes on: (1) complete sets of hereditary information (the chromosomes) and (2) cytoplasmic materials essential for survival. The **cell cycle** includes cell activities from one cell division to the next.

2) What Are the Events of the Prokaryotic Cell Cycle?

The DNA of bacteria is found in a single, circular chromosome within the cytoplasm. Cell division in bacteria is called **binary fission**, and can occur in 30 minutes or less. The cell absorbs nutrients, grows, replicates its DNA, and divides. Steps in the prokaryotic cell cycle: (1) one point on the chromosome attaches to plasma membrane; (2) binary fission begins when the chromosome replicates, each identical chromosome attaching to a separate point on the plasma membrane; (3) the cell elongates, pushing the chromosomes apart; (4) the plasma membrane around the middle of the cell grows inward; and (5) two new daughter cells form, each receiving one of the replicated chromosomes and about half the cytoplasm.

3) What is the Structure of the Eukaryotic Chromosome?

Human cells each have a total of 200 cm (6.5 feet) of DNA in its chromosomes. Most chromosomes have two arms extending from a **centromere**. Before cells divide, each chromosome replicates to form a duplicated chromosome, with its two copies (identical sister chromatids) attached at their centromeres. During cell division, the two sister chromatids physically separate, and each chromatid becomes an independent daughter chromosome.

Eukaryotic chromosomes usually occur in **homologous pairs** with similar sizes, staining patterns, and genetic information in nonreproductive **diploid cells** ($2n$). So human body cells have 46 chromosomes that can be organized into 23 homologous pairs. At some point in the life cycle of sexually reproducing organisms, **meiosis** produces **haploid cells** (n) called **gametes** (sperm and egg), with one copy of each type of chromosome.

4) What Are the Events of the Eukaryotic Cell Cycle?

Growth, replication of chromosomes, and most cell functions occur during **interphase**, the period between cell divisions. During the G_1 *phase* of interphase, the cell acquires nutrients, performs specialized functions, and grows. Cells no longer dividing enter the G_0 *phase*. Alternately, cells enter the *S phase*, during which chromosome replication occurs. Next, cells enter the G_2 *phase*, during which cells make molecules required for cell division.

  Mitotic cell division consists of nuclear division (**mitosis**) and cytoplasmic division (**cytokinesis**). The two cells produced by mitotic cell division are essentially identical to each other cytoplasmically and genetically identical to the parent cell. Different types of body cells (brain and liver, for example) are genetically identical but undergo **differentiation**, the process whereby cells assume specialized functions because they use different genes as they develop.

5) What Are the Phases of Mitosis?

Mitosis has four phases within a continuous process: (1) *prophase* (chromosomes condense and microtubular **spindle fibers** form from the **centrioles** and attach to the chromosomes); (2) *metaphase* (chromosomes become aligned along the equator of the cell); (3) *anaphase* (sister chromatids separate and are pulled to opposite poles of the cell); and (4) *telophase* (nuclear envelopes form around both groups of chromosomes). Each sister chromatid has a **kinetochore** (attachment point for spindle fibers) located at the centromere.

6) What Are the Events of Cytokinesis?

In most cells, cytokinesis occurs during telophase, enclosing each daughter nucleus into a separate cell. In animal cells microfilaments attached to the plasma membrane around the equator contract and constrict the equator region, pinching the cell in two. In plant cells, the Golgi complex buds off carbohydrate-filled vesicles along the equator. The vesicles fuse, forming the **cell plate** that expands to fuse with the plasma membrane. Cytokinesis is followed by interphase.

7) What Are the Functions of Mitotic Cell Division?

The functions of mitotic cell division are: (1) growth (from zygote to adult), maintenance, and repair of body tissues; and (2) **asexual reproduction** (formation of genetically identical offspring without fusion of eggs and sperm) in simple organisms and plants.

8) What Are the Advantages of Sexual Reproduction?

The reshuffling of genes among individuals to create genetically unique offspring results from **sexual reproduction**. DNA mutations are the ultimate source of genetic variability and the raw material for evolution. Mutations form **alleles**, alternate forms of a gene that confer variability on individuals (brown or blue eyes, for instance). Reshuffling genes may combine different alleles in beneficial ways through sexual reproduction (for instance, parent with good camouflage mated with parent with motionless behavior when predators are nearby may produce some offspring with good behavior and camouflage).

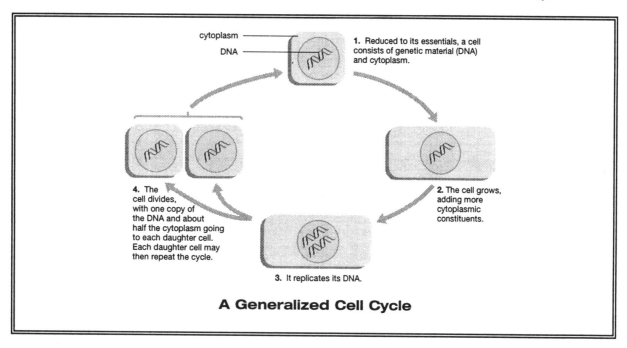

A Generalized Cell Cycle

9) What Are the Events of Meiosis?

Diploidy probably evolved when a temporarily $2n$ cell delayed meiosis for a while, remaining $2n$ for a significant portion of the life cycle. Eventually, some of the $2n$ cells reproduced mitotically, forming a diploid multicellular body. Sperm and egg fused to yield a $2n$ cell, the **zygote** (fertilized egg) that divides mitotically to form the adult body.

Meiosis separates homologous chromosomes in a diploid nucleus, producing haploid daughter nuclei. In **meiotic cell division** (meiosis followed by cytokinesis), each new cell receives one member from each pair of homologous chromosomes. Meiosis involves two nuclear divisions (meiosis I and II) and the following events occur: (1) chromosomes replicate before meiosis begins; (2) during meiosis I, duplicated homologous chromosomes separate, one duplicated chromosome moving into each of the two daughter cells ($2n$ becomes a pair of n cells); (3) in meiosis II, sister chromatids in each daughter cell separate and cytokinesis may occur to produce four haploid cells, each with one set of unduplicated chromosomes (each haploid cell with duplicated chromosomes becomes a pair of n cells with unduplicated chromosomes).

Meiosis I separates homologous chromosomes into two daughter nuclei. During prophase I, homologous chromosomes pair up and exchange DNA through **crossing over** and the formation of **chiasmata** (regions of exchange), resulting in **genetic recombination** (formation of new combinations of different alleles on a chromosome). During metaphase I, paired homologous chromosomes move to the equator of the spindle. Different pairs of chromosomes align themselves randomly at the equator, allowing independent assortment to occur. During anaphase I, homologous chromosomes separate. During telophase I, two haploid clusters of chromosomes are formed. Meiosis II (very similar to mitosis) separates sister chromatids. During anaphase II, the centromeres holding sister chromatids together split, and spindle fibers pull each chromatid (daughter chromosome) to opposite poles.

10) What Are the Roles of Mitosis and Meiosis in Eukaryotic Life Cycles?

The life cycles have a common overall pattern: (1) two haploid cells fuse a diploid cell with new gene combinations; (2) meiosis produces haploid cells; (3) mitosis of either haploid or diploid cells, or both, produces multicellular bodies and/or asexual reproduction. Many fungi and unicellular algae have haploid life cycles. Animals have a diploid life cycle. Plants have an **alternation of generations** life cycle: multicellular 2*n* body by mitosis→ haploid **spores** by meiosis → multicellular *n* body with gametes by mitosis → fusion of gametes to form 2*n* zygote → multicellular 2*n* body by mitosis.

11) What Are the Roles of Meiosis and Sexual Reproduction in Producing Genetic Variability?

During metaphase I, random alignment of homologous chromosomes at the equator creates novel combinations of chromosomes. Crossing over creates chromosomes with novel combinations of genes. Fusion of gametes adds further genetic variability to the offspring. In humans there are 2^{23} or about 8 million different types of gametes based on random alignment of homologous chromosomes at metaphase I. Fusion of gametes from two people makes 8 million x 8 million (64 trillion) possible genetically different children. Crossing over increases this number substantially.

KEY TERMS AND CONCEPTS

Fill-In: From the following list of key terms, fill in the blanks below.

asexual	crossing over	interphase
binary fission	cytokinesis	meiosis
chiasmata	gametes	mitosis
chromatin	genetic recombination	sexual
chromosomes	homologous	zygote

Eukaryotic cellular DNA is packaged in (1)_____ (darkly staining nuclear rods of condensed protein and DNA). In nondividing cells, chromosomal material is called (2)_____ (thin, hard to see chromosomes).

Cell division in bacteria is called (3)_____, and can occur in 30 minutes or less.

Eukaryotic chromosomes usually occur in (4)_____ pairs with similar sizes, staining patterns, and genetic information in nonreproductive diploid cells (2*n*).

At some point in the life cycle of sexually reproducing organisms, (5)_____ produces haploid cells (*n*) called (6)_____ (sperm and egg), with one copy of each type of chromosome.

Growth, replication of chromosomes, and most cell functions occur during (7)_____ , the period between cell divisions.

Mitotic cell division consists of nuclear division or (8)_____and cytoplasmic division called (9)_____.

(10)_____ reproduction (formation of genetically identical offspring without fusion of eggs and sperm) occurs in simple organisms and plants.

The reshuffling of genes among individuals to create genetically unique offspring results from (11)_____ reproduction.

Sperm and egg fused to yield a 2*n* cell, the (12)_____ (fertilized egg) that divides mitotically to form the adult body.

(13)_____ separates homologous chromosomes in a diploid nucleus, producing haploid daughter nuclei.

Meiosis I separates homologous chromosomes into two daughter nuclei. During prophase I, homologous chromosomes pair up and exchange DNA through (14)_____ and the formation of (15)_____ (regions of exchange), resulting in (16)_____ (formation of new combinations of different alleles on a chromosome).

Key Terms and Definitions

allele: an alternate form of a specific gene.

alternation of generation: a life cycle typical of plants in which a diploid spore-producing generation alternates with a haploid gamete-producing generation

anaphase: the stage of mitosis and meiosis II in which the sister chromatids of each chromosome separate from one another and are moved to opposite poles of the cell. In meiosis I, the stage in which homologous chromosomes are separated.

asexual reproduction: reproduction that does not involve the fusion of haploid gametes.

binary fission: cell division in prokaryotes.

cell cycle: the sequence of events in the life of a cell, from one division to the next.

cell plate: in plant cell division, a series of vesicles that fuse to form the new plasma membranes and cell wall separating the daughter cells.

centriole: a microtubular structure involved in spindle formation during cell division.

centromere: the region of a replicated chromosome at which the sister chromatids are held together.

chiasma (pl. chiasmata): during prophase I of meiosis, a point at which a chromatid of one chromosome crosses with a chromatid of the homologous chromosome. Exchange of chromosomal material between chromosomes takes place at a chiasma.

chromatid: one of the two identical strands of DNA and protein forming a replicated chromosome. The two sister chromatids are joined at the centromere.

chromatin: the complex of DNA and proteins that makes up eukaryotic chromosomes.

chromosome: in eukaryotes, a linear strand composed of DNA and protein, found in the nucleus of a cell, that contains the genes; in prokaryotes, a circular strand composed solely of DNA.

clone: an individual created from the body cells of another, so that both are genetically identical.

cloning: the process of artificially creating a new individual that is genetically identical to an existing individual.

cytokinesis: division of the cytoplasm and organelles into two daughter cells during cell division. Usually, cytokinesis occurs during telophase of mitosis.

crossing over: the exchange of corresponding segments of the chromatids of two homologous chromosomes during meiosis.

differentiation: the process whereby relatively unspecialized cells become specialized into particular tissue types.

diploid: referring to a cell with pairs of homologous chromosomes.

gamete: a haploid sex cell formed in sexually reproducing organisms.

haploid: referring to a cell that has only one member of each pair of homologous chromosomes.

homologue: a chromosome that is similar in appearance and genetic information to another chromosome with which it pairs during meiosis. Also called homologous chromosome.

interphase: the stage of the cell cycle between cell divisions. During interphase, chromosomes are replicated, and other cell functions occur, such as growth, movement, and nutrient acquisition.

kinetochore: a protein structure that forms at the centromere regions of chromosomes; attaches the chromosomes to the kinetochore microtubules of the spindle.

life cycle: the events in the life of an organism from one generation to the next.

meiosis: a type of cell division found in eukaryotic organisms, in which a diploid cell divides twice to produce four haploid cells.

meiotic cell division: a series of two cell divisions in which the diploid chromosome number is reduced to a haploid number and genetic variation is produced in the four resulting cells.

metaphase: the stage of mitosis or meiosis in which the chromosomes, attached to kinetochore microtubules, line up along the equator of the cell.

mitosis: a type of nuclear division found in eukaryotic cells. Chromosomes are duplicated during interphase before mitosis. During mitosis, one copy of each chromosome moves into each of two daughter nuclei. The daughter nuclei are therefore genetically identical to each other.

mitotic cell division: cell division involving prophase, metaphase, anaphase, and telophase; the resulting daughter cells are genetically identical to each other and to the parental cell.

prophase: the first stage of mitosis or meiosis, in which the chromosomes first become visible with a light microscope as thickened, condensed threads and the spindle begins to form. In meiosis I, the homologous chromosomes pair up and exchange parts at chiasmata.

recombination: creation of new combinations of alleles within chromosomes during meiosis through the process of crossing over.

sexual reproduction: a form of reproduction in which genetic material from two parental organisms is combined. In eukaryotes, two haploid gametes fuse to form a diploid zygote.

spindle fibers: elongated microtubular fibers making up the mitotic and meiotic spindle apparatus; some attach to the kinetochore regions of the centromeres of chromosomes and aid in chromosome movement.

spore: in the alternation of generation life cycle of plants, a haploid cell formed by meiosis that undergoes repeated mitotic cell divisions to form a multicellular haploid body.

telophase: the last stage of meiosis and mitosis, in which a nuclear envelope reforms around each new daughter nucleus, the spindle disappears, and the chromosomes relax from their condensed form.

zygote: in sexual reproduction, a diploid cell formed by the fusion of two haploid cells.

THINKING THROUGH THE CONCEPTS

True or False: Determine if the statement given is true or false. If it is false, change the underlined word so that the statement reads true.

17._____   Asexually produced organisms are similar but not identical to their parents.

18._____   Prokaryotes have many chromosomes per cell.

19._____   The somatic or body cells of a human are haploid.

20._____   A zygote is diploid.

21._____   Polyploid organisms contain more than two copies of each chromosome type.

22._____   DNA replication occurs during the G_2 portion of interphase.

23._____   Cell plates form during metaphase of the cell cycle.

24._____   Diploid cells produce haploid cells by the process of meiosis.

25._____   Crossing over occurs during meiosis II.

26._____   In the flowering plants, the diploid stage of the life cycle is greatly reduced.

27._____   Reduction of chromosome number occurs during <u>meiosis I</u>.
28._____   DNA <u>replicates</u> between meiosis I and meiosis II.
29._____   <u>Meiosis II</u> resembles mitosis.
30._____   Sister chromatids become daughter chromosomes during <u>anaphase I</u> of meiosis.

Matching: Mitosis and meiosis.

31.____   production of haploid cells from haploid   Choices:
         cells   a. mitosis
32.____   the mechanism by which unicellular   b. meiosis
         organisms reproduce   c. both of these
33.____   produces genetically variable cells   d. neither of these
34.____   produces sperm and egg cells in animals
35.____   allows multicellular organisms to grow
36.____   produces sperm and egg cells from
         haploid body cells in plants
37.____   ensures that each body cell gets a
         complete set of genes
38.____   can produce haploid cells
39.____   produces spores in plants and fungi
40.____   produces genetically identical cells
41.____   production of haploid cells from diploid
         ones
42.____   occurs in the human body
43.____   maintains the same number of
         chromosomes
44.____   doubles the number of chromosomes
45.____   reduces the number of chromosomes by
         half
46.____   chromosomes replicate once
47.____   cells divide twice

Matching: Interphase

48.____ phase in cells that will no longer divide
49.____ period after DNA synthesis occurs
50.____ period before DNA synthesis occurs
51.____ period of DNA replication or synthesis
52.____ spindle fiber proteins are made
53.____ period of most cell growth and
metabolic activity
54.____ nearly eliminated in embryonic cells
that divide rapidly

Choices:
a. S phase
b. G_0 phase
c. G_1 phase
d. G_2 phase

Matching: Stages of mitosis.

55.____ Chromosomes have reached the opposite
ends of the spindle
56.____ Chromosomes migrate to the cell's equator
57.____ Replicated chromosomes coil up and
condense
58.____ The centromeres divide
59.____ Daughter chromosomes move to the poles
60.____ The spindle breaks down
61.____ The nuclear envelope disintegrates
62.____ Sister chromatids become daughter
chromosomes
63.____ The spindle forms
64.____ Cytokinesis occurs

Choices:
a. telophase
b. anaphase
c. metaphase
d. prophase

Matching: Meiosis.

65.____ Individual chromosomes migrate to the
equator
66.____ Chiasmata form
67.____ Chromosomes each with 2 chromatids
move towards the poles
68.____ Centromeres divide
69.____ Homologous chromosomes pair up
70.____ Daughter chromosomes migrate towards
opposite poles
71.____ Homologous pairs of chromosomes move
together to the equator
72.____ Crossing over occurs
73.____ Homologous chromosomes move towards
opposite poles

Choices:
a. prophase I   e. prophase II
b. metaphase I   f. metaphase II
c. anaphase I   g. anaphase II
d. telophase I   h. telophase II

Multiple Choice: Pick the most correct choice for each question.

74. The genetic material in bacteria consists of
    a. several circular DNA molecules
    b. one circular RNA molecule
    c. many rod-like DNA molecules with protein
    d. one circular DNA molecule
    e. DNA in mitochondria

75. The daughter cells of binary fission are
    a. structurally identical
    b. chromosomally different
    c. genetically identical
    d. structurally similar and genetically identical
    e. not genetically the same as the parent cell

76. A region of attachment for two sister chromatids is the
    a. centriole
    b. centromere
    c. equator
    d. microtubule
    e. spindle fiber

77. Cell reproduction in prokaryotic cells <u>differs</u> from eukaryotic cells in that
    a. prokaryotic cells reproduce asexually but eukaryotic cells do not.
    b. each prokaryotic cell has a circular chromosome but the chromosomes of eukaryotic cells are linear.
    c. prokaryotic cells lack nuclei and do not replicate their DNA before dividing but eukaryotic cells have nuclei and replicate their DNA before dividing.
    d. prokaryotic chromosomes have DNA and protein but eukaryotic chromosomes are made of only DNA.
    e. they do not differ significantly.

78. Which of the following does not occur during prophase?
    a. the nuclear membrane disintegrates
    b. nucleoli break up
    c. the spindle apparatus forms
    d. the chromosomes condense
    e. DNA replicates

79. In sexually reproducing organisms, the source of chromosomes in the offspring is
    a. almost all from one parent, usually the father
    b. almost all from one parent, usually the mother
    c. half from the father and half from the mother
    d. a random mixing of chromosomes from both parents

80. Meiosis results in the production of
    a. diploid cells with unpaired chromosomes
    b. diploid cells with paired chromosomes
    c. haploid cells with unpaired chromosomes
    d. haploid cells with paired chromosomes
    e. none of the above choices is correct

81. A diploid cell contains six chromosomes. After meiosis I, each of the cells contains:
    a. three maternal and three paternal chromosomes each time
    b. a mixture of maternal and paternal chromosomes totaling three
    c. six maternal or six paternal chromosomes each time
    d. a mixture of maternal and paternal chromosomes totaling six
    e. three pairs of chromosomes

82. Which occurs in meiosis I but not in meiosis II?
    a. diploid daughter cells are produced
    b. chromosomes without chromatids line up at the equator
    c. centromeres divide
    d. pairing of homologous chromosomes occurs
    e. the spindle apparatus forms

CLUES TO APPLYING THE CONCEPTS

These practice questions are intended to sharpen your ability to apply critical thinking and analysis to biological concepts covered in this chapter.

83. Suppose you are working in a lab that grows various types of human body cells in petri dishes, a technique known as "cell culturing." While you are looking at one particular cell culture under the microscope, you decide to count up the numbers of cells you see in the various phases of the mitotic cell cycle. You count 45 cells in anaphase, 34 cells in prophase, 23 cells in telophase, 11 cells in metaphase, and 1008 cells in interphase. Could you use these numbers to determine the relative amounts of time this cell type spends in each phase of the mitotic cell cycle? Explain.

84. Suppose that the cloning process used to make the ewe named Dolly becomes widely used to clone human beings. What do you suppose the effect of widespread cloning might be on human populations?

ANSWERS TO EXERCISES

1.	chromosomes	22.	false, S	42.	c	63.	d
2.	chromatin	23.	false, cytokinesis	43.	a	64.	a
3.	binary fission	24.	true	44.	d	65.	f
4.	homologous	25.	false, meiosis I	45.	b	66.	a
5.	meiosis	26.	false, haploid	46.	c	67.	c
6.	gametes	27.	true	47.	b	68.	g
7.	interphase	28.	false, does not	48.	b	69.	a
8.	mitosis		replicate	49.	d	70.	g
9.	cytokinesis	29.	true	50.	c	71.	b
10.	asexual	30.	false, anaphase II	51.	a	72.	a
11.	sexual	31.	a	52.	d	73.	c
12.	zygote	32.	a	53.	c	74.	d
13.	meiosis	33.	b	54.	c	75.	c
14.	crossing over	34.	b	55.	a	76.	b
15.	chiasmata	35.	a	56.	c	77.	b
16.	genetic recombination	36.	a	57.	d	78.	e
17.	false, identical	37.	a	58.	b	79.	c
18.	false, one	38.	c	59.	b	80.	c
19.	false, diploid	39.	b	60.	a	81.	b
20.	true	40.	a	61.	d	82.	d
21.	true	41.	b	62.	b		

83. If the cells are dividing randomly in the culture dish, you can relate the percentage of time the cells spend in each phase of the cell cycle with the percentage of cells observed at each stage of the cell cycle. For instance, if cells spend 50% of the time in interphase, you would expect to find about half of the actively dividing cells in interphase at any particular time. So, if you saw 1121 cells and 34 (34/1121 = 3.0%) were in prophase, 11 (1.0%) were in metaphase, 45 (4.0%) were in anaphase, 23 (2.1%) were in telophase, and 1008 (89.9%) were in interphase, the percentages would indicate the relative amounts of time the cells spend in each phase of the cell cycle. If the average time for a complete mitotic division is known for the cells growing in the culture, these percentages could be used to determine the amounts of time cells spend in each phase. For instance, if it takes 24 hours (1140 minutes) for a complete cell cycle, these cells spend about 34 minutes (3% of 1140) in prophase, 11 minutes in metaphase, 46 minutes in anaphase, 24 minutes in telophase, and 1025 minutes in interphase.

84. One effect might be a reduction in genetic variation in human populations as fewer types of people produce more and more copies of themselves. There is a danger of eugenic manipulation of human traits, since those with "superior" traits might be cloned and those with "inferior" traits might be prohibited from being cloned (or even from reproducing at all). Another effect might be a skewing of sex ratios as more and more males are cloned and fewer and fewer females, especially in countries like China where males are considered more "valuable" to society then females. A much higher male sex ratio could, in turn, lead to other societal changes such as legalized prostitution.

Chapter 12: Patterns of Inheritance

OVERVIEW

This chapter presents Mendel's concepts of inheritance: the segregation and independent assortment of genes, the dominance and recessiveness of different alleles, and the randomness of fertilization. The authors cover the relationship of genes to chromosomes, especially sex-linkage, sex determination, linkage, and crossing over. They also present variations on the Mendelian theme, including mutation, incomplete dominance, multiple alleles, codominance, polygenic inheritance, gene interactions, and environmental influences. Chromosomal nondisjunction as a cause of humans with abnormal numbers of chromosomes is discussed, and the more common results of nondisjunction (XO, XXX, XXY, XYY, and Down syndrome) are presented.

1) How Did Gregor Mendel Lay the Foundation for Modern Genetics?

A gene's specific location on a chromosome is its **locus**. Homologous chromosomes carry similar genes at similar loci. Slightly different DNA nucleotide sequences at the same gene locus on two homologous chromosomes produce alternate forms of the gene, called **alleles**. If both homologous chromosomes in an organism have identical alleles, the organism is **homozygous**; if the alleles are different, the organism is **heterozygous**.

Mendel was successful because he: (1) chose the right organism to work with (garden peas which reproduce by **self-fertilization** and are thus **true-breeding** or homozygous and not by **cross-fertilization**); (2) designed and performed the experiment correctly, choosing easy to distinguish traits and counting up all the offspring from each mating; and (3) analyzed the data properly, using simple statistics such as determining the ratios of offspring with differing traits.

2) How Are Single Traits Inherited?

The inheritance of dominant and recessive alleles on homologous chromosomes can explain the results of Mendel's crosses. Example: purple flowers (*PP*) crossed with white flowers (*pp*) → purple flowers (*Pp*) → 75% purple flowers (*PP*) and (*Pp*) and 25% white flowers (*pp*), a 3:1 ratio. His hypothesis in modern terms is: (1) each trait is determined by a pair of genes, found at corresponding loci in homologous chromosomes; (2) the members of a gene pair separate from each other during meiosis so that each gamete gets one copy (**law of segregation**); (3) which member of a pair gets into which gamete, relative to other pairs, is a matter of chance (**law of independent assortment**); and, (4) in heterozygous individuals, the **dominant allele** may mask the expression of the **recessive allele** without changing its structure. The actual combination of alleles carried by an organism is its **genotype** (*PP*, *Pp*, and *pp*), while the organism's traits (things that can be seen or measured) are its **phenotype** (purple or white flower color).

Mendel's hypothesis can be used to predict the outcome of new types of single-gene crosses. For instance, cross-fertilization of an individual with dominant phenotype but unknown genotype (purple flower with either *PP* or *Pp* genotype) with a homozygous recessive individual (white flower = *pp*) is called a **test cross**, and can be used to determine the exact genotype of the dominant parent (*Pp* if recessive offspring occur, *PP* if no recessive offspring occur).

3) How Are Multiple Traits on Different Chromosomes Inherited?

Multiple traits may be controlled by genes on different chromosomes,. Such traits are inherited independently of each other (the **law of independent assortment**). Example: smooth yellow seeds (*SSYY*) x wrinkled green seeds (*ssyy*) → smooth yellow → 9/16 smooth yellow (*S-Y-*) + 3/16 smooth green (*S-yy*) + 3/16 wrinkled yellow (*ssY-*) + 1/16 wrinkled green (*ssyy*), a 9:3:3:1 ratio. Genes for seed color and seed shape are inherited independently of each other (move independently of each other during meiosis) since they are located in different (non-homologous) chromosomes, which align themselves randomly during meiotic metaphase I.

4) How Are Genes Located on the Same Chromosome Inherited?

Chromosomes contain many hundreds of gene loci. Genes on the same chromosome tend to be inherited together, a situation called **linkage**. However, **crossing over** may separate linked genes, since exchange of corresponding segments of DNA forms new gene combinations on both homologous chromosomes. This **genetic recombination** is one way that genetic variability occurs in gametes.

5) How Is Sex Determined, and How Are Sex-Linked Genes Inherited?

Females have two identical **sex chromosomes**, called X chromosomes, whereas males have one X and one Y chromosome. All other chromosomes are called **autosomes**. The sex chromosomes carried by males determine the sex of offspring (X from egg + X from sperm = daughter, X from egg + Y from sperm = son). **Sex-linked genes** are found on one sex chromosome (X or Y) but not on the other. Males express any genes found on the X chromosome, and each male inherits his X chromosome from his mother. So, half of the sons of a mother heterozygous for a sex-linked recessive condition (like hemophilia or colorblindness) will be affected.

6) What Are Some Variations on the Mendelian Theme?

In incomplete dominance, heterozygotes have a phenotype intermediate between the homozygotes. In snapdragons, red flowers (*RR*) crossed with white flowers (*R'R'*) yield pink heterozygotes (*RR'*), which yield when self-fertilized 1/4 red + 2/4 pink + 1/4 white. **Incomplete dominance** occurs when the degree of color depends on the number of active genes present. In snapdragons, *R* makes 50 units of red pigment and *R'* makes 0 units, so *RR* makes 100 units of red in the homozygote, *RR'* makes 50 units of red pigment in the pink heterozygote, and *R'R'* flowers are white.

There may be multiple (more than two) alleles of a gene present in a population of individuals. An individual diploid can have at most two different alleles for a given gene, even though more than two alleleic forms of the gene (**multiple alleles**) are present in the population. For example. in the ABO blood type alleles in humans there are three alleles (*A* and *B* are **codominant** [both are expressed in heterozygotes] and *O* is recessive) and six possible genotypes (type A = *AA* and *AO*, type B =*BB* and *BO*, type AB = *AB*, and type O = *OO*).

Many traits (like height, skin color, and eye color) are influenced by several genes. Some traits show a continuous variation of phenotypes in a population and cannot be split up into convenient, easily defined categories. These show **polygenic inheritance** since the effects of two or more pairs of functionally similar genes add up to produce the phenotype. For instance, as the amount of melanin pigment increases in irises, eye color darkens from blue to green to brown to black. The more genes that contribute to a trait, the greater the number of phenotypes and the finer the distinction between them.

Single genes typically have multiple effects on phenotype, a situation called **pleiotropy**. For example, the SRY (sex-determining region of the Y chromosome) gene produces the many anatomical and hormonal differences between males and females.

The environment influences the expression of genes. The phenotype of an organism is influenced by both its genotype and its environment. For example, the cooler body areas of a himalayan rabbit produce hair pigment while the warmer regions do not; human height and weight are products of genes and nutrition; human skin color is a product of genes and degree of sun exposure; and human IQ is a product of genes and educational environment.

7) How Are Human Disorders Caused by Single Genes Inherited?

Most human genetic disorders are caused by homozygous recessive alleles since the conditions occur because the recessive genes do not generate enough enzymes. Albinism results from a defect in melanin production due to homozygosity for recessive alleles. **Sickle-cell anemia** is caused by homozygosity for a defective hemoglobin synthesis allele. A few human genetic disorders, like **Huntington disease**, are caused by dominant alleles. Some human disorders, like red-green color blindness and **hemophilia**, are sex-linked.

8) How Do Errors in Chromosome Number Affect Humans?

Nondisjunction, defined as errors involving chromosome distribution during meiosis, produce gametes with one (or a few) extra or missing sex chromosomes or autosomes. Most embryos formed from such gametes miscarry, but a few types survive. Some genetic disorders among liveborn humans are caused by abnormal numbers of sex chromosomes. These include: (1) **Turner syndrome** (XO females); (2) **Trisomy X** (XXX females); (3) **Klinefelter syndrome** (XXY males); and (4) **XYY males**. Some genetic disorders among liveborn humans are caused by abnormal numbers of autosomes, the most common of which is **trisomy 21 (Down syndrome)**.

KEY TERMS AND CONCEPTS

Fill-In: From the following list of key terms, fill in the blanks in the following statements.

allele	gene	locus
carrier	genotype	nondisjunction
crossing over	heterozygote	phenotype
Down syndrome	homozygote	recessive
dominant		

One of several alterative forms of a particular gene is an (1)_____.

(2)_____ is a genetic disorder caused by the presence of three copies of chromosome 21.

An organism carrying two copies of the same allele of a gene is a (3)_____, while an organism carrying two different alleles of a gene is a (4)_____.

A unit of heredity containing the information for a particular characteristic is called a (5)_____.

(6)_____ is an error in meiosis in which chromosomes fail to segregate properly into the daughter cells.

The genetic composition of an organism is its (7)_____, while the physical properties of an organism are called its (8)_____.

A (9)_____ is an individual heterozygous for a recessive condition.

The physical location of a gene on a chromosome is its (10)_____.

The exchange of corresponding segments of the chromatids of two homologous chromosomes during meiosis is called (11)_____.

A (12)_____ allele determines the phenotype of heterozygotes completely, whereas a (13)_____ allele is expressed only in homozygotes.

Key Terms and Definitions

allele: one of several alterative forms of a particular gene.

autosome: a chromosome found in homologous pairs in both males and females, and which does not bear the genes determining sex.

carrier: an individual who is heterozygous for a recessive condition. Carriers display the dominant phenotype but can pass on their recessive allele to their offspring.

codominance: the relation between two alleles of a gene, such that both alleles are phenotypically expressed in heterozygous individuals.

cross-fertilization: union of sperm and egg from two different individuals of the same species.

crossing over: the exchange of corresponding segments of the chromatids of two homologous chromosomes during meiosis.

dominant: an allele that can determine the phenotype of heterozygotes completely, so that they are indistinguishable from individuals homozygous for the allele. In the heterozygotes, the expression of the recessive allele is completely masked.

Down syndrome: a genetic disorder caused by the presence of three copies of chromosome 21. Common characteristics include mental retardation, abnormally shaped eyelids, a small mouth with protruding tongue, short fingers, heart defects, and unusual susceptibility to infectious diseases.

gene: a unit of heredity containing the information for a particular characteristic. A gene is a segment of DNA located at a particular place on a chromosome.

genetic recombination: the recombining of alleles on homologous chromosomes, due to exchange of DNA during crossing over.

genotype: the genetic composition of an organism; the actual alleles of each gene carried by the organism.

hemophilia: a recessive, sex-linked disease in which the blood fails to clot normally.

heterozygous: an organism carrying two different alleles of the gene in question; sometimes called a hybrid.

homozygous: an organism carrying two copies of the same allele of the gene in question; also called a true-breeding organism.

Huntington disease: a dominant genetic condition causing degeneration of the central nervous system with symptoms typically beginning around age 40 or later.

hybrid: an organism that is the offspring of parents differing in at least one genetically determined characteristic; also used to refer to the offspring of parents of different species.

incomplete dominance: a pattern of inheritance in which heterozygotes have a phenotype intermediate between those of the two homozygotes.

inheritance: the genetic transmission of characteristics from parents to offspring.

Klinefelter syndrome: a set of characteristics typically found in individuals who have two X chromosomes and one Y chromosome. These individuals are phenotypically males, but sterile, and have several female-like traits, including narrow shoulders, broad hips, and partial breast development.

law of independent assortment: a pattern of inheritance of multiple traits, in which the distribution of alleles for one trait into the gametes does not affect the distribution of alleles for other traits. Occurs with genes that are located on different chromosomes.

law of segregation: a principle of inheritance, that the two allele of a given gene separate from each other during gamete formation, with the result that each gamete receives one allele; occurs because of the separation of homologous chromosomes during meiosis.

linkage: the inheritance of certain genes as a group because they are parts of the same chromosome. Linked genes do not show independent assortment.

locus: the physical location of a gene on a chromosome.

multiple alleles: more than two forms of a gene existing within the individuals of a species.

nondisjunction: an error in meiosis in which chromosomes fail to segregate properly into the daughter cells.

pedigree: a diagram showing genetic relationships among a set of individuals, usually with respect to a specific genetic trait.

phenotype: the physical properties of an organism. Phenotype can be defined as outward appearance (e.g., flower color), as behavior, or in molecular terms (e.g., ABO glycoproteins on red blood cells).

pleiotropy: a situation in which a single gene influences more than one phenotypic characteristic.

polygenic inheritance: a pattern of inheritance in which the interactions of two or more genes determine phenotype.
Punnett square method: an intuitive way to predict the

genotypes and phenotypes of offspring.

recessive: an allele expressed only in homozygotes, and which is completely masked in heterozygotes.

self-fertilization: union of sperm and egg from the same individual.

sex chromosome: one of the pair of chromosomes that differ between the sexes and usually determine the sex of an individual; e.g., human females have similar sex chromosomes (XX) while males have dissimilar ones (XY).

sex linked: a pattern of inheritance characteristic of genes located on one type of sex chromosome (e.g., X) and not found on the other type (e.g., Y).

sickle-cell anemia: a recessive disease caused by a single amino acid substitution in the hemoglobin molecule. Sickle-cell hemoglobin molecules tend to cluster together, distorting the red blood cell shape and causing them to break and clog the capillaries.

test cross: a breeding experiment in which an individual showing the dominant phenotype is mated with an individual that is homozygous recessive for the same gene. The ratio of offspring with dominant versus recessive phenotypes can be used to determine the genotype of the phenotypically dominant individual.

trisomy 21: see *Down syndrome*.

trisomy X: a condition of females who have three X chromosomes instead of the normal two. Most of these women are phenotypically normal and are fertile.

true-breeding: pertaining to an individual all of whose offspring produced through self-fertilization are identical to the parental type. True-breeding individuals are homozygous for the trait in question.

Turner syndrome: a set of characteristics typical of a woman with only one X chromosome. These women are sterile, with a tendency to be very short and to lack normal female secondary sexual characteristics.

THINKING THROUGH THE CONCEPTS

True or False: Determine if the statement given is true or false. If it is false, change the underlined word so that the statement reads true.

14._____   Each trait is determined by a pair of discrete units called underlined chromosomes.

15._____   Alternate forms of a gene are called chromatids.

16._____   In a heterozygote, the gene that is not expressed is recessive.

17._____   An *AabbDdEeGg* individual will produce 16 different types of gametes.

18._____ Linkage modifies Mendel's law of <u>segregation</u>.
19._____ New combinations of traits controlled by linked genes occur due to <u>crossing over</u>.
20._____ When red snapdragons are crossed with white snapdragons to produce pink snapdragons, this is an example of <u>polygenic inheritance</u>..
21._____ Human <u>females</u> determine the sex of their children.
22._____ Males are <u>haploid</u> for sex-linked genes.
23._____ A son inherits sex-linked traits from his <u>father</u>.
24._____ A person with Turner syndrome is <u>XXY</u>.
25._____ A person with Klinefelter syndrome is a <u>male</u>.
26._____ People who have Turner syndrome have <u>two</u> Barr bodies.

Matching: Chromosome anomalies in humans.

27.____ have an abnormal number of autosomes
28.____ sterile males with some breast development
29.____ females with three X chromosomes per nucleus
30.____ may be male or female
31.____ fertile females with normal phenotypes
32.____ short sterile females.
33.____ have a normal number of sex chromosomes
34.____ much more common among the babies of older mothers
35.____ males with more than one X chromosome
36.____ sterile females with less than two X chromosomes
37.____ have a possible predisposition towards violence
38.____ the most common chromosome anomaly among newborns
39.____ body cells have 45 chromosomes
40.____ trisomy 21
41.____ have 46 chromosomes in their body cells

Choices:
a. Turner syndrome
b. Klinefelter syndrome
c. trisomy X syndrome
d. XYY syndrome
e. Down syndrome
f. all of the above
g. none of the above

Multiple Choice: Pick the most correct choice for each question.

42. Which of these is not a genetic disorder?
    a. sickle-cell anemia
    b. hemophilia
    c. albinism
    d. Huntington disease
    e. malaria

43. Traits controlled by sex-linked recessive genes are expressed more often in males because
    a. Males inherit these genes from their fathers.
    b. Males are always homozygous.
    c. All male offspring of a female carrier get the gene.
    d. The male has only one gene for the trait.

44. Which of the following could be detected by counting up the number of chromosomes in a cell of the affected person?
    a. hemophilia
    b. albinism
    c. Huntington disease
    d. color-blindness
    e. trisomy 21

45. Each normal human possesses in his or her body cells
    a. 2 pairs of sex chromosomes and 46 pairs of autosomes
    b. 2 pairs of sex chromosomes and 23 pairs of autosomes
    c. 1 pair of sex chromosomes and 46 pairs of autosomes
    d. 1 pair of sex chromosomes and 23 pairs of autosomes
    e. 1 pair of sex chromosomes and 22 pairs of autosomes

46. A colorblind woman marries a noncolorblind man. Which of the following is true of their children?
    a. All will be colorblind.

    b. All daughters will be normal and all sons will be carriers.
    c. All daughters will be colorblind and all sons will be normal.
    d. All daughters will be heterozygous and all sons will be colorblind.
    e. It is impossible to predict with any reasonable degree of certainty

47. A recessive allele on the X chromosome causes colorblindness. A noncolorblind woman (whose father is colorblind) marries a colorblind man. What is the chance their son will be colorblind?
    a. 0
    b. 25%
    c. 50%
    d. 75%
    e. 100%

48. Hemophilia is an X-linked recessive gene causing a blood disorder. What are the chances that the daughter of a normal man and a heterozygous woman will have hemophilia?
    a. 0
    b. 25%
    c. 50%
    d. 75%
    e. 100%

49. A colorblind boy has a noncolorblind mother and a colorblind father. From which parent did he get the colorblind gene?
    a. father
    b. mother
    c. either parent could have given him the gene

50. A man who carries a harmful X-linked gene will pass the gene on to
    a. all of his daughters
    b. half of his daughters
    c. half of his sons
    d. all of his sons
    e. all of his children

Short Answer Genetics Problems: Monohybrid Crosses.

51. When two plants with red flowers are mated together, the offspring always are red, but if two purple-flowered plants are mated together, sometimes some of the offspring have red flowers. Which flower color is dominant?

52. In sheep, white (B) is dominant to black (b). Give the F_2 phenotypic and genotypic ratios resulting from the cross of a pure-breeding white ram with a pure-breeding black ewe.

53. If you found a white sheep and wanted to determine its genotype, what color animal would you cross it to and why?

54. Squash may be either white or yellow. However, for a squash to be white, at least one of its parents must also be white. Which color is dominant?

55. In peas, yellow seed color is dominant to green. Give the expected proportion of each color in the offspring of the following crosses: a heterozygous yellow with a heterozygous yellow; a heterozygous yellow with a green; and a green with a green.

56. If tall (D) is dominant to dwarf (d), give the genotypes of the parents that produce 3/4 tall plants and 1/4 dwarf plants among their progeny.

Short Answer Genetics Problems: Dihybrid Crosses.

57. In pigs, mule hoof (fused hoof) is dominant (C) while cloven foot is recessive (c). Belted coat pattern (S) is dominant to solid color (s). Give the F_2 genotype and phenotype ratios expected from the cross $CCSS$ x $ccss$.

58. In the F_2 generation of the previous question, what proportion of the cloven-hoofed, belted pigs would be homozygous?

59. Flat tail (F) is dominant to fuzzy tail (f), and toothed (T) is dominant to toothless (t). Give the results of a cross between two completely heterozygous parents.

60. In rabbits, black (B) is dominant to brown (b), and spotted coat (S) is dominant to solid coat (s). Give the genotypes of the parents if a black, spotted male is crossed with a brown, solid female and all the offspring are black and spotted.

61. In the preceding problem, give the genotypes of the parents if some of the offspring were brown and spotted.

62. In cattle, having horns (p) is recessive to hornless or polled (P). Coat color is controlled by incompletely dominant genes RR for red, rr for white, and Rr for roan. If two heterozygous roan-polled cattle are mated, what kinds of offspring are expected?

63. If a yellow guinea pig is crossed with a white one, the offspring are cream-colored. What is the simplest explanation for this result? What kinds of offspring are expected if two cream-colored guinea pigs mate?

64. In carnations, red or white phenotypes are dependent on homozygous genotypes, while the heterozygotes are pink. Give the F_1 and F_2 genotypic and genotypic ratios expected from a cross: red x white.

Short Answer Genetics Problems: Sex-Linked.

65. A normal woman whose father was a hemophiliac marries a normal man. What are the chances of hemophilia occurring in their children?

66. Another woman with no history of hemophilia in her family marries a normal man whose father was a hemophiliac. What are the chances of hemophilia occurring in their children?

67. Colorblindness (c) is a sex-linked recessive trait, while normal color vision (C) is dominant.

    a.   If two normal-visioned parents have a colorblind son, what are the parents' genotypes?
    b.   What are the chances that their daughter will be colorblind?

68. In cats, yellow is due to gene B, and black to its allele b. These genes are sex-linked. The heterozygous condition results in tortoiseshell. What kinds of offspring (sex and color) are expected from a cross of a black male with tortoise-shell female?

69. In fruit flies, normal long wings are dominant (V) and vestigial (shortened) wings are recessive (v). These genes are autosomal. The sex-linked gene controlling red eye color (W) is dominant to white eyes (w). A male with red eyes and normal wings mates a white-eyed vestigial-winged female. Give the expected ratio of phenotypes in the F_2 generation.

Short Answer Genetics Problems: Gene Interactions.

70. In poultry, there are two independently assorting gene loci, each with two alleles that affect the shape of a chicken's comb. One locus has a dominant allele (R) for rose comb while its recessive allele (r) produces single comb. The other locus has a dominant gene (P) for pea comb while its recessive allele (p) also produces single combs. When the two dominant genes occur together (R-P-), a walnut comb is produced. So, R-P- = walnut, R-pp = rose, rrP- = pea, and $rrpp$ = single. Give the expected phenotypic ratios of offspring from the following matings:

    a.   *RRPP* x *rrpp*   d.   *RrPP* x *RrPp*
    b.   *RrPp* x *rrpp*   e.   *rrPp* x *RrPP*
    c.   *Rrpp* x *rrPp*

71. In humans, deafness can be the result of a recessive allele affecting the middle ear (dd = deaf), or another recessive allele (ee = deaf) that affects the inner ear. Suppose two deaf parents have a child that can hear. Give the genotypes of all three individuals.

72. If two hearing people, heterozygous at both loci (*DdEe*) for deafness marry, what are the chances that their first child would be normal hearing? What is the chance of deafness in this child?

Short Answer Genetics Problems: Multiple Alleles.

73. Mallard ducks show a multiple allele pattern of inheritance in which M^R produces "restricted mallard" coloring and is dominant over M for mallard coloring, and both of these alleles are dominant over m for "dusky mallard" coloring. Give the phenotypic ratios expected among offspring from the following crosses:

    a. $M^R M \times M^R m$ b. $M^R M \times Mm$ c. $Mm \times mm$

Short Answer Genetics Problems: Multiple Genes.

74. If there are two pairs of genes involved in producing skin color in black x white crosses, and if the five phenotypic classes are black, dark, medium (mulatto), light, and white, give the expected F_2 results of a white x black mating.

75. Give the darkest phenotype possible among the offspring of the following matings:

    a. black x dark   e. dark x white
    b. black x medium   f. medium x light
    c. black x white   g. light x light
    d. dark x medium   h. light x white

CLUES TO APPLYING THE CONCEPTS

This practice question is intended to sharpen your ability to apply critical thinking and analysis to biological concepts covered in this chapter.

76. Occasionally, a family occurs in which both parents have recessive albinism but all of their children have normal amounts of skin pigmentation. Propose a genetic explanation for the inheritance of albinism in these families.

ANSWERS TO EXERCISES

1. allele
2. Down syndrome
3. homozygote
4. heterozygote
5. gene
6. nondisjunction
7. genotype
8. phenotype
9. carrier
10. locus
11. crossing over
12. dominant
13. recessive
14. false, genes
15. false, alleles
16. true
17. true
18. false, independent

    assortment
19. true
20. false, incomplete
    dominance
21. false, males
22. true
23. false, mother
24. false, XO
25. true
26. false, no
27. e
28. b
29. c
30. e
31. c
32. a
33. e

34. e
35. b
36. a
37. d
38. e
39. a
40. e
41. g
42. e
43. d
44. e
45. e
46. d
47. c
48. a
49. b
50. a

51. Purple is dominant.
52. Genotypic ratio: 1 *BB*: 2 *Bb*: 1 *bb*; Phenotypic ratio: 3 white: 1 black
53. A test cross with a black (*bb*) sheep
54. White is dominant.
55. a.   3/4 yellow: 1/4 green;
    b.   ½ yellow: ½ green;
    c.   all green
56. *Dd* x *Dd*
57. 9 *C-S-* mule-foot, belted pigs: 3 *C-ss* mule-foot, solid pigs: 3 *ccS-* cloven-foot, belted pigs: 1 *ccss* cloven-foot, solid pig
58. There are 3 cloven-foot, belted pigs, but only one is homozygous (*ccSS*); the other two are heterozygous (*ccSs*).
59. 9 *F-T*-flat and toothed: 3 *F-tt* flat and toothless: 3 *ffT-* fuzzy and toothed: 1 *fftt* fuzzy and toothless
60. black spotted male (*BBSS*) mated with brown solid female (*bbss*)
61. *BbSS* male mated with *bbss* female
62. 3 *P-RR* polled red: 6 P-Rr polled roan:
    3 P-rr polled white: 1 ppRR horned red:
    2 *ppRr* horned: roan: 1 *pprr* horned white
63. a.   Incomplete dominance with cream being heterozygous
    b.   1 yellow: 2 cream: 1 white
64. F$_1$: *Rr* pink;
    F$_2$: 1/4 *rr* red: 2/4 *Rr* pink: 1/4 *rr* white
65. Half of the sons will inherit the defective gene from mom and get Y from dad. Half of the daughters will also inherit the defective gene from mom but they will be heterozygous since they also inherit a normal gene from dad.
66. None is expected to have hemophilia or inherit the gene.

67. a.   *Cc* (normal but carrier woman) x *CY* (normal man) [$X^C X^c$ x $X^C Y$]
    b.   No chance: she inherits a normal C gene from dad as well as either C or c a gene from mom.
68. Black female, tortoiseshell female, black male, yellow male
69. F_2 for both sexes: 3/8 red long, 3/8 white long, 1/8 red vestigial, 1/8 white vestigial
70. a.   all walnut
    b.   1/4 walnut: 1/4 rose: 1/4 pea: 1/4 single
    c.   1/4 walnut: 1/4 rose: 1/4 pea: 1/4 single
    d.   3/4 walnut: 1/4 pea
    e.   ½ walnut: ½ pea
71. *DDee* x *ddEE* → *DdEe*
72. 9/16 *D-E-* normal: 3/16 *D-ee* deaf:
    3/16 *ddE-* deaf: 1/16 *ddee* deaf;
    so, 9/16 normal: 7/16 deaf
73. a.   3/4 restricted mallard: 1/4 mallard
    b.   ½ restricted mallard: ½ mallard
    c.   ½ mallard: ½ dusky mallard.
74. F_2: 1/16 white: 4/16 light: 6/16 medium: 4/16 dark: 1/16 black
75. a.   black
    b.   black
    c.   medium
    d.   black
    e.   medium
    f.   dark
    g.   medium
    h.   light

76. Like recessive deafness, there are several different genetic situations leading to albinism. In one type of albinism, the gene necessary for making an enzyme needed to convert a substrate into melanin pigment is missing (genotype *aa*), while in another type of albinism the gene necessary for transporting the substrate for melanin into the pigment-producing cells is lacking (genotype *bb*). So, both *aaBB* and *AAbb* people would have albinism. However, if *aaBB* married *AAbb*, all their children would be *AaBb* and would have normal pigmentation because the children would make the enzyme necessary to produce melanin and the enzyme necessary to transport the substrate for melanin into the pigment cells.

Chapter 13: Biotechnology

OVERVIEW

This chapter begins by describing natural examples of genetic recombination and then discusses several techniques used in recombinant DNA technology, including building DNA libraries, identifying and making copies of genes, and using the genes to modify living organisms. The methods used to locate and sequence genes also are covered.

1) What Is Bioechnology?

Biotechnology is any use or alteration of organisms, cells, or molecules for practical purposes. Common examples are the domestication of plants and animals. Modern technology uses **genetic engineering** (modification of DNA) to achieve specific goals to: (1) understand more about inheritance and gene expression; (2) better treat various genetic diseases; and (3) generate economic benefits, including better agricultural organisms and valuable biological molecules. One important tool is **recombinant DNA** (DNA altered by the incorporation of genes from other organisms) transferred into animals or plants using **vectors** (carriers such as bacteria or viruses) to make **transgenic** organisms.

2) How Does DNA Recombination Occur in Nature and in the Laboratory?

DNA recombination occurs naturally through processes such as sexual reproduction, bacterial **transformation** (gene transfer between different bacteria as they pick up either free DNA from the environment or tiny circular DNA **plasmids** that often carry genes for antibiotic resistance), and viral infections, which may transfer DNA between bacteria (via viruses called **bacteriophages**) or between eukaryotic species.

3) What Are Some of the Methods of Biotechnology?

A **DNA library** is produced by inserting the DNA from the entire **genome** of a selected organism into bacterial plasmids, making the DNA readily accessible and easy to duplicate. Researchers use **restriction enzymes**, produced by bacteria, to cut DNA at specific nucleotide sequences that often are palindromes (mirror image sequences like "madam"), producing small pieces of DNA with "sticky ends" that can hydrogen-bond with other pieces of DNA (from different species) with complementary sticky ends. Restriction enzymes are used to insert DNA into plasmids to build a DNA library. For example, using a restriction enzyme, bacterial plasmids and DNA from white blood cells (wbc) are cut open to produce fragments with sticky ends, the molecules are mixed together so that plasmids and wbc DNA hydrogen-bond at their sticky ends, and each recombinant plasmid is inserted into a separate bacterium; the resulting population of bacteria with recombinant plasmids constitute a human DNA library.

Researchers find genes of interest in a library by using **DNA probes** (sequences complementary to the gene) to identify bacteria in the library carrying plasmids with the gene of interest. Once the gene of interest has been located, it must be amplified. This can be done by either culturing the bacteria containing the gene to create multiple copies or using the **polymerase chain reaction (PCR)** that makes billions of copies of the gene using heat-stable DNA polymerase enzymes. The steps in PCR include:

(1) separating double-stranded DNA into single strands by heating to 90°C; (2) using specific primers to start the reaction at appropriate places; (3) using heat-resistant enzymes to construct new DNA copies; (4) cooling the mixture to stop the reaction; repeating step (1); etc. Each cycle takes a few minutes, so that billions of genes can be made in a single afternoon starting with a single copy.

4) What Are Some of the Applications of Biotechnology?

Amplified genes provide enough DNA to determine the exact base sequence of the gene. From the nucleotide sequence of the gene, the amino acid sequence of the protein it makes can be determined, which can lead to the determination of the protein's normal function in cells. This was done using the gene for cystic fibrosis. Information about a gene's base sequence can lead to rapid tests to determine the presence or absence of defective genes in fetuses (prenatal diagnosis), newborns, and adults.

DNA fingerprinting facilitates genetic detection. If DNA is cut into **restriction fragments** of different sizes, these can be separated by size by **gel electrophoresis** since smaller fragments move faster through a gel through which an electrical current is flowing. If the restriction fragments from a defective gene and a normal gene are cut into different-sized pieces (called **restriction fragment length polymorphism** or **RFLP**), comparing the fragments on a gel can determine which genes are present. RFLPs can be used to locate mutated genes that cause genetic disorders. Researchers collect DNA from a number of genetically related people. They look for a uniquely sized DNA fragment whose presence correlates with the presence of a particular disease. This was done for the genes causing cystic fibrosis and Huntington disease. RFLP analysis can then be used for the prenatal diagnosis of individuals carrying the gene.

Genetic engineering is rapidly replacing traditional breeding techniques to produce improved varieties of crops. Recombinant plants can be made using the bacterial Ti plasmid to carry in genes for cold tolerance or to make insecticides against insect pests. From clumps of cultured cells, individual cells can become **clones** (genetically identical organisms), each carrying the same inserted genes. "Gene guns" also are used to insert foreign DNA directly into plant seedlings that don't grow well in culture, although the success rates are low. The largest agricultural application of genetic engineering lies in using genes to improve resistance to various pests such as insects and weeds and to the chemicals used to kill pests. Genetic engineering may improve other qualities of food plants and animals. For example, the Flavr-Savr® tomato ripens on the vine but remains firm longer than typical tomatoes so that they can be harvested later (to taste better) and shipped without damage.

5) What Are Some Medical Uses of Biotechnology?

Knockout mice (with mutated genes similar to those in humans) provide models of human genetic diseases. Studies of such mice and their defects help researchers clarify the roles of human genes (like that for cystic fibrosis) and allow scientists to study the effects of different treatments for the disorder without risking human patients.

Genetic engineering allows the production of therapeutic proteins like insulin and human growth hormone (both grown in *E. coli* bacteria). Using transgenic "pharm" animals (into which human genes have been introduced using viral vectors or DNA injected into fertilized eggs), products such as alpha-1-antitrypsin (to treat emphysema) can be harvested from milk. The two potential types of human gene therapy are (1) replacing defective genes in body cells (being tested for cystic fibrosis and for SCID, using vectors such as liposomes and retroviruses); and (2) altering genes in fertilized eggs, thus permanently repairing the genetic defect. The Human Genome Project aims to sequence the entire human genome by 2005.

6) What Are Some Ethical Implications of Human Biotechnology?

Since 850 babies are born each year in the US with cystic fibrosis (CF), scientists have recommended that all couples be tested to find carriers of CF. Will unbiased and affordable genetic counseling be available for all those couples who are both heterozygous for CF? Should society bear the expenses of treating a child born with CF to parents who knew they were carriers? Should insurance companies be allowed to deny coverage to such couples? The increasing availability and use of tests for genetic diseases raise concerns about genetic discrimination by employers and insurance companies. The potential of cloning humans raises ethical issues. Are there valid reasons to clone or not to clone humans?

KEY TERMS AND CONCEPTS

From the information in this chapter and previous chapters, fill in the crossword puzzle based on the following clues.

Across

6. A procedure for sampling the fluid surrounding a fetus.
8. A method of producing large numbers of copies of a specific piece of DNA.
9. Type of nucleic acid containing uracil.
10. A procedure for sampling cells from the fetal chorionic villi.
11. A sequence of complementary nucleotides that can be used to identify a DNA segment that carries a gene.
12. A technique that separates DNA fragments on the basis of size.
14. The process of producing a new organism genetically identical to an existing organism.
16. Transgenic "_____" animals used to produce human gene products.
17. A virus specialized to infect bacteria.

Down

1. A bacterium, plasmid, or virus that carries DNA between different organisms.
2. A palindrome with the letters A, D, and M.
3. The use of DNA to identify an individual on the basis of a unique set of restriction fragment length polymorphisms.
4. A protein, normally isolated from bacteria, that cuts double-stranded DNA at a specific nucleotide sequence.
5. A readily accessible, easily duplicable, assemblage of all the DNA of a particular organism, normally cloned into bacterial plasmids.
7. Genetic material altered by the incorporation of genes from a different organism.
13. A small, circular piece of DNA found in the cytoplasm of many bacteria.
15. Differences in the lengths of fragments produced by cutting DNA from different individuals of the same species with the same restriction enzymes.

Key Terms and Definitions

amniocentesis: a procedure for sampling the amniotic fluid surrounding a fetus. A sterile needle is inserted through the abdominal wall, uterus, and amnion of a pregnant woman, into the amniotic fluid, and 10 to 20 milliliters of fluid are withdrawn. various tests may be performed on the fluid and the fetal cells suspended in it to provide information on the developmental and genetic state of the fetus.

bacteriophage: a virus specialized to infect bacteria; also called a phage.

biotechnology: the use or modification of organisms, cells, or biological molecules to achieve specific practical goals. Modern biotechnology commonly involves grenetic engineering.

chorionic villus sampling (CVS): a procedure for sampling cells from the chorionic villi produced by a fetus. A tube is inserted into the uterus of a pregnant woman, and a small sample of villi are suctioned off for genetic and biochemical analysis.

cloning: the process of producing a clone, a new organism genetically identical to an existing organism; also refers to the production of identical copies of a gene.

DNA fingerprinting: the use of DNA to identify an individual on the basis of a unique set of restriction fragment length polymorphisms.

DNA library: a readily accessible, easily duplicable, assemblage of all the DNA of a particular organism, normally cloned into bacterial plasmids.

DNA probe: a sequence of nucleotides, complementary to a specific gene, that can be used to identify a DNA segment that carries the gene.

gel electrophoresis: a technique that separates DNA fragments on the basis of size according to the different rates of migration through a gel through which an electric current is passing.

genetic engineering: the modification of genetic material to achieve specific goals.

genome: the entire set of genes carried by a member of any given species.

plasmid: a small, circular piece of DNA found in the cytoplasm of many bacteria; most do not carry genes required for the normal functioning of the bacterium, but they may carry genes that assist bacterial survival in certain environments, such as those for antibiotic resistance.

polymerase chain reaction (PCR): a method of producing virtually unlimited numbers of copies of a specific piece of DNA, starting with as little as one copy of the desired DNA.

recombinant DNA: genetic material that has been altered by the incorporation of genes from a different organism, typically from another species.

restriction enzyme: an enzyme, normally isolated from bacteria, that cuts double-stranded DNA at a specific nucleotide sequence; the cleavage sequence differs for different restriction enzymes.

restriction fragment: a piece of DNA that has been isolated by cutting up a larger piece of DNA with restriction enzymes.

restriction fragment length polymorphisms (RFLPs): differences in the lengths of restriction fragments, produced by cutting samples of DNA from different individuals of the same species with the same set of restriction enzymes; occur because individuals of the same species have different nucleotide sequences.

Tay-Sachs disease: a recessive disease caused by a deficiency of enzymes that regulate lipid breakdown in the brain.

transformation: the process by which bacteria pick up foreign DNA from other bacteria or viruses that may, in turn, be transmitted to their offspring.

transgenic: animals or plants that express DNA derived from another species.

vector: in the context of genetic engineering, a bacterium, plasmid, or virus that carries DNA between different organisms.

THINKING THROUGH THE CONCEPTS

True or False: Determine if the statement given is true or false. If it is false, change the underlined word so that the statement reads true.

18. _____   DNA recombination does not occur in nature.

19. _____   Sexual reproduction between humans is an example of DNA recombination in nature.

20. _____   Bacteria pick up free DNA and incorporate it into their chromosomes during sexual reproduction.

21. _____   A recessive condition caused by inability to break down fatty materials in nerve cells is called sickle cell anemia.

22. _____   A readily accessible, easy to duplicate collection of all the DNA of a particular organism is a DNA library.

23. _____   An example of a palindrome is the word madman.

24. _____   An enzyme that cuts palindromic DNA open to form sticky ends is called ligase enzyme.

25. _____   Bacterial DNA is protected from the action of restriction enzymes by methylation.

26. _____   A method for making many copies of a small amount of DNA is the restriction fragment length polymorphism reaction.

27._____   RFLPs are helpful in <u>gene mapping</u>.

Fill-In: Using the figure below, fill in the answers to the following questions.

28. Describe what is happening in step 1.

29. Describe what is happening in step 2.

30. Describe what is happening in step 3.

31. Describe what is happening in step 4.

Matching: Recombinant DNA.

32.____ accessory chromosomes in bacteria
33.____ defends bacteria against viral infection
by cutting apart the viral DNA
34.____ self-replicating tiny loops of DNA
35.____ used to identify marker genes in
chromosomes
36.____ readily accessible, easy to duplicate
collection of all the DNA of a
particular organism
37.____ bacteria protect themselves from this
by methylating their DNA
38.____ a bacterium may contain hundreds of
these "parasites"
39.____ can cut apart DNA to create single-
stranded ends
40.____ restriction fragment length
polymorphisms
41.____ cut at palindromic DNA sequences
42.____ often contains genes for antibiotic-
digesting enzymes
43.____ cuts up DNA into fragments of various
sizes that can be separated by gel
electrophoresis

Choices:
a. DNA library
b. restriction enzymes
c. plasmids
d. RFLPs

Multiple Choice: Pick the most correct choice
for each question.

44. Small accessory chromosomes found in
bacteria and useful in recombinant DNA
procedures are called
a. plasmids
b. palindromes
c. centrioles

45. Which of the following is not a goal of
biotechnology?
a. generating economic benefits
b. efficiently producing biologically
important molecules
c. improving agriculturally important
food plants
d. more effectively treating disease
e. creating humans with higher

intelligence levels

46. In biotechnology research, DNA fragments
created by restriction enzyme action are
separated from one another by
a. crossing over
b. gel electrophoresis
c. centrifugation
d. filtering
e. the polymerase chain reaction

47. The enzymes used to cut genes in
recombinant DNA research are called
a. DNA polymerases
b. RNA polymerases
c. spliceosomes
d. replicases
e. restriction enzymes

48. DNA recombinations controlled by scientists in the laboratory
    a. are random and undirected
    b. involve specific pieces of DNA moved between deliberately chosen organisms
    c. use natural selection to determine their usefulness
    d. are of little practical use to humans
    e. usually cause harmful mutations

49. Which of the following is a palindrome?
    a. <u>CCGTA</u>
       GGCAT

    b. <u>GAATTC</u>
       CTTAAG

    c. <u>CATTG</u>
       GTAAC

    d. <u>GGAATC</u>
       CCTTAG

    e. <u>AAAAA</u>
       TTTTT

50. The polymerase chain reaction (PCR) is useful in
    a. analyzing a person's fingerprints
    b. cutting DNA into many small pieces
    c. allowing restriction enzymes to cut DNA at palindromes
    d. creating recombinant plasmids
    e. making many copies of a small amount of DNA

CLUES TO APPLYING THE CONCEPTS

This practice question is intended to sharpen your ability to apply critical thinking and analysis to biological concepts covered in this chapter.

51. What do you think are some of the potential risks of releasing genetically engineered organisms into the environment, and what are some of the ethical issues raised by recombinant DNA technology?

ANSWERS TO EXERCISES

1. vector
2. MADAM
3. DNA fingerprinting
4. restriction enzyme
5. library
6. amniocentesis
7. recombinant DNA
8. PCR
9. RNA
10. CVS

11. DNA probe
12. electrophoresis
13. plasmid
14. cloning
15. RFLP
16. pharm
17. phage
18. false, does
19. true

20. false, transformation
21. false, Tay-Sachs disease
22. true
23. false, madam
24. false, restriction enzyme
25. true
26. false, polymerase chain reaction
27. true

28. A plasmid vector is removed from a bacterial cell and cut open with a restriction enzyme.

29. DNA from another organism is cut with the same restriction enzyme.

30. The plasmid DNA and a piece of DNA from the other organism are joined by complementary base pairing to form a recombinant molecule.

31. The recombinant plasmid is taken up by a host cell.

32. c
33. b
34. c
35. d
36. a

37. b
38. c
39. b
40. d
41. b

42. c
43. b
44. a
45. e
46. b

47. e
48. b
49. b
50. e

51. Recombinant organisms could compete with existing organisms and possible replace them in nature, leading to undesirable consequences. Genetic engineering could produce organisms that act as hazardous pathogens, infecting plants, animals, and/or humans to cause hard-to-treat diseases. Newly engineered organisms also could transfer genes to other species, harming them or causing them to become harmful. Some ethical questions that should be considered: Are the potential benefits worth the potential risks? Who decides what kinds of alterations in a species are acceptable? Should ecosystems be altered by introducing altered species? Do scientists or governments have the right to stop or forbid recombinant DNA research if there is a potential for it to solve medical or environmental problems? Who has the right to decide whether human genes should be altered and who should decide whose genes and which genes should be changed? What uses should be made of genetic information and who should decide what these are?

Chapter 14: Principles of Evolution

OVERVIEW

This chapter introduces the concept of evolution. The authors present an historical account of pre-Darwinian thought, followed by the Darwin-Wallace theory of evolution by natural selection. Finally, various proofs of evolution are given.

1) How Did Evolutionary Thought Evolve?

Pre-Darwinian science, heavily influenced by theology, held that all organisms were simultaneously created by God and that each distinct life form remained unchanged from the moment of creation. Plato said that each object on Earth is an imperfect reflection of an "ideal form" and Aristotle categorized all organisms into a linear hierarchy (the "ladder of creation"). These ideas went unchallenged for nearly 2000 years. However, as European explorers noted in the 1700s, the numbers of **species** (different kinds of organisms) in newly discovered lands was greater than expected and led to thoughts that similar species might have developed from a common ancestor.

    Fossils (preserved remains of organisms that lived long ago) found in rocks resembled parts of living organisms. The organization of fossils is consistent: (1) older fossils are found in rock layers beneath younger fossils; (2) the resemblance to modern forms of life gradually increased as increasingly younger fossils are examined, like a ladder of nature stretching back in time; and (3) many fossils are of species now extinct. Scientists concluded that different types of organisms have lived at various times in the past.

    Geology provided evidence that the Earth is exceedingly old. Biblical calculations suggest the Earth is 4000 to 6000 years old. The theory of **catastrophism** claims that successive catastrophes on Earth (like the biblical Great Flood) produced layers of rock and caused many species to become extinct in short time periods, perhaps with the creation of more species after each event. Actually, Earth is very old, having been formed from the forces of wind, water, earthquakes, and volcanoes in much the same way then as now (the theory of **uniformitarianism**). Modern geologists estimate the Earth to be about 4.5 billion years old.

    The French biologist Lamarck proposed in 1801 that organisms evolved through the **inheritance of acquired characteristics**: Through an innate drive for perfection (never scientifically demonstrated), living organisms can modify their bodies through the use or disuse of parts (actually true) and these modifications can be inherited by their offspring (actually false). Though Lamarck's theory was abandoned, by the mid-1800s biologists began to realize that the fossil record suggested that present-day species had evolved from preexisting ones. But how?

SUMMARY OF THE DARWIN-WALLACE THEORY OF EVOLUTION

Observation 1: All natural populations have the potential to increase geometrically in size due to reproductive abilities.

Observation 2: Most natural populations maintain a relatively constant size.

> **Conclusion 1**: Thus, many organisms must die young, producing few or no offspring each generation.

Observation 3: Individuals in a population differ in many abilities that affect survival and reproduction (some are "better adapted").

> **Conclusion 2**: The most well adapted organisms probably reproduce the most, since they survive the best. This differential reproduction is due to **natural selection**.

Observation 4: Some of the variation in adaptiveness among individuals is genetic and is passed on to the offspring.

> **Conclusion 3**: Over many generations, differential reproduction among individuals with different genotypes changes the overall frequencies of genes in populations, resulting in evolution.

> 1) Darwin did not know the mechanism of heredity (to explain Observation 4) and could not prove Conclusion 3.
> 2) In desperation, Darwin resorted to a version of Lamarck's inheritance of acquired characteristics and this nearly destroyed his entire theory.

---

2) How Do We Know That Evolution Has Occurred?

The fossil record provides evidence of evolutionary change over time. Giraffes, elephants, horses, and other types of organisms show a progressive series of fossils leading from ancient primitive organisms, through several intermediary stages, to the modern forms.

Comparative anatomy provides structural evidence of evolution. Through **convergent evolution**, unrelated species in similar environments evolve similar body functions from dissimilar underlying structures, called **analogous structures** (for example, wings of birds and butterflies). Also, closely related species in dissimilar environments evolve dissimilar body functions from similar underlying structures, called **homologous structures** for example, among mammals, the forelimbs of apes, seals, dogs, and bats). Some species of organisms have **vestigial structures** (structures with no apparent purpose), which are homologous to functional structures in other species.

Embryological stages of animals can provide evidence of common ancestry. All vertebrate embryos look similar to one another early in their development (all have gill slits and tails, even humans), indicating that all vertebrate species have similar genes.

Modern biochemical and genetic analyses reveal relatedness among diverse organisms. All cells have DNA, RNA, ribosomes, similar genetic codes, similar amino acids in proteins, and similar chromosome structures.

3) What Is the Evidence That Populations Evolve by Natural Selection?

Artificial selection (breeding domestic organisms such as dogs to produce specific desired features) demonstrates that organisms may be modified by controlled breeding. Also, evolution by natural selection occurs today, as illustrated by selection for dark-colored moths in industrially polluted areas and the evolution of populations of roaches in Florida for which the poison roach bait called "Combat ®" is ineffective.

KEY TERMS AND CONCEPTS

Fill-In: From the following list of key terms, fill in the blanks in the following statements.

analogous   homologous
artificial   inheritance of acquired characteristics
catastrophism   uniformitarianism
convergent   vestigial
fossils

(1)_____, the preserved remains of organisms that lived long ago, are often found in rocks and typically resemble parts of living organisms.

The theory of (2)_____ claims that successive worldwide events, like the biblical Great Flood, produced layers of rock and caused many species to become extinct in short periods of time.

According to the theory of (3)_____, Earth is very old, having been formed from the forces of wind, water, earthquakes, and volcanoes in much the same way then as now.

In 1801, the French biologist Lamarck proposed that organisms evolved through the (4)_____.

Through (5)_____ evolution, unrelated species in similar environments evolve similar body functions from dissimilar underlying structures, called (6)_____ structures.

Closely related species living in dissimilar environments, evolve dissimilar body functions from similar underlying structures, called (7)_____ structures.

Some species of organisms have (8)_____ structures with no apparent purpose, which are homologous to functional structures in other species.

The wing of a butterfly and the wing of a bird are (9)_____ structures, while the forelimb of a human and the forelimb of a whale are (10)_____ structures.

(11)_____ selection, the breeding of domestic organisms such as dogs to produce specific desired features, demonstrates that organisms may be modified by controlled breeding.

Key Terms and Definitions

analogous structures: structures that have similar functions and superficially similar appearance but very different anatomy, such as the wings of insects and birds. The similarities are due to similar selective pressures rather than common ancestry.

artificial selection: the breeding of domestic plants and animals to produce specific desirable features.

catastrophism: the hypothesis that Earth has experienced a series of geological catastrophes, probably imposed by a supernatural being, which accounts for the multitude of species, both extinct and modern, while preserving creationism.

convergent evolution: the independent evolution of similar structures among unrelated organisms, due to similar selective pressures. See *analogous structures*.

evolution: the descent of modern organisms from preexisting life forms; strictly speaking, any change in the proportions of different genotypes in a population from one generation to the next.

fossil: the remains of an organism, usually preserved in rock. Fossils include petrified bones or wood; shells; impressions of body forms such as feathers, skin, or leaves; and markings made by organisms such as footprints.

homologous structures: structures that may differ in

function but that have similar anatomy, presumably because of descent from common ancestors.

inheritance of acquired characteristics: the hypothesis that organisms' bodies change during their lifetimes by use and disuse, and that these changes are inherited by their offspring.

natural selection: the unequal survival and reproduction of organisms due to environmental forces (for example, physical factors, such as climate, and living organisms, such as predators or prey) that act differently upon genetically different members of a population.

population: a group of individuals of the same species, found in the same time and place, and actually or potentially interbreeding.

species: all of the populations of organisms that are potentially capable of interbreeding under natural conditions but that usually do not interbreed with members of other species.

uniformitarianism: the hypothesis that Earth developed gradually through natural processes similar to those at work today occurring over long periods of time.

vestigial structures: structures that serve no apparent purpose, but which are homologous to functional structures in related organisms and evidence of evolution.

THINKING THROUGH THE CONCEPTS

True or False: Determine if the statement given is true or false. If it is false, change the underlined word so that the statement reads true.

12. _____ Before Darwin, most people thought species were capable of change.
13. _____ The idea that God created some new species after every catastrophe was proposed by Agassiz.
14. _____ The remains of organisms preserved in rock are called vestigial structures.
15. _____ Lamarck proposed that an internal drive toward complexity within cells is the driving force in evolution.
16. _____ The similarity in the bones making up a bird's wing and a horse's foot are due to convergent evolution.
17. _____ Amino acid sequences in proteins of different animals tend to support evolution.
18. _____ Aristotle's "ladder of creation" was considered immutable.
19. _____ In convergent evolution, the two forms being modified are closely related.
20. _____ The many different varieties of dogs are the result of natural selection.
21. _____ Analogous structures arise due to convergent evolution.

Matching: Theories about life.

22. _____   uniformitarianism
23. _____   ladder of creation
24. _____   "ideal forms"
25. _____   inheritance of acquired
              characteristics
26. _____   catastrophism
27. _____   multiple creations
28. _____   natural selection
29. _____   Essay on Population
30. _____   coevolution

Choices:

a. Lamarck   f. Agassiz
b. Aristotle   g. Cuvier
c. Wallace   h. Lyell
d. Plato   i. Darwin
e. Malthus

Multiple Choice: Pick the most correct choice for each question.

31. Which of the following proposes that living organisms inherited body parts modified through use or disuse?
    a. natural selection
    b. catastrophism
    c. inheritance of acquired characteristics
    d. evolution

32. Fossils resulted from a successive series of geological upheavals according to the theory of
    a. natural selection
    b. catastrophism
    c. uniformitarianism
    d. independent assortment

33. The evolution of adaptations between different species as a result of extensive interactions with each other is termed
    a. analogous evolution
    b. divergent evolution
    c. convergent evolution
    d. coevolution

34. Which of the following is homologous to the human arm?
    a. wing of insect
    b. wing of bird
    c. body of snake
    d. fin of fish
    e. tail of salamander

35. Supportive evidence for evolution is found in studies of
    a. biochemistry
    b. embryos
    c. comparative anatomy
    d. domestication of plants and animals
    e. all of the answers are correct

36. Fossils provide direct evidence for
    a. behavioral adaptations
    b. physiological characteristics
    c. habitat preference
    d. structural similarities and differences
    e. catastrophism

CLUES TO APPLYING THE CONCEPTS

This practice question is intended to sharpen your ability to apply critical thinking and analysis to biological concepts covered in this chapter.

37. A species of moth has a very long proboscis (tubular mouth part) that is used to suck nectar from the inner base of a particular type of long, trumpet-shaped flower. Closely related moth species, however, have much shorter mouth parts and feed off nectar from plants with shorter tubular flowers. How would Darwin and Lamarck each explain the evolution of the species of moth with the exceptionally long proboscis?

ANSWERS TO EXERCISES

1. fossils
2. catastrophism
3. uniformitarianism
4. inheritance of acquired characteristics
5. convergent
6. analogous
7. homologous
8. vestigial
9. analogous

10. homologous
11. artificial
12. false, incapable of change
13. true
14. false, fossils
15. true
16. false, common ancestry
17. true
18. true

19. false, unrelated
20. false, artificial
21. true
22. h
23. b
24. d
25. a
26. g
27. f
28. c, i

29. e
30. i
31. c
32. b
33. d
34. b
35. e
36. d

37. Lamarck would explain the long proboscis by the "inheritance of acquired characteristics." The ancestral species was uniform for a short proboscis, but the moths began to habitually stretch their probosci deep into the long trumpet flowers to get nectar, stretching their probosci in the process. These slightly stretched probosci were in turn passed down to the next generation who repeated the process of stretching their probosci into the flowers and passing the longer probosci on. After many generations of this, all the moths of this species were born with very long probosci. Darwin would explain the long probosci by natural selection. Individuals in the ancestral species displayed genetic variation in proboscis length and those with longer probosci could get more nectar from the flowers, allowing them to live longer and out-reproduce those with shorter probosci. This produced the next generation with a higher percentage of moths with longer probosci. After many generations of natural selection favoring moths with longer probosci, the entire species showed this trait.

Chapter 15: How Organisms Evolve

OVERVIEW

This chapter covers the basic principles of population genetics, including how to calculate gene and genotype frequencies and the Hardy-Weinberg Principle. The authors examine the major forces causing evolution (mutation, migration, small population size, nonrandom mating, and natural selection), and discuss the three ways that natural selection acts on populations (stabilizing, disruptive, and directional). The results of natural selection are enumerated, and the ultimate fate for most species (extinction) is discussed.

1) How Are Populations, Genes, and Evolution Related?

Evolutionary changes occur from generation to generation so that descendants are different from their ancestors. Evolution is a property not of individuals but of **populations** (all individuals of a species living in a given area). Inheritance provides the link between the lives of individuals and evolution of populations. **Population genetics** is the study of the frequency, distribution, and inheritance of alleles in populations. The **gene pool** is the sum of all the genes in a population. The relative frequencies of various alleles in a population are the **allele frequencies**. Evolution is changes in gene frequencies that occur in a gene pool over time.

An **equilibrium population** is a hypothetical population in which evolution does not occur. The **Hardy-Weinberg principle** states that under certain conditions, allele and genotype frequencies in a population will remain constant over time (evolution will not occur). Such an evolution-free population is an equilibrium population that will remain in **genetic equilibrium** as long as: (1) there is no mutation; (2) there is no **gene flow** (migration) between populations; (3) the population is extremely large; (4) all mating is random; and (5) there is no **natural selection** (all genotypes reproduce equally well). Few natural populations are in genetic equilibrium.

2) What Causes Evolution?

There are five major causes of evolutionary change: mutation, gene flow, small population size, nonrandom mating, and natural selection. Mutations (changes in DNA sequence) are the ultimate source of genetic variability.

Mutations occur at rates between 1 in 10,000 and 1 in 100,000 genes per individual per generation. Although mutation by itself is not a major evolutionary force, without mutations there would be no evolution and no diversity among life forms. Mutation is not goal-directed, but random.

Gene flow (migration) between populations changes allele frequencies. It spreads advantageous alleles throughout a species, and helps maintain all the organisms over a large area as one species since the populations cannot become very different in allele frequencies as long as gene flow occurs.

Small populations are subject to random changes in allele frequencies (**genetic drift**). In large populations, chance events are unlikely to significantly alter the overall gene frequencies, but in small populations chance events could reduce or eliminate alleles, greatly altering in a random way its genetic makeup. Genetic drift tends to reduce genetic variability within small populations, and genetic drift tends to increase genetic variability between or among populations. A **population bottleneck** (drastic reduction in numbers followed by expansion in numbers from the few survivors) may cause both changes in allele frequencies and reduction in genetic variability, not allowing the population to evolve in response to environmental changes. In the **founder effect**, some isolated populations are founded by a small number of individuals who may have different allele frequencies than the larger parent population due to chance, and this may lead to a sizable new population that differs greatly from the original. Populations occasionally become very small, and these may contribute significantly to major evolutionary changes.

Random mating rarely occurs within populations, and all genotypes are not equally adaptive. Any time an allele confers a slight advantage to some individuals, **natural selection** will favor the enhanced reproduction of those individuals. Four important points about natural selection and evolution are: (1) natural selection does not cause genetic changes in individuals; (2) natural selection acts on individuals (unequal reproduction) but evolution occurs in populations; (3) the **fitness** of an organism is a measure of its reproductive success; and (4) evolutionary changes are not progressive in an absolute sense, just relative to the environmental circumstances present at any particular time and place.

3) How Does Natural Selection Work?

Natural selection is primarily an issue of **differential reproduction**: Organisms with favorable alleles leave more offspring (who inherit those alleles) than do other individuals with less favorable alleles. Natural selection acts on the phenotype, which reflects the underlying genotype. Natural selection can influence populations in three major ways. **Directional selection** shifts character traits in a specific direction: it favors individuals at one end of a distribution range for a trait and selects against average individuals and those at the opposite extreme of the distribution. **Stabilizing selection** acts against individuals who deviate too far from the average: it favors individuals having an average value for a trait and selects against individuals with extreme values due to opposing environmental pressures. Opposing environmental pressures may produce **balanced polymorphisms**, in which two or more alleles are maintained in a population because each is favored by a separate environmental force. **Disruptive selection** adapts individuals within a population to different habitats: it favors individuals at both ends of the distribution of a trait and selects against average individuals.

A variety of processes can cause natural selection. Organisms with reproductively successful phenotypes have the best **adaptations** (characteristics that help an individual survive and reproduce) to their particular environments. **Competition** for scarce resources favors the best-adapted individuals. When two species compete, as seen with predators and their prey, each exerts strong selection pressure on the other. When one evolves a new feature or modifies an old one, the other typically evolves new adaptations in response, a constant mutual feedback situation called **coevolution**. **Predation** includes any situation where one organism (the predator) eats another (the prey). **Symbiosis** (individuals of different species live in direct contact for long periods) leads to the most intricate coevolutionary adaptations. **Sexual selection** favors traits that help an organism mate. Kin selection favors altruistic behaviors. **Altruism** is any behavior that endangers an individual or reduces its reproductive success but benefits other members of the species. If the altruistic individual helps relatives who possess the same alleles, this is called **kin selection**.

4) What Causes Extinction?

Natural selection may lead to **extinction** (death of all members of a species). Two characteristics predispose a species to extinction when the environment changes: extremely limited ranges (localized distribution), and very narrow structural or behavioral requirements (overspecialization). Wide-ranging species do not succumb to local environmental catastrophes, and species which feed on a variety of foods do not die off if one food supply vanishes. In addition, interactions with other organisms may drive a species to extinction, as happened when the Panama land bridge allowed North American species to migrate into South America, causing the extinction of most native South American species due to competition. Finally, habitat change and habitat destruction are the leading causes of extinction. **Mass extinctions** are disappearances of many varied species in a short time over a large area and may be caused by traumatic environmental events such as the effects of the impact of a large meteorite.

KEY TERMS AND CONCEPTS

From the key terms in this chapter and previous chapters, fill in the crossword puzzle based on the following clues.

Across

3. (Across) a trait that helps an organism to survive and reproduce in a particular environment
5. ____ extinction: disappearances of many varied species in a short time over a large area
6. _____ polymorphism: prolonged maintenance of two or more alleles in a population
7. genes are made of this
9. movement of alleles from one population to another due to migration of organisms
10. genetic _____: change in the allele frequencies of a small population purely by chance
11. nucleic acid containing ribose sugar
13. ___ selection: favors a certain allele because it increases the survival or reproductive success of relatives bearing the same allele
15. unit of inheritance
16. evolution of adaptations in different species due to extensive interactions with each other
18. _____ selection: both extremes are favored over the average phenotype
20. function as enzymes in cells
21. alternate forms of the same gene
22. death of all members of a species.
23. wrote "On the Origin of Species" in 1859

24. an isolated population founded by a small number of individuals may develop allele frequencies that are very different from those of the parent population
25. significant changes in allele frequencies over time in a population

Down

1. _____-_____ principle: under certain conditions, allele and genotype frequencies in a population will remain constant over time
2. group of individuals of the same species found in the same time and place
3. (Down) behavior that endangers an individual but benefits other members of its species
4. the relative proportion of each allele of a gene found in a population
8. the total of all alleles of all genes in the population
12. measure of the reproductive success of an organism
14. relationship between individuals in which both require the same resource not available in sufficient quantity to satisfy the needs of all
17. organisms eaten by predators
19. any situation where one organism eats another

Key Terms and Definitions

adaptation: a characteristic of an organism that helps it to survive and reproduce in a particular environment; also the process of acquiring such characteristics.

allele frequency: for any given gene, the relative proportion of each allele of that gene found in a population.

altruism: a behavior that endangers an individual or its reproductive success but that benefits other members of its species.

balanced polymorphism: the prolonged maintenance of two or more alleles in a population, usually because each allele is favored by a separate selective force.

coevolution: the evolution of adaptations in two different species due to their extensive interactions with one another, so that each acts as a major force of natural selection upon the other.

competition: a relationship between individuals or species in which both require the same resource, which is not available in sufficient quantity to satisfy the needs of all users.

differential reproduction: differences in reproductive output among individuals of a population, usually as a result of genetic differences.

directional selection: a type of natural selection in which one extreme phenotype is favored over all others.

disruptive selection: a type of natural selection in which both extremes are favored over the average phenotype.

equilibrium population: a population in which allele frequencies and the distribution of genotypes do not change from generation to generation.

extinction: the death of all members of a species.

fitness: the reproductive success of an organism, usually expressed in relation to the average reproductive success of all individuals in the same population.

founder effect: a type of genetic drift in which an isolated population founded by a small number of individuals may develop allele frequencies that are very different from those of the parent population, because of chance inclusion of disproportionate numbers of certain alleles in the founders.

gene flow: the movement of alleles from one population to another owing to migration of individual organisms.

gene pool: the total of all alleles of all genes in the population; for a single gene, the total of all the alleles of

that gene that occur in a population.

genetic drift: a change in the allele frequencies of a small population purely by chance.

genetic equilibrium: situation when the frequencies of alleles in a population are essentially equal to the values expected from the Hardy-Weinberg principle, and no evolution is occurring in that population.

Hardy-Weinberg principle: under certain conditions, allele and genotype frequencies in a population will remain constant over time (evolution will not occur).

inclusive fitness: the reproductive success of all organisms bearing a given allele, usually expressed in relation to the average reproductive success of all individuals in the same population. Compare with fitness.

kin selection: selection favoring a certain allele because it increases the survival or reproductive success of relatives bearing the same allele.

mass extinction: disappearances of many varied species in a short time over a large area and may be caused by traumatic environmental events such as the effects of the impact of a large meteorite.

natural selection: the differential survival or reproduction of organisms due to environmental forces that act differently upon genetically different members of a population.

population: a group of individuals of the same species, found in the same time and place, and actually or potentially interbreeding.

population bottleneck: a form of genetic drift in which a population becomes extremely small, which may lead to differences in allele frequencies as compared with other populations of the species, and to a loss in genetic variability.

population genetics: the study of the frequency, distribution, and inheritance of alleles in a population of organisms.

predation: any situation where one organism (the predator) eats another (the prey)

sexual selection: a type of natural selection in which the choice of mates by one sex is the selective agent.

stabilizing selection: a type of natural selection in which those organisms displaying extreme phenotypes are selected against.

symbiosis: a relationship in which individuals of different species closely interact with one another for an extended time.

THINKING THROUGH THE CONCEPTS

True or False: Determine if the statement given is true or false. If it is false, change the underlined word so that the statement reads true.

26._____   Individual plants or animals change in response to selection.

27._____   For a population to remain at equilibrium, it must be large.

28._____   Mutation is the factor that controls the direction of evolution.

29._____   Genetic drift is a characteristic of large populations.

30._____   Natural selection acts on genotypes directly.

31._____   Stabilizing selection results in change.

32._____   Gene flow tends to decrease differences between populations.

33._____   Both large-beaked and small-beaked varieties of a single species of birds can be maintained in an area by disruptive selection.

34._____   Behavior that endangers an organism but benefits its close relatives is symbiosis.

35._____   Most species eventually give rise to new species.

Short Answer: A population of 600 plants contains 294 *AA*, 252 *Aa*, and 54 *aa* individuals. The *AA* and *Aa* plants produce purple flowers while the *aa* plants produce white flowers. What are the frequencies of the following?

36._____   allele *A*

37._____   allele *a*

38._____   genotype *AA*

39._____   genotype *Aa*

40 _____   genotype *aa*.

41._____   purple-flowered plants

42._____   white-flowered plants

Short Answer.

43. Give the 5 conditions that Hardy and Weinberg stated were necessary to keep a population in genetic equilibrium.

_____   _____

_____   _____

_____

44. Explain why mutation by itself is not a major force in evolution but that mutation is necessary for evolution to occur.

45. Explain why population size greatly influences the potential for chance events to change allele frequencies.

Matching: Evolutionary mechanisms.

46._____ only a few males fertilize all females in a population.

47._____ the frequencies of white and brown guinea pigs in a population change because a large number of white animals enters the population.

48._____ a small number of organisms begins a new colony that becomes large and has gene frequencies very different from the parent population and other neighboring populations.

49._____ provides "genetic potential" to a population, upon which natural selection acts.

50._____ chance loss of genetic variation from a population due to its small size.

51._____ temporary restriction in population size due to short term unfavorable environmental conditions.

52._____ causes the evolution in males of elaborate structures and behaviors related to reproduction.

53._____ average individuals survive and reproduce best since the population is well-adapted to a stable environment.

54._____ selection involving sickle cell anemia in Africans.

55._____ selecting a mate with traits similar to your own.

56._____ selection favoring both extreme phenotypes and not favoring average individuals.

57._____ a change in a prey species forces a change in a predator species.

58._____ selection of one extreme phenotype in populations living in a rapidly changing environment.

59._____ selection for flowering times in New England wildflowers.

60._____ behavior that endangers an organism or reduces its reproductive success but benefits others in the population.

61._____ selection within a bird species encountering only large and small seeds.

62._____ selecting causing industrial melanism.

63._____ the total of all populations of organisms that interbreed under natural conditions.

Choices:

a. sexual selection
b. population bottleneck
c. gene flow
d. mutation
e. altruism
f. assortative mating
g. founders effect
h. species
i. genetic drift
j. harem breeding
k. coevolution
l. disruptive selective
m. stabilizing selection
n. directional selection

Multiple Choice: Pick the most correct choice for each question.

64. All of the following are true **except**
   a. differential reproduction leads to a change in allele frequency
   b. evolution occurs in populations
   c. evolution is a change in allele frequency
   d. natural selection causes genetic changes in individuals
   e. with stabilizing selection, population averages do not change.

65. All of the following meet the Hardy-Weinberg requirements for equilibrium in a population **except**
   a. no random mating may occur
   b. no mutations may occur
   c. no migrations may occur
   d. no natural selection may occur
   e. populations must be very large

66. A population that meets the Hardy-Weinberg requirements
   a. evolves
   b. is small and usually isolated
   c. has allele frequency in equilibrium
   d. changes genotypic distribution from generation to generation

   e. always has 75% A and 25% a allele frequencies

67. In a hypothetical population, the frequency of the dominant A allele is 80% and the frequency of the recessive allele a is 20%. What percentage of the population would you expect to be heterozygous (Aa) in genotype?
   a. 4%
   b. 16%
   c. 32%
   d. 50%
   e. 25%

68. Which of the following is more likely to occur in a small population than in a large population?
   a. gene flow
   b. immigration
   c. genetic drift
   d. nonrandom mating
   e. natural selection

69. A characteristic that better enables an organism to survive and reproduce is
   a. a mutation
   b. an adaptation
   c. a bottleneck gene
   d. a stabilizing factor

CLUES TO APPLYING THE CONCEPTS

This practice question is intended to sharpen your ability to apply critical thinking and analysis to biological concepts covered in this chapter.

70. In certain parts of Africa, the frequency of sickle cell anemia among newborn African infants is about 2% and holding steady, while in the United States, the frequency of sickle cell anemia among newborn African-Americans is about 0.2% and slowly declining. Why are these populations behaving differently in evolutionary terms regarding sickle cell anemia?

ANSWERS TO EXERCISES

1. Hardy-Weinberg
2. population
3. (across) adaptation
3. (down) altruism
4. allele frequency
5. mass
6. balanced
7. DNA
8. gene pool
9. gene flow
10. drift
11. RNA

12. fitness
13. kin
14. competition
15. gene
16. coevolution
17. prey
18. disruptive
19. predation
20. protein
21. allele
22. extinction
23. Darwin

24. founder effect
25. evolution
26. false, do not change
27. true
28. false, selection
29. false, small
30. false, phenotypes
31. false, lack of change
32. true
33. true
34. false, altruism
35. false, become extinct

36. 588 (294 *AA* x 2) + 252 (252 *Aa* x 1) = 840/1200 (600 plants x 2 alleles each) = 0.70 = 70%.
37. 108 (54*aa* x 2) + 252 (252 *Aa* x 1) = 360/1200 = 0.30 = 30%.
38. 294 *AA* /600 total plants = 0.49 = 49%.
39. 252/600 = 0.42 = 42%.
40. 54/600 = 0.09 = 9%.
41. 294 *AA* + 252 *Aa* /600 total plants = 546/600 = 0.91 = 91%.
42. 54 *aa* /600 = 0.09 = 9%.

43. A population will remain in genetic equilibrium as long as: (1) there is no mutation; (2) there is no gene flow (migration) between populations; (3) the population is extremely large; (4) all mating is random; and (5) there is no natural selection (all genotypes reproduce equally well).

44. Because the rate of mutation is very low (less than 1 in 100,000 for most genes), mutation alone will not change gene frequencies quickly enough to account for observed evolutionary changes. However, mutation is important in the process of evolution by natural selection because mutation is the ultimate source of the genetic variation upon which natural selection acts in natural populations.

45. Small populations are subject to relatively large random changes in allele frequencies (genetic drift). In large populations, chance events are unlikely to significantly alter the overall gene frequencies, but in small populations chance events could reduce or eliminate alleles, greatly altering in a random way its genetic makeup. Genetic drift tends to reduce genetic variability within small populations, and genetic drift tends to increase genetic variability between or among small populations.

46. j
47. c
48. g
49. d
50. i

51. b
52. a
53. m
54. m
55. f

56. l
57. k
58. n
59. m
60. e

61. l
62. n
63. h
64. d
65. a

66. c
67. c
68. c
69. b

70. In Africa, stabilizing selection is at work with sickle cell anemia: Normal individuals (*AA*) are at a disadvantage due to susceptibility to malaria and those with sickle cell anemia (*SS*) are at a disadvantage as well. Heterozygotes (*AS*) have highest fitness since they are somewhat resistant to malaria and do not suffer from sickle cell disease. Thus, both alleles are maintained at relatively high frequencies in those areas of Africa. In the United States, normal individuals (*AA*) have the highest fitness since there is no malaria and their children will not have sickle cell disease. Heterozygotes (*AS*) are healthy but can produce offspring with sickle cell disease (*SS*) if two heterozygotes mate; hence, their fitness is slightly reduced. Thus, in the United States, the *A* allele is slowly rising and the *S* allele is slowly declining due to directional selection favoring normal homozygotes.

Chapter 16: The Origin of Species

OVERVIEW

This chapter covers speciation. After defining biological species, the authors describe the general mechanisms of allopatric and sympatric speciation. They cover premating and postmating mechanisms for maintaining reproductive isolation, the distinctions between divergent and phyletic speciation, and the concepts of gradualism and punctuated equilibrium.

1) What Is a Species?

Biologists define **species** as groups of actually or potentially interbreeding natural populations, which are reproductively isolated from other such groups. So, two organisms are the same species if they interbreed in nature and have normal, vigorous, fertile offspring. This definition does not apply to asexually reproducing species, however.

2) How Do New Species Form?

Speciation is the process by which new species form. Speciation depends on two factors: (1) isolation (gene flow between diverging populations must be small or nonexistent); and (2) genetic divergence of two populations (they must evolve large genetic differences so that they cannot interbreed or produce normal offspring).

Hypothetical mechanisms of speciation are **allopatric speciation** (the populations are geographically separated from each other and thus isolated from gene flow) and **sympatric speciation** (the populations share the same area but are isolated from gene flow). There is some debate as to whether genetic drift or natural selection normally plays the major role in allopatric speciation, but geographic isolation is involved in most cases of speciation in animals.

With sympatric speciation, two likely mechanisms can reduce gene flow: **ecological isolation** (if a geographical area contains two distinct types of habitat, different members of the same species might begin to specialize in one habitat or the other) and **chromosomal aberrations** (changes in chromosome numbers can cause immediate reproductive isolation of a population). A common speciation mechanism in plants is **polyploidy**, the acquisition of multiple copies of each chromosome. If a fertilized egg duplicates its chromosomes but doesn't divide into daughter cells, the chromosome number can rise from diploid ($2n$, or pairs of chromosomes) to become tetraploid ($4n$, or 4 doses of each chromosome). Healthy $4n$ plants produce $2n$ gametes. Tetraploids breed successfully with other tetraploids. But if $4n$ breeds with 2n, sterile $3n$ (triploid) offspring result. So, tetraploid plants and their diploid parents form distinct reproductive communities that cannot interbreed successfully. Speciation by polyploidy is common in plants (which can reproduce by self-fertilization and asexually) but rare in animals.

3) How Is Reproductive Isolation Between Species Maintained?

Speciation occurs through the evolution of mechanisms that prevent interbreeding. Genetic divergence during a period of isolation is necessary for new species to arise, but speciation will occur only if mechanisms ensuring **reproductive isolation** also develop. **Isolating mechanisms** are structural and/or behavioral modifications that prevent interbreeding.

Premating isolating mechanisms include geographical isolation, ecological isolation, temporal isolation, behavioral isolation, and mechanical incompatibility. **Geographical isolation** prevents members of different species from meeting each other and usually is considered to be a mechanism that allows new species to form. **Ecological isolation** involves populations having different resource requirements involving the use of different local habitats within the same general area: thus, they are not likely to meet during the mating season. **Temporal isolation** occurs between species that breed at different times of the year. Even if two species occupy similar habitats, they cannot interbreed if they have different breeding seasons. **Behavioral isolation** involves species with different courtship rituals. Courtship rituals involve recognition and evaluation signals between males and females and also aid in distinguishing among species. Colors and songs of birds, frog croaks, cricket chirping patterns, and firefly flashing colors and frequencies are examples of signals used in courtship rituals. **Mechanical incompatibility** occurs when physical barriers between species prevent fertilization. For example, male and female sex organs of different species may not fit together properly for sperm transfer in animals, or different flower sizes or structures of different species may prevent pollen transfer in plants.

Postmating isolating mechanisms prevent production of vigorous, fertile offspring, and include gametic incompatibility, hybrid inviability, and hybrid inferiority. **Gametic incompatibility** occurs when sperm from one species are unable to fertilize eggs of another. **Hybrid inviability** occurs if hybrid offspring survive poorly. Hybrids may die during development or display behaviors that are mixtures of the two parental types and be unable to attract mates. **Hybrid infertility** occurs if hybrid offspring are unable to produce normal sperm or eggs. Most animal hybrids such as mules (horse mating with donkey) or ligers (lion mating with tiger) are sterile since their chromosomes do not pair properly during meiosis. Crosses between tetraploid ($4n$) and diploid ($2n$) plant species usually result in sterile triploid ($3n$) offspring.

4) At What Rate Does Speciation Occur?

The rate of speciation varies considerably over evolutionary time, and bursts of speciation are seen in both the fossil record and in the distribution of modern organisms. In animals, instantaneous speciation is unlikely, although it can happen in plants due to polyploidy. During periods of **adaptive radiation** (where populations of a single species invade a variety of new habitats and evolve in response to the differing environmental pressures in these habitats), one species gives rise to many in a relatively short time.

The fossil record is ambiguous regarding the tempo of evolution. If fossils are arranged in chronological order, we see a series of rapid jumps from one stage of evolution of the group to the next stage. These rapid jumps may be due to gaps in the fossil record when no intermediate forms happened to be preserved ("missing links") or may really represent periods of very rapid evolution interspersed between long periods of very little change (**punctuated equilibrium**). According to this theory, the "rapid" periods of change (often 50,000 years or more) represent periods of speciation. This theory does not contradict any of the accepted mechanisms of evolution. The pace of evolution depends on the circumstances and on the species involved, and few biologists doubt that a pattern of punctuated equilibrium could occur in some situations.

KEY TERMS AND CONCEPTS

From the following list of key terms, fill in the blanks in the following story.

adaptive radiation
allopatric
behavioral isolation
ecological isolation
geographic isolation
hybrid infertility
mechanical incompatibility

polyploidy
postmating isolating mechanisms
premating isolating mechanisms
reproductive isolation
speciation
species
temporal isolation

Michael took his son Zachary to the zoo. Zachary is taking a high school biology class and was loaded with questions for his dad. "How do we know that African lions and Asian tigers are different (1)_____?" he asked. Mike replied: "Zach, they have different physical characteristics, but more importantly they show (2)_____ isolation since they don't try to interbreed in nature or even in zoos." This led to another question: "Dad, since both lions and tigers are mammals and cats, how could (3)_____ have occurred in such closely related organisms?" "Well, son, when a single group of organisms, like mammals, give rise to many closely related species, we call this (4)_____. In the case of lions and tigers, first of all there was (5)_____ isolation since they evolved in different parts of the world, leading to (6)_____ speciation. And even if they lived for a time in the same general area, there may have been (7)_____ isolation since they would occupy different habitats there. These are examples of (8)_____ isolating mechanisms, son, since they prevent lions and tigers from mating."

They then entered the amphibian and reptile house, where many different species of frogs were living. "Dad, the wood frogs and green frogs breed in the same areas each year, but are different species. How can this be?" "Son, they don't interbreed because wood frogs mate during early spring and green frogs mate during late spring, a situation called (9)_____ isolation. Also, the males of the two species have different sounding croaks and females are attracted only to males of her own kind. This is an example of (10)_____ isolation because they have different courtship rituals." When they looked at the lizards, Zach said: "Dad, I just thought of something weird. Suppose a large lizard of one species tried to mate with a small lizard of another species. Could they succeed?" "Probably not, son," replied Mike as he chuckled at the thought of such an unlikely liaison, "since their sex organs couldn't fit together properly. This is a situation called (11)_____ isolation."

Moving to the building housing hoofed animals, Zach noticed the horses, donkeys and mules. "Dad, since horses and donkeys produce mules when farmers force them to mate, aren't they all members of the same species?" "Son, horses and donkeys would never interbreed under natural conditions, and besides, the mules they produce are sterile, an example of (12)_____. This is an example of a (13)_____ isolating mechanism since it acts after mating between different species takes place."

"Dad, this is all way cool, but now I have a really important question. When do we eat?" "Soon, son, but just eat your banana right now. By the way, Zach, bananas don't have seeds and are sterile because they have three copies of each chromosome, a condition called (14) _____." "Banana, smanana!" exclaimed Zach. "I want a burger with fries!"

Key Terms and Definitions

adaptive radiation: the rise of many new species in a relatively short time as a result of a single species invading different habitats and evolving under different selective pressures in those habitats.

allopatric speciation: speciation that occurs when two populations are separated by a physical barrier that prevents gene flow between them (geographical isolation).

behavioral isolation: lack of mating between species of animals that differ substantially in courtship and mating rituals.

ecological isolation: lack of mating between organisms belonging to different populations that occupy distinct habitats within the same general area.

gametic incompatibility: the inability of sperm from one species to fertilize eggs of another species.

geographical isolation: the separation of two populations by a physical barrier.

hybrid infertility: reduced fertility (often complete sterility) in hybrid offspring of two different species.

hybrid inviability: the failure of a hybrid offspring of two different species to survive to maturity.

isolating mechanism: any structural and/or behavioral modification that prevent interbreeding between different species.

mechanical incompatibility: the inability of male and female organisms to exchange gametes, usually because of incompatibility of the reproductive structures.

polyploidy: the acquisition by cells of multiple copies of each chromosome usually by some disturbance of cell division after chromosomal replication; a mechanism of speciation common in plants.

postmating isolating mechanism: any structure, physiological function, or developmental abnormality that prevents organisms of two different populations, once mating has occurred, from producing vigorous, fertile offspring.

premating isolating mechanism: any structure, physiological function, or behavior that prevents organisms of two different populations from exchanging gametes.

punctuated equilibrium: a model of evolution, stating that morphological change and speciation are rapid, simultaneous events (the "punctuation"), separated by long periods during which a species remains unchanged (the "equilibrium").

reproductive isolation: the failure of organisms of one population to breed successfully with members of another population; may be due to premating or postmating isolating mechanisms.

speciation: the process whereby two populations achieve reproductive isolation.

species: groups of actually of potentially interbreeding natural populations, which are reproductively isolated from other such groups.

sympatric speciation: speciation of populations that are not physically divided; usually due to ecological isolation or chromosomal aberrations (such as polyploidy).

temporal isolation: the inability of organisms to mate if they have significantly different breeding seasons.

THINKING THROUGH THE CONCEPTS

True or False: Determine if the statement given is true or false. If it is false, change the underlined word so that the statement reads true.

15. _____ Temporal isolation is isolation by <u>distance</u>.
16. _____ Polyploidy is most common in <u>plants</u>.
17. _____ Mechanical incompatibility is a <u>postmating</u> isolating mechanism.
18. _____ The most valid way to determine whether two organisms belong to different species is to look at <u>mating behavior</u>.
19. _____ Speciation depends on <u>lack of isolation</u> between populations.
20. _____ Populations physically separated are <u>allopatric</u>.
21. _____ Sympatric speciation can occur if <u>chromosome aberrations</u> occur.

22._____   Geographic isolation is a <u>premating</u> isolating mechanism.
23._____   Hybrid infertility is a <u>postmating</u> isolating mechanism.
24._____   Adaptive radiation causes <u>convergent</u> evolution.

Matching: Speciation.

25.____ causes "instant speciation"   Choices:
26.____ causes most animal speciation   a.   gene flow
27.____ populations are separated by a physical   b.   geographical isolation
            barrier   c.   polyploidy
28.____ causes much plant speciation but little animal d.   mutation
            speciation   e.   adaptive radiation
29.____ reduces genetic differences between
            populations, retarding animal speciation
30.____ one species gives rise to many new species in
            a short time
31.____ acquisition of more than 2 copies of each
            chromosome in the nucleus of cells

Matching: Maintaining reproductive isolation.

32.____ Interbreeding does not occur in nature   Choices: a.   Geographical isolation
            between British peppered moths and   b.   Ecological isolation
            Canadian peppered moths.   c.   Temporal isolation
33.____ Interbreeding does not occur between closely   d.   Behavioral isolation
            related species of fruit flies with slightly   e.   Mechanical isolation
            different courtship rituals.   f.   Gamete incompatibility
34.____ When horses and donkeys are forced to   g.   Hybrid inviability
            interbreed, sterile mules are produced.   h.   Hybrid infertility
35.____ Leopard frogs and pickerel frogs that live in
            the same area with similar mating seasons do
            not interbreed because one species breeds in
            swamps and the other breeds in clear lakes.
36.____ Closely related species of katydid insects do
            not interbreed because the male and female
            sex organs cannot fit together properly to
            allow sperm transfer to occur.
37.____ Wood frogs and green frogs breed in the same
            lakes but do not interbreed because one
            species breeds in April while the other
            species breeds in May.
38.____ Two closely related species of fruit flies
            sometimes mate but the female's immune
            system kills the male's sperm as though it
            were a foreign invading microbe.

Short Answer.

39. Name and briefly describe two genetic models for speciation.

    _____

    _____

40. Name the two general conditions necessary for speciation in animals.

    _____   _____

41. Briefly describe three postmating reproductive isolating mechanisms.

    _____

    _____

    _____

Multiple Choice: Pick the most correct choice for each question.

42. Among animals in particular, which of the following must occur for speciation to happen?
    a. geographic isolation
    b. adaptive radiation
    c. reproductive isolation
    d. ecological isolation
    e. migration

43. In plants a common method of sympatric speciation is
    a. ecological isolation
    b. geographical isolation
    c. adaptive radiation
    d. polyploidy
    e. nondisjunction

44. Two species of pines releasing pollen at separate times in the same habitat is an example of
    a. geographical isolation
    b. ecological isolation
    c. temporal isolation
    d. behavioral isolation
    e. mechanical incompatibility

45. The Everglades kite is almost extinct because of
    a. fire in the Everglades
    b. saltwater intrusion in the Everglades
    c. invasion of the walking catfish
    d. disappearance of the apple snail
    e. industrial melanism

Short Answer: Answer the following question, based on the figure below.

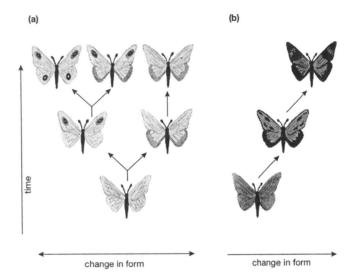

(a)   (b)

time

change in form   change in form

46. Compare parts a and b in the figure and comment on which evolutionary pattern is the result of speciation and which is not. Please briefly explain your answer.

CLUES TO APPLYING THE CONCEPTS

This practice question is intended to sharpen your ability to apply critical thinking and analysis to biological concepts covered in this chapter.

47. Dogs (*Canis familiaris*) and coyotes (*Canis latrans*) are given different species names by biologists. Interestingly, dogs and coyotes will eagerly mate with each other in captivity and produce perfectly healthy, fertile offspring. What sorts of criteria do you think biologists have used to determine that dogs and coyotes are different species? Do you personally think that dogs and coyotes should be considered different species? Do you think that dogs and coyotes are closely related types of animals?

ANSWERS TO EXERCISES

1.	species	13.	postmating	26.	b
2.	reproductive	14.	polyploidy	27.	b
3.	speciation	15.	false, time	28.	c
4.	adaptive radiation	16.	true	29.	a
5.	geographic	17.	false, premating	30.	e
6.	allopatric	18.	true	31.	c
7.	ecological	19.	false, isolation	32.	a
8.	premating	20.	true	33.	d
9.	temporal	21.	true	34.	h
10.	behavioral	22.	true	35.	b
11.	mechanical	23.	true	36.	e
	incompatibility	24.	false, divergent	37.	c
12.	hybrid infertility	25.	c	38.	f

39. Two hypothetical mechanisms of speciation are allopatric speciation, in which the populations are geographically separated from each other and thus isolated from gene flow, and sympatric speciation, in which the populations share the same area but are isolated from gene flow due to different behaviors.

40. Speciation is the process by which new species form. Speciation in animals generally depends on two factors: isolation, since gene flow between diverging populations must be small or non-existent; and the genetic divergence of two populations, since they must evolve large genetic differences so that they cannot interbreed or produce normal offspring.

41. Postmating isolating mechanisms prevent production of vigorous, fertile offspring, and include gametic incompatibility, hybrid inviability, and hybrid inferiority. Gametic incompatibility occurs when sperm from one species are unable to fertilize eggs of another. Hybrid inviability occurs if hybrid offspring survive poorly. Hybrids may die during development, or display behaviors that are mixtures of the two parental types and be unable to attract mates. Hybrid infertility occurs if hybrid offspring are unable to produce normal sperm or eggs due to problems in chromosome pairing during meiosis.

42. a   43. d   44. c   45. d

46. Speciation has occurred twice in part a to produce three contemporary species at the top of the figure. The branched arrows indicate the speciation events which resulted in reproductive isolation between the new species formed. In part b, a large amount of phenotypic change has occurred over an extended period of time, but no reproductive isolation has occurred so that the entire lineage has remained as a single species.

47. Since dogs and coyotes successfully interbreed in captivity, they must be closely related, having diverged from common ancestors in the not too distant past. Biologists consider them different species because they hardly ever interbreed under natural conditions in the wild. Also, recent molecular genetics studies have shown that dogs are much more closely related to wolves than they are to coyotes.

Chapter 17:   The History of Life on Earth

OVERVIEW

This chapter traces the evolution of life from the beginning of the universe through prebiotic evolution, spontaneous generation of the first living prokaryotic cells, metabolic evolution, the rise of eukaryotes and multicellularity, the invasion of land, and finally human evolution.

1) How Did Life Begin?

In the 1600s, biologists thought life arose through **spontaneous generation** from nonliving matter and unrelated life forms (like trees giving rise to fish and birds). This was not substantially refuted until the mid-1800s by Louis Pasteur. In the 1930s, Oparin and Haldane proposed **prebiotic** (or chemical) **evolution**: evolution before life, when the atmosphere of the Earth contained hydrogen, methane, carbon dioxide, nitrogen, hydrogen sulfide, hydrochloric acid, water vapor, and ammonia gases but no free oxygen gas. In 1953, Miller and Urey did lab experiments to show that under prebiotic atmospheric conditions, organic molecules like amino acids, nucleotides, and ATP could be produced. Since no organisms or oxygen gas existed in prebiotic times, organic molecules could accumulate in shaded shallow pools not subjected to ultraviolet solar radiation, forming a "primordial soup."

In the 1980s, Cech and Altman discovered that certain small RNA molecules called **ribozymes** act as enzymes to make more RNA. Possibly, molecular evolution began in the primordial soup when ribozymes began to copy themselves and make other molecules. Cellular life requires self-replicating molecules enclosed within membranes. If water containing proteins and lipids is agitated, hollow membrane-like balls called **microspheres** form. If microspheres formed around ribozymes, nonliving **protocells** formed, possibly evolving into living cells.

2) What Were the Earliest Organisms Like?

Life began about 3.5 to 3.9 billion years ago, according to fossils and chemical data. The first cells were prokaryotic and obtained nutrients and energy by absorbing organic molecules from the environment and breaking down these molecules anaerobically (no oxygen gas was present) and gaining a little energy. These cells were primitive anaerobic bacteria. Eventually, some cells evolved photosynthesis, the ability to use solar energy to make their own complex, energy-rich molecules from water and carbon dioxide. The cyanobacteria evolved, and through photosynthesis, released oxygen gas into the atmosphere. About 2.2 billion years ago, free oxygen gas began to accumulate. This, in turn, allowed the evolution of microbes capable of aerobic respiration, breaking down organic molecules completely using oxygen gas to release significant amounts of chemical energy.

Predation soon evolved, with larger prokaryotic cells engulfing bacteria. About 1.7 billion years ago, eukaryotic cells having membrane bound nuclei and cytoplasmic organelles evolved from predatory bacteria. According to Lynn Margulis' **endosymbiotic hypothesis**, primitive cells acquired the precursors of mitochondria and chloroplasts by engulfing certain types of bacteria and forming a symbiotic (mutually supportive) relationship with them. The fact that these organelles retain their own bacterial-like DNA supports the hypothesis. Cilia, flagella, centrioles, and microtubules may have evolved from a symbiosis between spiral-shaped bacteria and a primitive eukaryotic cell. The origin of the nucleus is obscure.

3) How Did Multicellularity Arise?

Increased size was an advantage, but large unicellular organisms couldn't survive due to the slowness of diffusion. Multicellular organisms evolved about one billion years ago. Multicellular plants developed specialized structures (roots and leaves) that helped them invade diverse habitats.

    Multicellular animals developed specializations to help them capture prey, feed, and escape their enemies. Primitive jellyfish had a single opening serving as mouth and anus where food taken in was digested, and wastes excreted. More efficient means of feeding evolved: separate mouth and anal openings for one-way flow of food and wastes seen in most animals today. Animals evolved muscular movement so that predators could chase prey and prey could escape. First there were hydrostatic skeletons (water-filled tubes) for locomotion in some worms and external skeletons in the arthropods, then finally internal skeletons in the vertebrates developed. Greater sensory capabilities and more-sophisticated nervous systems evolved.

4) How Did Life Invade the Land?

Terrestrial organisms must find adequate water, protect their gametes from drying out, and resist the effects of gravity without a buoyant watery environment. The plants that first colonized the land, however, had ample sunlight, rich nutrient sources in the soil, and no predators. Some plants developed specialized structures that adapted them to dry land. Waterproof coatings on the aboveground parts reduced water loss, rootlike structures were anchored in the soil, mining water and nutrients, and extra-thick cell walls enabled stems to stand erect.

    Primitive land plants (mosses and ferns) retained swimming sperm and required water to reproduce, but the **conifers** (cone-bearing plants) retained their eggs internally and encased sperm within pollen grains blown around by the wind, allowing the conifers to flourish in dry habitats.   Landing on a female cone near the egg, the pollen released sperm cells directly into living tissue, eliminating the need for a surface film of water. As the moist climate dried up, conifers flourished. Flowering plants enticed animals (mainly insects) to carry pollen from flower to flower, thus wasting much less pollen than conifers do. Flowering plants also reproduce more rapidly and grow more quickly than do conifers.

    Some animals evolved specialized structures that adapted them to life on dry land. Some animals were **preadapted** for land life (they already had structures suitable for life on land, such as exoskeletons in the arthropods).   Amphibians evolved from lobed fishes. Lobefins had two preadaptations for land: (1) stout fleshy fins for crawling; and (2) a pouch off the digestive tract that could act as a primitive lung or swim bladder for buoyancy. With improvements in lungs and legs, lobefins evolved into amphibians. But amphibians depended on water to keep the skin moist (for gas exchange) and for egg laying.

    Reptiles, which evolved from amphibians, developed several adaptations to dry land: (1) internal fertilization; (2) shelled, waterproof eggs containing a supply of water and food; (3) scaly, waterproof skin; and (4) improved lungs. Two groups of smaller reptiles developed insulation to retain body heat: feathers and hair. Reptiles gave rise to both birds (feathers) and mammals (hair). Unlike birds which lay eggs, mammals evolved live birth and mammary (milk-producing) glands to feed the young. When the reptilian dinosaurs became extinct (65 million years ago), the mammals adaptively radiated out into the vast array of modern forms.

5) How Did Humans Evolve?

Primate evolution is linked to: (1) grasping hands for powerful (club swinging) and precise (writing, sewing) manipulations; (2) binocular vision (forward-looking eyes with overlapping fields of vision) for accurate depth perception and color vision for finding ripe fruit; and (3) a large brain with high intelligence. Dryopithecine primates evolved into the **hominids** (humans) and pongids (great apes) some 5 to 8 million years ago. The earliest hominids, called the *australopithecines* (southern apes of Africa) could stand and walk upright. Upright posture freed the hands to carry weapons and manipulate tools. The genus *Homo* diverged from the *australopithecines* about 2.5 million years ago. The evolution of *Homo* was accompanied by advances in tool technology. *Homo neanderthalensis* (Neanderthals) appeared about 150,000 years ago and had large brains and ritualistic behaviors but did not lead to modern humans based on DNA analysis. Modern humans (*Homo sapiens*) evolved about 150,000 years ago in Africa and spread into the Near East, Europe, and Asia, supplanting all other hominids. Humans and Neanderthals coexisted in Europe until humans overran and displaced the Neanderthals. The evolution of human behavior is highly speculative. Human **cultural evolution** (learned behavior passed down from previous generations) now far outpaces biological evolution. There have been three major surges of human population growth, each associated with a cultural revolution: (1) development of tools (ending 10,000 years ago, 5 million humans worldwide); (2) agricultural revolution (the past 8000 years, 500 million humans worldwide); and (3) scientific and technological revolution (the past 300 years, 5.6 billion humans worldwide).

KEY TERMS AND CONCEPTS

Fill-In: From the following list of terms, fill in the following statements.

aquatic	ferns	preadapted
conifers	hominids	ribozymes
Cro-Magnon	hypothesis	spontaneous generation
cultural evolution	microspheres	terrestrial
endosymbiotic	Neanderthals	

In the 1600s, biologists thought life arose through (1)_____ from nonliving matter.

Certain small RNA molecules called (2)_____ act as enzymes to make more RNA. Possibly, molecular evolution began in the primordial soup when they began to copy themselves and make other molecules.

 If water containing proteins and lipids is agitated, hollow membrane-like balls called (3)_____ form.

According to Lynn Margulis' (4)_____, primitive cells acquired the precursors of mitochondria and chloroplasts by engulfing certain types of bacteria and forming a symbiotic (mutually supportive) relationship with them.

(5)_____ organisms must find adequate water, protect their gametes from drying out, and resist the effects of gravity without a buoyant watery environment .

Primitive land plants (mosses and ferns) retained swimming sperm and required water to reproduce, but the (6)_____ retained their eggs internally within cones and encased sperm within pollen grains blown around by the wind, allowing them to flourish in dry habitats.

Some animals were (7)_____ for land life since they already had structures suitable for life on land, such as exoskeletons in the arthropods.

Dryopithecine primates evolved into the (8)_____ (humans) and the pongids (great apes) about 5 to 8 million years ago.

Human (9)_____ (learned behavior passed down from previous generations) now far outpaces biological evolution.

(10)_____ appeared about 150,000 years ago and had large brains and ritualistic behaviors, but did not lead to modern humans based on DNA analysis.

Key Terms and Definitions

conifer: a plant group that reproduces using seeds formed inside cones and retains its leaves throughout the year.

cultural evolution: changes in the behavior of a population of animals, especially humans, by learning behaviors acquired by members of previous generations.

endosymbiotic hypothesis: the hypothesis that certain organelles, especially chloroplasts and mitochondria, evolved from bacteria captured by ancient predatory prokaryotic cells.

exoskeleton: a rigid external skeleton that supports the body, protects the internal organs, and has flexible joints to allow for movement.

hominid: a general term applied to all humans and their prehistoric relatives, beginning with the Australopithecines, whose fossils date back at least 4.4 million years.

microsphere: a small, hollow sphere formed from proteins or proteins complexed with other compounds.

preadaptation: a feature evolved under one set of environmental conditions that, purely by chance, helps an organism adapt to new environmental conditions.

prebiotic evolution: evolution before life existed; especially abiotic synthesis of organic molecules.

primate: a group of mammals including the lemurs, monkeys, apes, and humans.

protocell: a nonliving microscopic structure consisting of a microsphere of protein and lipid enclosing a mixture of ribozymes.

ribozyme: an RNA molecule that can catalyze certain chemical reactions, especially those involved in synthesis and processing of RNA itself.

spontaneous generation: the proposal that living organisms can arise from nonliving matter.

THINKING THROUGH THE CONCEPTS

True or False: Determine if the statement given is true or false. If it is false, change the underlined word so that the statement reads true.

11._____   Primitive Earth was characterized by an <u>abundance</u> of free oxygen.

12._____   The first living organisms were most likely <u>prokaryotic</u>.

13._____   The first living organisms were <u>autotrophic</u>.

14._____   Photosynthesis arose first in the <u>green algae</u>.

15._____   Separate mouth and anal openings is <u>less</u> efficient than a single common opening.

16._____   In plants, spores are <u>asexual</u> reproducing agents.
17._____   Spores germinate and grow into <u>sporophytes</u> .
18._____   Gymnosperms, as a rule, produce <u>more</u> pollen than angiosperms.
19._____   Feathers probably evolved first for <u>insulation</u>.
20._____   Internal skeletons are called <u>exoskeletons</u>.

Short Answer:

21. Define spontaneous generation. Explain why scientists assert that it cannot happen today. Explain why scientists say that it probably happened many years ago.

22. Arrange the following events into the sequence that scientists assert occurred during the history of Earth.
    A. evolution of terrestrial organisms
    B. oxygen gas begins to accumulate in the atmosphere
    C. spontaneous formation of simple organic molecules which, in the absence of O_2, accumulated in the seas
    D. evolution of anaerobic prokaryotic cells
    E. evolution of mitochondria and chloroplasts (the Endosymbiotic theory)
    F. chance formation of ribozymes with the ability to make accurate and inaccurate copies of itself
    G. evolution of aerobic prokaryotic cells
    H. evolution of multicellular eukaryotic organisms
    I. evolution of primitive photosynthetic anaerobic cells
    J. by chance, primitive microspheres surround the proper mix of organic molecules and form primitive living cells

23. List four reptilian adaptations that help them cope successfully with a fully terrestrial lifestyle.

    _____   _____

    _____   _____

Matching: Terrestrial plants.

24.____ better adapted to colder climates

25.____ use insects to transport pollen

26.____ need water for sexual reproduction since sperm
must swim to the eggs

27.____ primarily use wind to passively transport pollen

28.____ evolved from conifer-like ancestors

29.____ dominant plants today

30.____ dominant plants 250 million years ago

31.____ dominant plants 325 million years ago

Choices:

a. flowering plants

b. ferns and mosses

c. conifers

Matching: Terrestrial animals. (Hint: there may be more than one correct answer for some questions.)

32.____ evolved from reptilian ancestors

33.____ evolved from lobefins whose fins evolved into
limbs for crawling

34.____ dinosaurs

35.____ use their lungs and moist skin to exchange
gases with the air

36.____ first land animals to evolve waterproof eggs

37.____ humans

38.____ first land animals

39.____ evolved from lobefins that developed swim
bladders from their lungs

40.____ land animals that shed their eggs and sperm into
the water

41.____ were preadapted to life on land due to their
exoskeletons

42.____ directly evolved from amphibian-like ancestors

Choices:

a. reptiles

b. amphibians

c. mammals

d. arthropods

e. birds

f. bony fish

g. jellyfish

Multiple Choice: Pick the most correct choice for each
question.

43. It is proposed that the primitive atmosphere
contained all of the following **except**

a. CO_2

b. O_2

c. NH_2

d. H_2O

44. The first living "cells" probably were hollow, ball-
shaped structures called

a. microspheres

b. protenoids

c. polypeptides

d. ribozymes

e. bacteria

45. The first organisms probably were
primitive

a. photosynthetic bacteria

b. cyanobacteria

c. anaerobic bacteria

d. aerobic microbes

e. viruses

46. The first cells probably
    a. produced their own food
    b. absorbed food from the environment
    c. engulfed food from the environment
    d. did not require food
    e. underwent sexual reproduction

47. Reptiles are more advanced than amphibians because of
    a. internal fertilization
    b. eggs with shells
    c. scaly skin
    d. improved lungs
    e. All of the choices are correct

48. Primate fossils are relatively rare because
    a. primates did not live in areas where fossils were likely to be preserved
    b. the bones of these small animals likely were destroyed by predators and scavengers
    c. early primates were not very numerous
    d. all of the above choices are correct
    e. they are all buried in graves

Short Answer: Based on the following figure, answer the questions below.

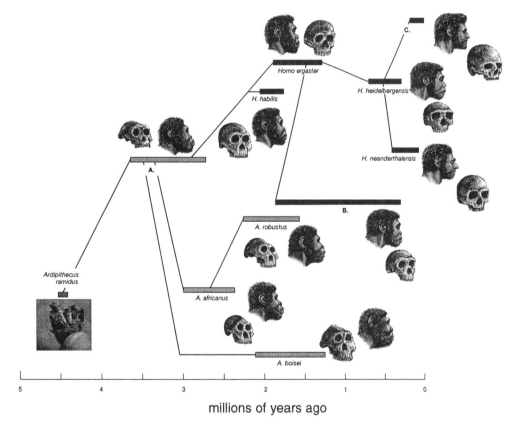

49. Identify member "a" of the human family tree: _____

50. Identify member "b" of the human family tree: _____

51. Identify member "c" of the human family tree: _____

CLUES TO APPLYING THE CONCEPTS

This practice question is intended to sharpen your ability to apply critical thinking and analysis to biological concepts covered in this chapter.

52. Which of the following discoveries, if true, would force scientists to revise drastically their thinking about the origin of life on Earth? (1) Earth is 5 billion years old, not 4.6 billion years old; (2) lipids can spontaneously form selectively permeable membranes without the presence of proteins; or (3) the atmosphere of the Earth 4.5 billion years ago contained about the same amount of oxygen gas as it does today. Explain your choice.

ANSWERS TO EXERCISES

1. spontaneous generation
2. ribozymes
3. microspheres
4. endosymbiotic hypothesis
5. terrestrial
6. conifers
7. preadapted
8. Neanderthals
9. cultural evolution
10. hominids

11. false, absence
12. true
13. false, heterotrophic
14. false, cyanobacteria
15. false, more
16. false, sexual
17. false, gametophytes
18. true
19. true
20. false, endoskeleton

21. Spontaneous generation is the generation of living cells from nonliving matter and unrelated life forms. This was not substantially refuted until the mid-1800s by Louis Pasteur. In the 1930s, Oparin and Haldane proposed prebiotic (or chemical) evolution: evolution before life, when the atmosphere of the Earth contained hydrogen, methane, carbon dioxide, nitrogen, hydrogen sulfide, hydrochloric acid, water vapor, and ammonia gases but no free oxygen gas. In 1953, Miller and Urey did lab experiments to show that under prebiotic atmospheric conditions, organic molecules like amino acids, nucleotides, and ATP could be produced. Since no organisms or oxygen gas existed in prebiotic times, organic molecules could accumulate in shaded shallow pools not subjected to ultraviolet solar radiation, forming a "primordial soup" and making spontaneous generation of the first living cells possible. Once living organisms and oxygen gas existed on Earth, spontaneous generation was no longer possible.

22. C, F, J, D, I, B, G, E, H, A

23. Reptiles, which evolved from amphibians, developed several adaptations to dry land: (1) internal fertilization; (2) shelled, waterproof eggs containing a supply of water and food; (3) scaly, waterproof skin; and (4) improved lungs.

24.	c	31.	b	37.	c	43.	b
25.	a	32.	c and e	38	d	44.	a
26.	b	33.	b	39.	f	45.	c
27.	c	34.	a	40.	b	46.	b
28.	a	35.	b	41.	d	47.	e
29.	a	36.	a	42.	a	48.	d
30.	c						

49. *Australopithecus afarensis*

50. *Home erectus*

51. *Homo sapiens*

52. The atmosphere of Earth 4.5 billion years ago contained about the same amount of oxygen gas as it does today. If this were true, the thinking of scientists regarding the possibility of spontaneous generation in prebiotic times would be erroneous since large molecules would be broken down by free oxygen gas, not allowing large quantities of such molecules to accumulate and serve as the raw materials for the chance development of living cells from nonliving materials.

Chapter 18: Systematics: Seeking Order Amidst Diversity

OVERVIEW

This chapter presents the basic principles used to classify organisms into discrete groups for the purpose of systematic study.

1) How Are Organisms Named and Classified?

Systematics is the science of reconstructing **phylogeny** (evolutionary history). **Taxonomy** is the science of naming organisms and placing them into categories based on their evolutionary relationships. The more characteristics two organisms share, the closer is their evolutionary relationship. The seven major taxonomic categories are (1) **kingdom**, (2) **division** (plants) or **phylum**, (3) **class**, (4) **order**, (5) **family**, (6) **genus**, and (7) **species**. These form a nested hierarchy in which each level includes all the levels below it. As we move down the hierarchy, smaller and smaller groups are included. Each category is increasingly narrow and specifies a group whose common ancestor is increasingly recent. The **scientific name** (always underlined or italicized) of an organism is formed from its genus (closely related species that do not interbreed) and species (populations of organisms that can interbreed under natural conditions). These names are recognized by biologists worldwide.

Numerous criteria are used for classification. Past evolutionary relationships are inferred based on features shared by living organisms who inherited them from common ancestors. Historically, the most important distinguishing characteristics have been anatomical (bones and teeth, for instance). Developmental stages also provide clues to common ancestry. Ultimately, evolutionary relationships among species must reflect genetic similarities (DNA nucleotide sequences and chromosome structures).

2) What Are the Kingdoms of Life?

Before 1970, taxonomists classified all life into two kingdoms (plants and animals), an oversimplification. In 1969, Whittaker proposed five kingdoms: (1) Monera (unicellular prokaryotes); (2) Protista (unicellular eukaryotes); (3) Plantae (multicellular eukaryotes that obtain nutrients by photosynthesis); (4) Fungi (multicellular eukaryotes that absorb externally digested nutrients); and (5) Animalia (multicellular eukaryotes that ingest food and digest it internally). Recent evidence based on rRNA sequences indicates the Monera are two distinct groups. Woese has proposed classifying life into three broad categories or **domains**: (1) Bacteria (or eubacteria); (2) Archaea (or archaebacteria); and (3) Eukarya (eukaryotic organisms). Each domain could contain a number of kingdoms (see Table 18-2 in the text).

3) Why Do Taxonomies Change?

Taxonomic categories are revised as taxonomists learn more about evolutionary relationships, particularly using molecular biology techniques. Asexually reproducing organisms pose a particular challenge to taxonomists since the criteria of interbreeding cannot be used to distinguish among species.

4) Exploring Biodiversity: How Many Species Exist?

Each year, between 7,000 and 10,000 new species are named, most of them insects, many from the tropical rain forests. Today, about 1.4 million species have been identified (5% are prokaryotes and protists, 22% are plants and fungi, and 73% are animals). However, scientists think that 7 to 10 million species may exist on Earth. This total range of species diversity, and the interrelationships among species, is known as **biodiversity**.

KEY TERMS AND CONCEPTS

From the key terms in this and previous chapters, fill in the crossword puzzle based on the following clues.

Across
3. Taxonomic category contained within a kingdom and consisting of related classes of plants.
5. Multicellular, eukaryotic organisms that absorb externally digested nutrients.
6 Taxonomic category composed of related orders.
8. Referring to unicellular prokaryotic organisms.
10. Taxonomic category contained within an order and consisting of related genera.
13. The science of reconstructing the evolutionary histories of life forms on Earth.
14. Multicellular, eukaryotic, photosynthetic organisms
15. Unicellular eukaryotic organisms.

Down
1. Broadest taxonomic category, consisting of phyla.
2. New taxonomic category based on molecular analysis indicating two groups of bacteria exist.
4. Domain containing all eukaryotic organisms.
7. A group of organisms within a genus that interbreed under natural conditions.
9. Taxonomic category of animals contained within a kingdom and consisting of related classes.
11. Multicellular, eukaryotic organisms that ingest food and digest it internally.
12. Taxonomic group containing closely related species.

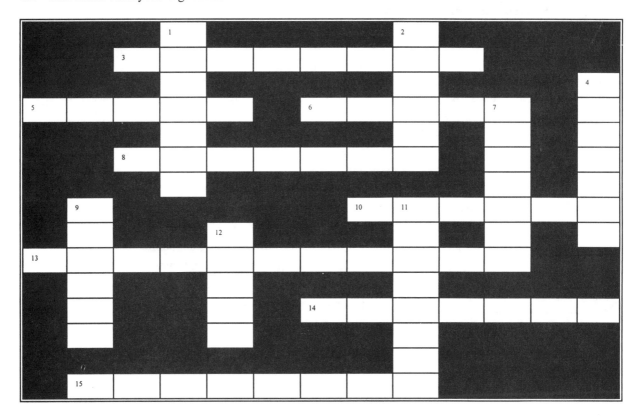

Key Terms and Definitions

class: the taxonomic category composed of related genera. Closely related classes form a division or phylum.

division: the taxonomic category contained within a kingdom and consisting of related classes of plants, fungi, bacteria, or plantlike protists.

domain: a new broad taxonomic category proposed after molecular analysis revealed that two quite distinct groups of bacteria exist.

DNA-DNA hybridization: a technique by which DNA from two species is separated into single strands and then allowed to reform. Hybrid double-stranded DNA from the two species can occur where the sequence of nucleotides is complementary. The greater the degree of hybridization, the closer the evolutionary relatedness of the two species.

DNA sequencing: determination of the nucleotide sequences within segments of DNA; used to compare DNA from different species.

family: the taxonomic category contained within an order and consisting of related genera.

genus: the taxonomic category consisting of very closely related species.

kingdom: the broadest taxonomic category, consisting of phyla or divisions. We recognize five kingdoms in this text.

order: the taxonomic category contained within a class and consisting of related families.

phylogeny: evolutionary history, deduced by comparison of genetically derived differences and similarities among organisms.

phylum: the taxonomic category of animals and animal-like protists contained within a kingdom and consisting of related classes.

scientific name: the genus and species designation of an organism.

species: a group of organisms within a genus that interbreed under natural conditions or, if asexually reproducing, are more closely related to one another than to other organisms within the genus.

systematics: the science of reconstructing the evolutionary histories of life forms on Earth.

taxonomy: the science by which organisms are classified into hierarchically arranged categories that reflect their evolutionary relationships.

THINKING THROUGH THE CONCEPTS

True or False: Determine if the statement given is true or false. If it is false, change the underlined word so that the statement reads true.

16._____   The science that places organisms into categories based on their evolutionary relationships is <u>classification</u>.

17._____   A genus contains several <u>classes</u>.

18._____   A kingdom is a more <u>general</u> category than a phylum.

19._____   The two-part name for a type of organism is its <u>genus and species</u>.

20._____   Organisms within the same <u>genus</u> interbreed in nature.

21._____   Eastern and mountain bluebirds have the same <u>genus</u> name.

22._____   <u>Darwin</u> developed the classification system we use today.

23._____   One important criterion used to classify organisms is <u>geographical location</u>.

24._____   All bacterial and protistan organisms are <u>prokaryotic</u>.

25._____   Plants are, but fungi are not, <u>multicellular</u>.

Multiple Choice: Pick the most correct choice for each question.

26. The classification of organisms is called
    a. taxonomy
    b. morphology
    c. ecology
    d. nomenclature
    e. phylogeny

27. Which hierarchical order goes from more specific to more general?

    a. kingdom ➙– division ➙ class
    b. genus ➙ family ➙ order
    c. order ➙ family ➙ genus
    d. order ➙ kingdom ➙ species
    e. family ➙ order ➙ species

28. In the scientific name, *Canis lupis*, "*Canis*" is the name of the
    a. genus
    b. species
    c. family
    d. wolf group only
    e. dog group only

Matching: The five kingdoms.

29.____ unicellular, eukaryotic organisms
30.____ multicellular, non photosynthetic organisms with extracellular digestion
31.____ prokaryotic organisms
32.____ organisms are unicellular and lack nuclei
33.____ all organisms are photosynthetic
34.____ bacteria
35.____ multicellular, non-photosynthetic organisms with digestion within the body
36.____ organisms are unicellular and have nuclei

Choices:
    a   Kingdom Plantae
    b   Kingdom Fungi
    c   Kingdom Monera
    d   Kingdom Animalia
    e   Kingdom Protista

CLUES TO APPLYING THE CONCEPTS

This practice question is intended to sharpen your ability to apply critical thinking and analysis to biological concepts covered in this chapter.

37. In what way did Darwin reinterpret the significance of Linnaeus' system of classification?

ANSWERS TO EXERCISES

1. kingdom
2. domain
3. division
4. Eukarya
5. Fungi
6. class
7. species
8. moneran
9. phylum
10. family
11. Animalia
12. genus
13. systematics
14. Plantae
15. Protista
16. false, taxonomy
17.. false, species
18. true

19. true
20. false, species
21 true
22. false, Linnaeus
23. false, anatomical similarities
24. false, unicellular
25. false, photosynthetic
26. a
27. b
28. a
29. e
30. b
31. c
32. c
33. a
34. c
35. d
36. e

37. Linnaeus placed organisms into a series of hierarchically arranged categories based on resemblances to other life forms, a system he thought was related to the creation of these organisms by God. Darwin proposed that Linnaeus' categories reflected the evolutionary relatedness of organisms.

Chapter 19:   The Hidden World of Microbes

OVERVIEW

This chapter describes the three major groups of microorganisms: viruses, prokaryotic unicellular microbes (Kingdom Monera), and eukaryotic unicellular microbes (Kingdom Protista).

1) What Are Viruses, Viroids, and Prions?

Viruses have no cells, membranes, ribosomes, cytoplasm, or energy source, cannot grow or move, and can reproduce only within **host** (virus-infected) cells.   A virus consists of a molecule of DNA or RNA surrounded by a protein coat and is too small to see under the light microscope. The protein coat is specialized to allow a virus to enter a specific host cell. Once inside, the viral genetic material takes command, forcing the host to make more viral protein and genetic material that assemble to form new viruses to burst out of the host cell and invade new host cells. Viral infections cause diseases that are difficult to treat. Viruses that infect bacteria are called **bacteriophages**. Within a particular organism, viruses specialize in attacking particular cell types. Antiviral agents may destroy host cells as well as virus. Antibiotics are useless against viruses.

Some infectious agents are even simpler than viruses. Some plant diseases are caused by **viroids**, pieces of RNA without protein coats only one-tenth the size of viruses. **Prions**, infectious "replicating" particles made of protein alone, cause degenerative brain diseases in humans (**kuru** and *Creutzfeldt-Jacob disease*) and livestock (*scrapie* and "mad cow disease"). Apparently, these proteins fold abnormally, inducing normal proteins to refold abnormally as well, causing nerve cell damage and degeneration. The origin of viruses, viroids and prions is obscure.

2) Which Organisms Compose the Prokaryotic Domains, Bacteria and Archaea?

Prokaryotes lack nuclei, chloroplasts, and mitochondria and are 10 times smaller than eukaryotic cells. Lack of sexual reproduction and fossil evidence make bacteria hard to classify. Taxonomists use criteria such as shape, means of movement, pigments, staining properties, nutrient requirements, DNA and RNA sequences, and the appearance of colonies to classify them. Bacterial cell walls afford protection from osmotic rupture in watery environments, and give bacteria their shapes (rodlike **bacilli**, corkscrew-shaped **spirilla**, and spherical **cocci**). Cell walls contain **peptidoglycan** (sugar chains attached by short peptides of amino acids). The **Gram staining** of cell walls reveals two types of bacteria: (1) *gram-positive* bacteria have peptidoglycan walls only and are often sensitive to penicillin; and (2) *gram-negative* bacteria have an additional outer membrane-like coating often toxic to mammals.

Outside the cell wall, sticky polysaccharide or protein **capsules** (escape detection by host immune systems) and **slime layers** (help bacteria adhere) aid in bacterial survival. Protein **pili** (hairlike projections) help bacteria attach to other cells. Some bacteria can move by using simple **flagella**, orienting toward various stimuli, a behavior called **taxis**: (1) **chemotactic** bacteria move toward food or away from toxins; (2) **phototactic** bacteria move to or from light; and (3) **magnetotactic** bacteria contain iron crystals and orient to the Earth's magnetic field. In hostile environments, many rod-shaped bacteria form internal **endospores** (thick-walled coating around the chromosome), which often are used for bacterial dispersal.

Bacteria reproduce rapidly by cell division called **binary fission** (Chapter 11). Some transfer genetic material in **plasmids** (small circular DNA molecules often containing drug-resistance genes) from donor to recipient bacteria during **bacterial conjugation** using hollow *sex pili*. Conjugation produces new genetic combinations in the recipient cell. Bacteria perform many functions important to other life forms: (1) **cyanobacteria** engage in plantlike photosynthesis; (2) **chemosynthetic** bacteria release sulfates (sulfur) and nitrates (nitrogen) into the soil; (3) some **anaerobes** (use no oxygen) release sulfur into the atmosphere during a special type of photosynthesis; (4) some **symbiotic** bacteria live in the digestive tracts of animals and help them break down food molecules, making nutrients like vitamins K and B$_{12}$; (5) **nitrogen-fixing bacteria** live in the roots of **legume** plants (beans and their relatives), converting nitrogen gas into ammonium nutrients for the plants. Some **pathogenic** bacteria cause disease and pose a threat to human health.

Archeae are fundamentally different from bacteria, especially in their membrane lipids, cell walls, and rRNA nucleotide sequences. The domain Archaea include: (1) **methanogens** that convert carbon dioxide to methane; (2) **halophiles** that live in very salty environments; and (3) **thermoacidophiles** that live in hot, acidic environments. Eukaryotes probably evolved from archaeal ancestors.

3) Which Organisms Make Up the Kingdom Protista?

Protists are a diverse group including funguslike water molds and slime molds, plantlike unicellular algae, and animal-like protozoa, among others.

Funguslike protists (absorb nutrients from the environment and typically decompose dead organisms). **Water molds** (Oomycota) cause powdery mildew of grapes and late blight of potatoes. **Slime molds** have a mobile feeding stage and a stationary reproductive stage (**fruiting body**) producing spores. The **acellular slime molds** (Myxomycota) consist of a mass of cytoplasm (**plasmodium**) containing thousands of nuclei. The **cellular slime molds** (Acrasiomycota) consist of masses of independent cells that move and feed by producing extensions (**pseudopods**). These cells release a chemical signal that causes the cells to form a dense aggregation (**pseudoplasmodium**) that forms a fruiting body.

Plantlike unicellular algae or photoplankton (gain energy by photosynthesis). **Dinoflagellates** (Pyrrophyta) have two flagella for movement and have red pigment, sometimes causing "red tides" when the water is warm and rich in nutrients. **Diatoms** (Chrysophyta) are called "pastures of the sea" since they are important as food in marine food webs. They encase themselves in silica (glassy) pillbox-like shells that accumulate as "diatomaceous earth." **Euglenoids** (Euglenophyta) each have a flagellum for movement and an eyespot for light detection.

Animal-like protozoa ingest food and digest it internally. **Zooflagellates** (Sarcomastigohora) have at least one flagellum for movement or food gathering. One parasitic type (*Trypanosoma*) causes African sleeping sickness while another (*Girardia*) causes dysentery in the United States. **Sarcodines** (Sarcomastigohora) are the freshwater **amoebas** (shell-less, one form causes amoebic dysentery) and **heliozoans** (glassy shells). The marine **foraminiferans** (chalky limestone shells) and **radiolarians** (glassy shells) move and capture food by forming pseudopods. **Sporozoans** (Apicomplexa) have no means of movement and live as parasites, causing malaria (*Plasmodium*). **Ciliates** (Ciliphora), which move by cilia, are the most complex in cell structure; they have well-developed nervous systems and prey-capturing explosive darts called **trichocysts**, as well as mouth openings, anal openings, and contractile vacuoles to excrete water.

KEY TERMS AND CONCEPTS

Fill-In: Using the terms below, fill in all the boxes in the following concept map.

bacilli	diatoms	protistans	spirilla
bacteria and	dinoflagellate	sarcodines	viruses
Archaea	DNA and RNA	slime molds	water molds
ciliates	euglenoids	sporozoans	zooflagellates
cocci	protein coats		

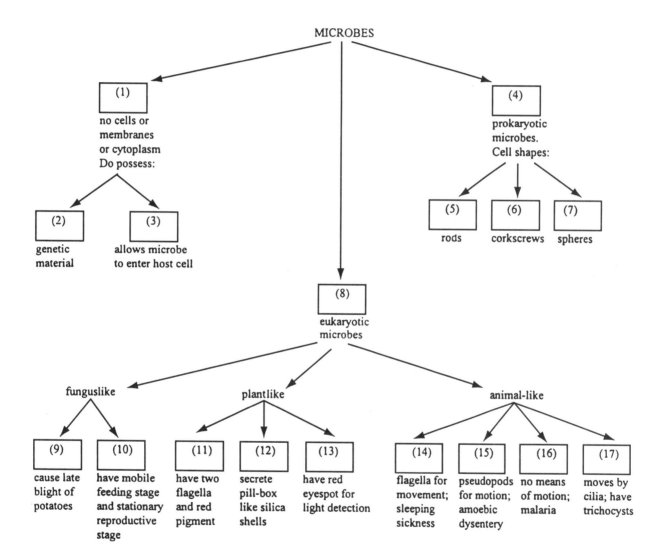

Key Terms and Definitions

acellular slime mold: a fungus-like protist consisting of a mass of cytoplasm containing many nuclei, which in turn forms a fruiting body.

amoeba: shell-less animal-like protists which move and capture food by forming pseudopods

anaerobe: prokaryotic cells that do not depend on oxygen to extract energy from the environment.

antibiotic resistance: the ability of a mutated pathogen to resist the effects of an antibiotic that normally kills it.

Archaea: a domain of prokaryotic organisms fundamentally different from bacteria in their lipids, cell walls, and ribosomal RNA.

bacillus (pl. bacilli): a rod-shaped bacterium.

bacteria: an organism consisting of a single prokaryotic cell surrounded by a complex polysaccharide coat.

bacterial conjugation: the exchange of genetic material between bacteria.

bacteriophage: a virus specialized to attack bacteria.

binary fission: the process by which a single bacterium divides in half, producing two identical offspring.

capsule: a polysaccharide or protein coating that surrounds the cell walls of some bacteria.

cellular slime mold: a fungus-like protist consisting of individual amoeboid cells that can aggregate to form a sluglike mass, which in turn forms a fruiting body.

chemosynthetic: capable of oxidizing inorganic molecules to obtain energy.

chemotactic: a behavior of some bacteria in regars to chemicals in the environment.

ciliate: a category of protozoan characterized by cilia and a complex unicellular structure, including harpoonlike organelles called trichocysts. Members of the genus *Paramecium* are well-known ciliates.

cilium (pl. cilia): a short, hairlike organelle projecting through the cell membrane, usually numerous and engaged in coordinated beating, which moves a cell through a fluid environment or moves the fluid over the surface of the cell.

coccus (pl. cocci): a spherical bacterium.

cyanobacteria: photosynthetic prokaryotic cells, utilizing chlorophyll and releasing oxygen as a photosynthetic by-product, sometimes called "blue-green algae."

diatom: a category of protist that includes photosynthetic forms with two-part glassy outer coverings. Diatoms are important photosynthetic organisms in fresh and salt water.

dinoflagellate: a category of protist that includes photosynthetic forms in which two flagella project through armorlike plates. Abundant in oceans, these sometimes reproduce rapidly, causing "red tides."

endospore: a resistant or resting structure that disperses readily and withstands unfavorable environmental conditions.

euglenoid: a category of protist characterized by one or more whiplike flagella used for locomotion and a photoreceptor for detecting light. Euglenoids are photosynthetic, but some are capable of heterotrophic nutrition if deprived of chlorophyll.

flagellum (pl. flagella): a motile hairlike organelle that propels a cell through a fluid.

foraminiferan: a type of marine animal-like protozoan sarcodine protist that form calcium carbonate (chalky or limestone) shell.

fruiting body: a spore-forming reproductive structure found in certain protists, bacteria, and fungi.

Gram stain: a stain which is selectively taken up by the walls of ceratin types of bacteria (gram positive bacteria) and rejected by the cell walls of others called (gram negative bacteria). The stain is used to distinguish bacteria based on their cell wall construction.

halophile: a type of archaebacterium that thrives in concentrated salt solutions.

heliozoan: a type of freshwater sarcodine animal-like protist that attaches to stalks of underwater plants or rocks and may produce a silica shell.

host: a cell or organism which is invaded by a parasitic organism or a virus.

kuru: a human degenerative brain disease caused by a type of prion passed on by eating diseases brains.

legume: a family of plants characterized by root swellings in which nitrogen-fixing bacteria are housed. Includes soybeans, lupines, alfalfa, and clover.

magnetotactic: behavior of some flagellated bacteria that orient in relation to the Earth's magnetic field due to the presence of iron crystals in their cytoplasm.

methanogen: a type of anaerobic archaebacterium capable of converting carbon dioxide to methane.

nitrogen-fixing bacteria: possessing the ability to remove nitrogen from the atmosphere and combine it with hydrogen to produce ammonia.

pathogen: an organism capable of producing disease.

peptidoglycan: material found in prokaryotic cell walls consisting of chains of sugars cross-linked by short chains of amino acids called peptides.

phototactic: a behavior of some bacteria in relation to moving toward or away from sunlight.

phytoplankton: a general term describing photosynthetic protists that are abundant in marine and freshwater environments.

pilus: a hairlike projection made of protein and found on the surface of certain bacteria, often used to attach the bacterium to another cell.

plasmid: a small circular DNA molecule in some bacteria, often containing genes for antibiotic resistance; passed fron donor to recipient cells during bacterial conjugation.

plasmodial slime mold: a funguslike protist consisting of a multinucleate mass or plasmodium that crawls and ingests decaying organic matter in amoeboid fashion and forms a fruiting body under adverse conditions.

plasmodium: a sluglike mass of cytoplasm containing thousands of nuclei that are not confined within individual cells.

prion: a protein which, in mutated form acts as an infectious agent causing certain neurodegenerative diseases, including kuru and scrapie.

Protista: a taxonomic kingdom including unicellular, eukaryotic organisms.

protozoan (pl. protozoa): a nonphotosynthetic or animal-like protist.

pseudoplasmodium: an aggregation of individual amoeboid cells that form a sluglike mass.

pseudopod: extension of the cell membrane by which certain cells, such as amoebae, locomote and engulf prey.

radiolarian: a type of marine sarcodine animal-like protozoan protist that forms shells of silica.

sarcodine: a category of nonphotosynthetic protist (protozoa) characterized by the ability to form pseudopodia. Some, such as amoebae, are naked, while others have elaborate shells.

slime layer: outer layer secreted by some bacteria enabling them to adhere in clusters to structures within hosts.

spirillum: a spiral shaped bacterium.

sporozoan: a category of parasitic protist. Sporozoans have complex life cycles often involving more than one host, and are named for their ability to form infectious spores. A well-known member (genus *Plasmodium*) causes malaria.

symbiotic: a mutually beneficial association between two organisms.

taxis: a behavior of a cell or organism toward or away from a particular stimulus in the environment.

thermoacidophile: a form of archaebacterium that thrives in hot, acidic environments.

trichocyst: a stinging organelle of protists.

viroid: a particle of RNA capable of infecting a cell and directing the production of more viroids; responsible for certain plant diseases.

virus: a noncellular parasitic particle consisting of a protein coat surrounding a strand of genetic material. Viruses multiply only within cells of living organisms.

water mold: a division (Oomycota) of filamentous fungal-like protists which produce large eggs during sexual reproduction; produce powdery mildew of grapes, late blight of potatoes and other plant diseases.

zooflagellate: a category of nonphotosynthetic protist that move using flagella.

zoospore: a nonsexual reproductive cell that swims using flagella, such as is formed by members of the fungal division Oomycota

THINKING THROUGH THE CONCEPTS

True or False: Determine if the statement given is true or false. If it is false, change the <u>underlined</u> word so that the statement reads true.

18. _____ Protistans organisms are <u>eukaryotic</u>.
19. _____ Viruses are <u>cellular</u>.
20. _____ Antibiotics are a <u>better</u> defense against viral infection than vaccines.
21. _____ Bacilli bacteria are <u>rod-shaped</u>.
22. _____ Bacteria reproduce asexually by <u>mitosis</u>.
23. _____ Cyanobacteria are <u>heterotrophic</u>.
24. _____ Unicellular algae are often called <u>phytoplankton</u>.
25. _____ <u>Dinoflagellates</u> are unicellular, eukaryotic, heterotrophic protists.
26. _____ The organism causing African sleeping sickness is a <u>sporozoan</u>.
27. _____ The organism causing malaria is a <u>zooflagellate</u>.

Matching: Archeae .

28.____ found in hot sulfur springs
29.____ produce "swamp gas" (methane)
30.____ thrive in bodies of concentrated salt water
31.____ thrive in very hot environments
32.____ found in the stomachs of cows
33.____ found in very acidic environments
34.____ found in the Dead Sea

Choices:

a. halophiles
b. thermoacidophiles
c. methanogens

Matching: Unicellular algae.

35.____ will absorb nutrients if maintained in darkness
36.____ produce glassy shells that fit together like shoe boxes
37.____ have pairs of flagella
38.____ lack cellulose cell walls
39.____ some are bioluminescent
40.____ have single flagella for locomotion
41.____ all live in fresh water
42.____ cause "red tides" that kill fish
43.____ have simple light-sensitive organelles
44.____ store reserve food as oils that help them stay buoyant
45.____ produce a toxic nerve poison
46.____ make food by photosynthesis

Choices:

a. euglenoids
b. dinoflagellates
c. diatoms
d. all these algae

Matching: Protozoans.

47.____ the "white cliffs of Dover" are made from their calcium shells

48.____ all are parasites

49.____ move by using flagella

50.____ use cilia for locomotion

51.____ some are symbiotic

52.____ use pseudopods for locomotion

53.____ have no means of locomotion

54.____ have trichocysts which aid in predation

55.____ one type digests cellulose in the guts of termites

56.____ one type causes malaria

57.____ one type causes amoebic dysentery

58.____ some have contractile vacuoles to excrete excess water

59.____ one type causes African sleeping sickness

60.____ some types form glassy shells

Choices:

   a.   sarcodines

   b.   zooflagellates

   c.   ciliates

   d.   sporozoans

   e.   all of these

Multiple Choice: Pick the most correct choice for each question.

61. Of the following, which are eukaryotic?
    a. bacteria
    b. blue-green algae
    c. protists
    d. viruses
    e. viroids

62. All of the following are true **except**
    a. viruses are prokaryotic
    b. viruses require a host cell
    c. viruses are intracellular parasites
    d. viruses possess proteins and hereditary material
    e. viruses may have DNA or RNA

63. Viral infections do <u>not</u> cause:
    a. AIDS
    b. herpes
    c. common cold
    d. flu
    e. malaria

64. Nitrogen-fixing bacteria:
    a. convert nitrates into nitrogen gas
    b. convert nitrogen gas into a usable form
    c. remove nitrates from the soil
    d. remove nitrogen gas from plants
    e. are found on the roots of corn and wheat plants

65. All protists are characterized by:
    a. single cell
    b. cell wall
    c. chlorophyll
    d. filamentous body form
    e. all of the choices are correct

66. Protozoans include all of the following **except**:
    a. dinoflagellates
    b. sporozoans
    c. zooflagellates
    d. sarcodines
    e. ciliates

67. "Red tides" are caused by a population explosion of:
    a. diatoms
    b. zooflagellates
    c. dinoflagellates
    d. euglenoids
    e. bacteria

68. The malarial parasite, *Plasmodium*, is a:
    a. sporozoan
    b. ciliate
    c. zooflagellate
    d. sarcodine
    e. blue-green alga

Short Answer: Based on the following figure, answer the questions below.

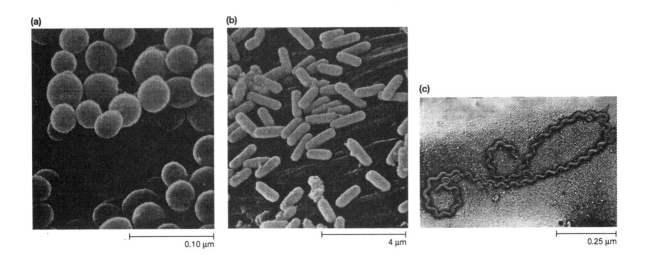

69. The shape of the bacteria of type "a" is _____.

70. The shape of the bacteria of type "b" is _____.

71. The shape of the bacteria of type "c" is _____.

CLUES TO APPLYING THE CONCEPTS

This question is intended to sharpen your ability to apply critical thinking and analysis to biological concepts covered in this chapter.

72. The common intestinal bacterium called *E. coli* is a typical bacterium in that it can reproduce rapidly under favorable growth conditions. Suppose you placed 10 bacteria in a very large flask containing the ideal culture medium for bacterial growth and reproduction. Under these conditions, the bacteria will divide every 20 minutes. How long would it take until the flask contained about 1 billion bacteria? How many bacteria would be present after 12 hours? If bacteria reproduce so rapidly, why aren't we neck-deep in bacteria all over the world?

ANSWERS TO EXERCISES

1. viruses
2. DNA and RNA
3. protein coats
4. bacteria and Archaea
5. bacilli
6. spirilla
7. cocci
8. protistans
9. water molds
10. slime molds
11. dinoflagellates
12. diatoms
13. euglenoids
14. zooflagellates
15. sarcodines
16. sporozoans
17. ciliates

18. true
19. false, non-cellular
20. false, worse
21. true
22. false, binary fission
23. false, autotrophic
24. true
25. false, protozoa
26. false, zooflagellate
27. false, sporozoan
28. b
29. c
30. a
31. b
32. c
33. b
34. a

35. a
36. c
37. b
38. a
39. b
40. a
41. a
42. b
43. a
44. c
45. b
46. d
47. a
48. d
49. b
50. c
51. b

52. a
53. d
54. c
55. b
56. d
57. a
58. c
59. b
60. a
61. c
62. a
63. e
64. b
65. a
66. a
67. c
68. a

69. cocci or spherical bacteria

70. bacilli or rod-shaped bacteria

71. spirilla or corkscrew-shaped bacteria

72. Since the bacteria double in number every 20 minutes, after one hour there would be 10 x 2 x 2 x 2 = 80 bacteria (an eight-fold increase every hour). At that rate of increase, it would take about 9 hours to accumulate 1.342 billion bacteria. After 12 hours, there would be approximately 700 billion bacteria in the flask (about 1.342 billion after 9 hours x 8 = about 10.74 billion after 10 hours x 8 = about 85.92 billion after 11 hours x 8 = about 687.36 billion after 12 hours). There are estimates that the total weight of all bacteria on Earth is much higher than the total weight of all other organisms combined. But we are not swamped with bacteria because seldom do they find the ideal conditions necessary to reproduce as rapidly as they can.

Chapter 20: The Fungi

OVERVIEW

In this chapter, the authors describe the general features and ecology of fungi, then give a brief description of the major fungal divisions.

1) What Are the Main Adaptations of Fungi?

Most fungi have a body called a **mycelium**, which is an interwoven mass of threadlike filaments called **hyphae**. The hyphae are either made of one cell with many nuclei, or made of many cells separated by porous partitions called **septa**. Fungal cell walls contain chitin. Fungi do not move about, but hyphae can grow rapidly in any direction within suitable environments.

Fungi obtain nutrients from other organisms. Some are **saprobes**, digesting the bodies of dead organisms, secreting enzymes outside their bodies to digest complex molecules, and then absorbing the smaller subunits. Others are parasitic, feeding on living organisms and causing disease. Still others live in mutually beneficial relationships (**symbiosis**) with other organisms. A few are predatory.

Most fungi can reproduce both sexually and asexually. In simple asexual reproduction, a mycelium breaks into pieces, each growing into a new mycelium. Some fungi produce sexual and asexual **spores** (small, spherical, thick-walled single cells that are dispersed by wind or water and develop into new mycelia). Mitotic division of haploid mycelium cells produces asexual spores. Fusion of haploid cells produces diploid zygotes, which undergo meiosis to make haploid sexual spores.

2) How Do Fungi Affect Humans?

Some fungi, such as mushrooms and truffles, serve as food. Some are decomposers of other organisms, releasing nutrients with carbon, nitrogen, and phosphorous, and minerals used by plants. Cycling of nutrients within ecosystems would cease without fungal and bacterial decomposers. Parasitic fungi can cause skin diseases such as ringworm and athlete's foot, lung diseases such as valley fever and histoplasmosis, and vaginal yeast infections. Fungi cause plant diseases such as chestnut blight, Dutch elm disease, and corn smut. Farmers use "fungal pesticides" to kill crop predators (termites, tent caterpillars, aphids, citrus mites, rice weevils). Fungi are used to make bread rise, produce wine from grapes, flavor cheeses, and make beer.

3) How Are Fungi Classified?

The **zygote fungi** (600 species, Zygomycota) reproduce by forming diploid **zygospores**. They cause soft fruit rot and bread mold. Haploid cells from hyphae of different mating types fuse sexually to produce diploid zygospores that disperse through the air, then undergo meiosis to produce new hyphae.

The **sac fungi** (30,000 species, Ascomycota) reproduce by forming sexual spores within sacs called **asci** (plural of **ascus**). Sac fungi are colorful molds, morels, and truffles, and cause Dutch elm disease and chestnut blight. Some yeasts cause vaginal infections, while others are used in the baking and brewing industries.

The **club fungi** (25,000 species, Basidiomycota) produce club-shaped reproductive structures (**basidia**) containing sexual spores (**basidiospores**). Common club fungi are mushrooms, puffballs, shelf fungi (monkey-stools), as well as parasitic rusts and smuts of grain crops. Germinating basidiospores form hyphae of two different mating types, which fuse to form underground mycelia. The filaments grow outward below ground from the original spore in a roughly circular pattern, occasionally sending up numerous mushrooms in a ringlike pattern called a **fairy ring**.

The **imperfect fungi** (25,000 species, Deuteromycota) have no known means of sexual reproduction. Some produce useful antibiotics (*Penicillium* makes penicillin). Some are used to flavor cheese. Others cause diseases such as ringworm and athlete's foot. Some are predators of roundworms.

Lichens are associations of fungi (sac fungi) with photosynthetic green algae or cyanobacteria. The fungus provides support and water, and the algal or bacterial partner provides food. They grow on bare rock, acting as original colonizers of new environments.

Mycorrhizae are fungi (club or sac fungi) associated symbiotically with the roots of about 80% of all rooted plants. The hyphae invade the root cells, digesting organic nutrients in the soil, absorbing the digested nutrients and water, and passing them directly into the root cells. The plant provides sugar to the fungal cells.

KEY WORDS AND CONCEPTS

From the information in this chapter, fill in the crossword puzzle based on the following clues.

Across
1. (Across)A symbiotic association between a fungus and a plant root.
3. The _____ fungi have no known mechanism of sexual reproduction; some cause ringworm and athlete's foot.
6. A saclike case in which sexual spores are formed.
8. A diploid cell, often club-shaped, which produces spores by meiosis.
10. The _____ fungi are molds of fruit and bread.
11. The _____ fungi are the morels, truffles, and yeasts.
12. A partition that separates the fungal hypha into individual cells.
13. The outward underground growth of mycelia from club fungi, with periodic emergence of numerous fruiting bodies in a ringlike pattern.

Down
1. (Down)The body of a fungus, consisting of a mass of hyphae.
2. A threadlike structure, many of which make up the fungal body, that consists of elongated cells.
4. The _____ fungi are the mushrooms, puffballs, shelf fungi, rusts, and smuts.
5. A sexual spore formed by members of the fungal division Basidiomycota.
7. A close relationship between two types of organisms.
9. An association between an alga and a fungus.

Key Terms and Definitions

ascus: a saclike case in which sexual spores are formed by members of the fungal division Ascomycota.

basidiospore: a sexual spore formed by members of the fungal division Basidiomycota.

basidium: a diploid cell, often club-shaped, formed on the basidiocarp of members of the fungal division Basidiomycota, which produces basidiospores by meiosis.

club fungi: members of the division Basidiomycota; reproduce sexually by forming club-shaped reproductive structures called basidia containing sexual spores called basidiospores; mushrooms, puffballs, shelf fungi, rusts, smuts.

fairy ring: outward underground growth of mycelia from club fungi basidiospores, with periodic emergence of numerous fruiting bodies in a ringlike pattern.

hypha (pl. hyphae): a threadlike structure, many of which make up the fungal body, that consists of elongated cells, often with many haploid nuclei.

imperfect fungi: members of the division Deuteromycota; no known mechanism of sexual reproduction; some produce antibiotics (penicillin), others are parasites causing ringworm and athlete's foot.

lichen: a symbiotic association between an alga or cyanobacterium and a fungus, resulting in a composite organism.

mycelium: the body of a fungus, consisting of a mass of hyphae.

mycorrhiza (pl. mycorrhizae): a symbiotic association between a fungus and a plant root. The fungus, often a basidiomycete or anascomycete, grows around and often into the roots of most vascular plants in a mutually beneficial relationship.

sac fungi: members of the division Ascomycota; produce sexual ascospores within saclike cases called asci; morels, truffles, yeasts.

saprobe: an organism that derives its nutrients from the bodies of dead organisms.

septum (pl. septa): a partition that separates the fungal hypha into individual cells. Pores in the septa allow transfer of materials between cells.

sporangium (pl. sporangia): a structure in which spores are produced.

spore: small, spherical, thick-walled single cells, produced

by either sexual and asexual reproduction in fungi, that are dispersed by wind or water and develop into new hyphae.

symbiosis: a close relationship between two types of organisms. Either or both may benefit from the association, or, in the case of parasitism, one of the participants is harmed.

zygospore: produced by the division Zygomycota, a fungal spore surrounded by a thick, resistant wall, which forms from a diploid zygote.

zygote fungi: members of the division Zygomycota; reproduce sexually by forming diploid zygospores; molds of fruit and bread.

THINKING THROUGH THE CONCEPTS

True or False: Determine if the statement given is true or false. If it is false, change the underlined word so that the statement reads true.

14._____ The single threadlike filaments that make up the body of a fungus are called mycelia.
15._____ The cell walls of most fungi are composed of chitin.
16._____ Fungi are autotrophic.
17._____ Most fungal nuclei are diploid.
18._____ Fungi digest food particles outside their bodies.
19._____ In fungi, haploid sexual spores are produced from diploid zygotes.
20._____ Fungi and protozoans are decomposers.
21._____ A fungus causes Dutch elm disease.
22._____ A fungus found growing beneficially with plant roots is called a lichen.
23._____ Yeast belong to the fungal division Zygomycota.

Matching: Symbiotic fungi.

24.____ hyphae often invade plant root cells
25.____ association between ascomycete fungi and cyanobacteria
26.____ harm the non-fungal partner
27.____ early colonizers of bare rocky volcanic islands
28.____ association between basidiomycetes fungi and the roots of vascular plants
29.____ association between ascomycete fungi and animals
30.____ digest organic compounds in the soil

Choices:
a. lichens
b. mycorrhizae
c. both of these
d. neither of these

Matching: Fungi.

31.____ mushrooms
32.____ sac fungi
33.____ egg fungi
34.____ causes ringworm
35.____ produce zygospores
36.____ no sexual reproduction
37.____ yeasts
38.____ black bread mold
39.____ late blight of Irish potatoes
40.____ makes penicillin
41.____ causes Dutch-elm disease
42.____ puffballs
43.____ truffles
44.____ have cellulose in their cell walls
45.____ used in making Roquefort cheese
46.____ water molds
47.____ club fungi
48.____ causes athlete's foot

Choices:
a. Zygomycota
b. Ascomycota
c. Deuteromycota
d. Basidiomycota
e. Oomycota

Multiple Choice: Pick the most correct choice for each question.

49. With few exceptions, the fungal plant body is composed of
a. hyphae
b. chitin
c. mycorrhizae
d. vascular tissue
e. cellulose

50. Fungal cell walls are characterized by the presence of
a. algin
b. cellulose
c. glucose
d. chitin
e. calcium carbonate

51. Ecologically, fungi as well as bacteria are important because they are
a. predators
b. the basis of food chain
c. producers of organic materials
d. decomposers of organic materials
e. capable of nitrogen-fixation

52. Spore formation occurs in sac-like structures within the
a. Zygomycetes
b. Basidiomycetes
c. Fungi imperfecti
d. Ascomycetes
e. Oomycetes

53. Lichens are a symbiotic relationship between
a. two different fungi
b. protist and alga
c. mold and mildew
d. protist and fungus
e. alga and fungus

CLUES TO APPLYING THE CONCEPTS

This practice question is intended to sharpen your ability to apply critical thinking and analysis to biological concepts covered in this chapter.

54. Be careful what you wish for, in case it comes true. The owner of a peach orchard was worried because his trees were attacked by a parasitic fungus. He wished to get rid of all fungi in his orchard, so he sprayed his trees and the ground with a powerful fungicide. The next season, no fungi were found in the orchard, but the trees grew even more poorly and produced even less fruit than the year before, when they were infected. Why did the orchard owner experience such poor results?

ANSWERS TO EXERCISES

1. (across) mycorrhiza	14. false, hyphae	28. b	41. b
1. (down) mycelium	15. true	29. d	42. d
2. hypha	16. false, heterotrophic	30. b	43. b
3. imperfect	17. false, haploid	31. d	44. e
4. club	18. true	32. b	45. c
5. basidiospore	19. true	33. e	46. e
6. ascus	20. false, bacteria	34. c	47. d
7. symbiosis	21. true	35. a	48. c
8. basidium	22. false, mycorrhiza	36. c	49. a
9. lichen	23. false, Ascomycota	37. b	50. d
10. zygote	24. b	38. a	51. d
11. sac	25. a	39. e	52. d
12. septum	26. d	40. c	53. e
13. fairy ring	27. a		

54. The orchard owner should not have wished for all the fungi to die, for his over-zealous spraying of fungicide killed not only the harmful parasites but also the beneficial mycorrhizae fungi. Without the mycorrhiza, the peach trees could not absorb water and minerals from the soil as efficiently and, therefore, grew poorly and yielded less fruit.

Chapter 21: The Plant Kingdom

OVERVIEW

This chapter focuses on the plants. After a discussion of evolutionary trends, the authors describe the multicellular algae, bryophytes, ferns, conifers, and flowering plants.

1) What Are the Key Features of Plants?

Most plants are multicellular and use photosynthesis to convert carbon dioxide and water into sugar and oxygen gas. They have an **alternation of generation** life cycle consisting of both a diploid **sporophyte** generation (that develops from a zygote and produces haploid spores by meiosis) and a haploid **gametophyte** generation (that develops from a spore and produces, by mitosis, gametes that fuse to produce a diploid zygote).

Plants produce separate diploid and haploid generations that alternate with each other. Generally, a diploid zygote develops into a diploid sporophyte plant by mitosis, which then produces haploid spores by meiosis. A haploid spore develops into a haploid gametophyte plant by mitosis, then produces haploid gametes by mitosis, followed by the fusion of sperm and egg to make a diploid zygote. Increasing prominence of the sporophyte generation, accompanied by decreasing size and duration of the gametophyte generation, is the general trend in plant evolution.

2) What Is the Evolutionary Origin of Plants?

Plant ancestors were most likely aquatic, photosynthetic protists, similar to present-day **algae**. Algae: (1) lack true roots, stems, leaves, and complex reproductive structures like flowers and cones; (2) shed gametes directly into the water; (3) have varied and complex life cycles, some with dominant haploid gametophyte stages, some with dominant diploid sporophyte stages, and others with nearly identical sporophyte and gametophyte stages; and (4) are colored by pigments that capture light energy for photosynthesis.

The red algae (4000 species, Rhodophyta) have red and green (chlorophyll) pigments and are mostly marine and multicellular forms in clear tropical waters. Some produce gelatinous carrageenan (used as stabilizers in paints, cosmetics, and ice cream) and agar (used in gelatins).

The brown algae (1500 species, Phaeophyta) are the multicellular seaweeds, have brownish-yellow and green pigments, and dominate cool marine coastal waters. The giant kelps provide food, shelter, and breeding areas for a wide variety of marine animals.

The green algae (7000 species, Chlorophyta) are mostly multicellular and colonial forms living in ponds and lakes. Evidence that land plants evolved from green algal ancestors include: (1) similar types of chlorophyll and accessory pigments in green algae and land plants; (2) green algae and land plants store food as starch and have cell walls made of cellulose; and (3) algae live in fresh water, leading to adaptations for life on land, such as fluctuating temperatures and periods of dryness.

3) How Did Plants Invade and Flourish on Land?

The plant body increased in complexity as plants invaded the land. Plants adapted to dry land by anchoring roots to absorb water and nutrients; developing conducting vessels to transport water and minerals upwards from roots to leaves, and move sugars from leaves to other body parts; and by producing stiffening polymer **lignin** in the conducting vessels to support the plant body. They developed a waxy waterproof **cuticle** on leaf and stem surfaces to limit water evaporation, and **stomata** (pores) in leaves and stems that open for gas exchange and close when water is scarce to reduce evaporation. Instead of flagellated gametes and spores (**zoospores**) that algae release into the water, plants evolved pollen, seeds, and later flowers and fruits to protect spores, gametes, and young embryos from dessication, attract pollinators, and aid in dispersion of offspring.

The bryophytes (16,000 species of mosses and liverworts) are nonvascular plants lacking true roots, stems, and leaves. Anchoring structures (**rhizoids**) bring water and nutrients into the plant body and they then diffuse throughout the body which must remain small (less than 1 inch tall). Enclosed reproductive structures are present to prevent dessication: **archegonia** (in which eggs develop) and **antheridia** (in which sperm develop). Sperm must swim to eggs through a film of water; thus, most bryophytes are confined to moist areas. The leafy gametophyte plant is larger than the leafless sporophyte, which develops from zygotes growing upward out of the archegonia in the gametophyte plants.

Adaptations that allowed plants to grow tall included support structures for the body, and vessels to conduct water and nutrients. Evolution of rigid conducting cells (**"vessels"**) in the **vascular plants** allowed plants to live on dry land. In vascular plants, and especially the seed plants, the diploid sporophyte plant body is dominant over the smaller, shorter-lived gametophyte plants.

Seedless vascular plants include the smaller divisions of club mosses (Lycophyta) and horsetails (Sphenophyta), and a large division of ferns (Pteridophyta, 12,000 species). Ferns are the only seedless vascular plants with broad leaves. In ferns, the small gametophyte plants lack conducting vessels, and the sperm must swim through water to reach the egg.

The seed plants dominate the land due to the evolution of **pollen grains** (to allow for sperm to find eggs without swimming through water) and **seeds** (to allow embryos to develop without being immersed in water). The pollen grain, containing sperm-producing cells, is all that remains of the male gametophyte generation in seed plants. The female gametophyte plant is a small group of haploid cells that produces an egg. Pollen grains are dispersed by wind or by animal pollinators like bees. Analogous to the eggs of birds and reptiles, seeds consist of: (1) an embryonic plant; (2) a supply of food for the embryo; and (3) a protective outer coat.

Gymnosperms evolved earlier than the flowering plants and include the conifers (500 species, Coniferophyta) and two smaller groups (the cykads, and gingkos). Conifers (pines, firs, spruces, hemlocks, and cypresses) are most abundant in the far north and at high elevations. They are adapted to dry, cold conditions due to: (1) retention of green leaves throughout the year (**evergreens**), allowing them to photosynthesize and grow all year long; (2) leaves that are thin needles covered with a thick cuticle to minimize evaporation; and (3) production of "antifreeze" in their sap to allow transport of nutrients in sub-freezing temperatures.

A pine tree is the sporophyte plant. It makes smaller male and larger female cones. Male cones release pollen (male gametophyte with sperm) carried by wind to female cones (female gametophyte with eggs). The pollen grain sends out a pollen tube that burrows into the female cone. After 14 months, it reaches an egg cell, releases sperm and fertilization occurs. The fertilized egg is enclosed in a seed and develops into an embryo. The seed is liberated when the cone matures and its scales separate.

Angiosperms (Anthophyta, 230,000 species) evolved from gymnosperm ancestors that formed an association with insects who carried pollen from tree to tree. Insects ate some pollen for food and carried pollen from plant to plant, and trees wasted less pollen than gymnosperms. Angiosperms are successful due to: (1) **flowers** (sporophyte structures containing the male and female gametophytes) where fertilizations occur within the flower ovary (where the eggs are and the seeds develop); (2) **fruits** (ripened flower ovaries containing seeds) that attract animals and entice them to disperse seeds; and (3) broad leaves to increase the amount of sunlight trapped for photosynthesis in warm, moist climates.

Most temperate climate angiosperms drop their leaves annually, during fall and winter, and become dormant. To discourage animals from eating tender leaves, angiosperms have evolved many defenses, including thorns, spines, resins, and chemical defenses now harvested by humans (used in such substances as taxol®, aspirin, nicotine, caffeine, mustard, and peppermint). Flowering plants may be grouped into the **monocots** (most grasses, grains, palms) and **dicots** (most trees, shrubs, herbs).

KEY TERMS AND CONCEPTS

From the following list of key terms, fill in the blanks in the following statements.

algae	ferns	roots
alternation of generation	gametophyte	seeds
angiosperms	green	sporophyte
brown	gymnosperms	stomata
bryophytes	lignin	vessels
cuticle	red	

Most plants have an (1)_____ life cycle consisting of both a diploid (2)_____ generation (that develops from a zygote and produces haploid spores by meiosis) and a haploid (3)_____ generation (that develops from a spore and produces, by mitosis, gametes that fuse to produce a diploid zygote).

Plant ancestors were most likely aquatic, photosynthetic protists, similar to present-day (4)_____.

The (5)_____ algae are mostly marine and multicellular forms in clear tropical waters. Some produce agar.

The (6)_____ algae are the multicellular seaweeds and dominate cool marine coastal waters.

The (7)_____ algae are mostly multicellular and colonial forms living in ponds and lakes.

Plant adaptations to dry land include: (8)_____ to absorb water and nutrients; (9) _____ to transport water and minerals upward from roots to leaves, and move sugars from leaves to other body parts; (10)_____ in the conducting vessels to support the plant body; waterproof (11)_____ on leaf and stem surfaces to limit water evaporation; (12)_____ in leaves and stems that open for gas exchange and close when water is scarce to reduce evaporation; and (13)_____ to protect young embryos from dessication.

The (14)_____ (mosses and liverworts) are nonvascular plants lacking true roots, stems, and leaves.

(15)_____ are seedless vascular plants with broad leaves.

(16)_____ include the conifers, the cykads, and the gingkos.

(17)_____ produce seeds, flowers, and fruits.

Key Terms and Definitions

algae (sing. alga): a general term for simple aquatic plants lacking vascular tissue.

alternation of generations: a life cycle typical of plants in which a diploid sporophyte (spore-producing) generation alternates with a haploid gametophyte (gamete-producing) generation.

angiosperm: a flowering vascular plant.

antheridium: a structure in which male sex cells are produced, found in the bryophytes and certain seedless vascular plants.

archegonium: a structure in which female sex cells are produced, found in the bryophytes and certain seedless vascular plants.

bryophyte: a division of simple nonvascular plants, including mosses and liverworts.

conifer: a class of tracheophyte that reproduces using seeds formed inside cones and retains its leaves throughout the year.

cuticle: a waxy or fatty coating on the exposed surfaces of epidermal cells of many land plants, which aids in the retention of water.

dicot: a class of angiosperm whose embryo has two cotyledons, or seed leaves.

evergreen: cone-bearing seed plants that do not shed their leaves.

flower: the reproductive structure of an angiosperm plant.

fruit: the mature ripened ovary of an angiosperm plant. This structure contains the seeds.

gametophyte: a multicellular haploid plant that produces haploid sex cells by mitosis.

gymnosperms: nonflowering seed plants such as conifers, cycads, and gingkos.

lignin: a stiffening polymer found in the conducting vessels of land plants to support the plant body.

monocot: a class of angiosperm plant in which the embryo has one cotyledon, or seed leaf.

ovule: a structure within the ovary of a flower, inside which the female gametophyte develops. After fertilization, the ovule develops into the seed.

pollen: the male gametophyte of gymnosperms and angiosperms.

rhizoid: a rootlike structure found in bryophytes that anchors the plant and absorbs water and nutrients from the soil.

seed: the reproductive structure of a seed plant. The seed is protected by a seed coat and contains an embryonic plant and a supply of food for it.

sporophyte: the diploid form of a plant that produces haploid, asexual spores through meiosis.

stoma (pl. stomata): an adjustable opening in a leaf that regulates the diffusion of carbon dioxide and water into and out of the leaf.

tracheophyte: a category of plant that has conducting vessels; a vascular plant.

vascular: conducting tissue in land plants, containing the polymer lignin for rigidity.

vessel: a tube of conducting cells, with perforated or missing end walls, leaving a continuous uninterrupted hollow cylinder.

zoospore: a nonsexual reproductive cell that swims using flagella.

zygote: a diploid cell resulting from the fusion of male and female sex cells.

THINKING THROUGH THE CONCEPTS

True or False: Determine if the statement given is true or false. If it is false, change the <u>underlined</u> word so that the statement reads true.

18._____   The gametophyte plant generation is <u>diploid</u>.
19._____   Spores develop into the <u>gametophyte</u> generation.
20._____   The gametophyte plant produces <u>gametes</u>.
21._____   The brown algae are commonly found in <u>tropical</u> waters.
22._____   The archegonium is a <u>male</u> reproductive organ.
23._____   The dominant generation of the bryophytes is the <u>gametophyte</u>.
24._____   A liverwort is an example of a <u>vascular</u> plant.
25._____   The <u>gametophyte</u> generation of ferns is the large, visible generation.
26._____   The <u>angiosperms</u> are the flowering plants.
27._____   The <u>anther</u> is the male part of a flower.

Matching: Alternation of generations.

28._____ haploid   Choices:
29._____ begins as a zygote   a.   sporophyte
30._____ multicellular   b.   gametophyte
31._____ produces sex cells by meiosis   c.   both of these
32._____ begins as a spore   d.   neither of these
33._____ produces spores by meiosis
34._____ grows by mitosis
35._____ dominant form in flowering plants

Matching: Algae.

36._____ shed gametes directly into water   Choices:
37._____ brown algae   a.   Rhodophyta
38._____ contribute limestone to coral reefs   b.   Phaeophyta
39._____ red algae   c.   Chlorophyta
40._____ much like land plants   d.   all of these
41._____ found in clear, tropical, marine habitats e.   none of these
42._____ stores food as starch
43._____ *Spirogyra*
44._____ seaweeds and kelps
45._____ photosynthesize
46._____ green algae
47._____ produce food emulsifiers
48._____ live in cool temperate oceans
49._____ live in fresh water environments
50._____ have vascular tissue

Matching: Land plants. (Hint: There may be more than one correct answer for sone questions.)

51.____ do not produce seeds
52.____ make pollen grains
53.____ lack true vascular tissues
54.____ some rely on insect pollinators
55.____ sperm must swim to eggs
56.____ produce fruits
57.____ seedless vascular plants
58.____ gametophytes larger than sporophytes
59.____ monocots and dicots
60.____ make cones
61.____ most successful land plants today
62.____ vascular plants
63.____ mosses
64.____ seed plants

Choices:

a.   angiosperms
b.   gymnosperms
c.   ferns
d.   bryophytes
e.   all of these
f.   none of these

Multiple Choice: Pick the most correct choice for each question.

65. In the alternation of generations life cycle, the haploid stage is called the
  a.   zygote
  b.   gametophyte
  c.   sporophyte
  d.   antheridium
  e.   archegonium

66. The haploid stage of an alternation of generations life cycle produces
  a.   seeds
  b.   spores
  c.   reproductive cells
  d.   zygotes
  e.   cones

67. During alternation of generations, meiosis results in the production of:
  a.   egg cells
  b.   sperm cells
  c.   spores
  d.   answers a and b are correct
  e.   answers a, b and c are correct

68. The ancestors of the terrestrial plants are considered to be:
  a.   bryophytes
  b.   green algae
  c.   red algae
  d.   brown algae

69. Which of the following is seen as a basic trend in plant evolution?
  a.   gametophyte generation increases in size
  b.   gametophyte generation decreases in size
  c.   sporophyte generation produces gametes
  d.   sporophyte generation decreases in size
  e.   water becomes more necessary for reproduction

70. Gymnosperms are characterized by all of the following **except**:
  a.   cones
  b.   seeds
  c.   needle-shaped leaves
  d.   evergreen
  e.   fruit

Short Answer: From the figure below, answer the following questions concerning the alternation of generations life cycle.

71. The type of plant body depicted by (1) is commonly called a _____.

72. The type of cell division (2) producing spores is called _____.

73. The type of plant bodies depicted by (3) are commonly called _____.

74. The fusion of gametes (4) is commonly referred to as _____.

75. Are zygotes typically haploid or diploid cells? _____.

76. Are spores typically haploid or diploid cells? _____.

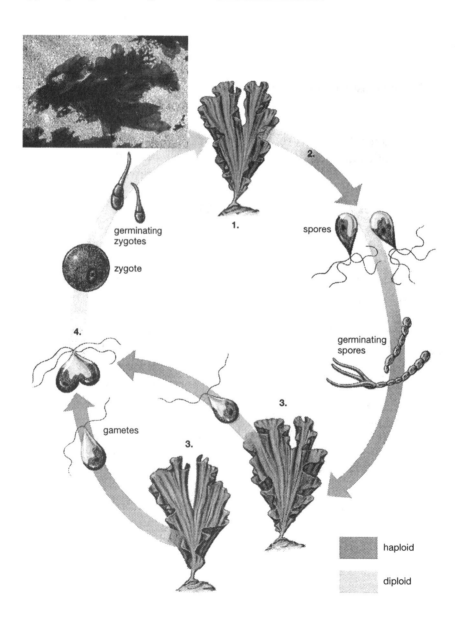

CLUES TO APPLYING THE CONCEPTS

This practice question is intended to sharpen your ability to apply critical thinking and analysis to biological concepts covered in this chapter.

77. In recent years, bees domestic to the United States have decreased in numbers for several reasons: the rise in bee parasites, unfavorable changes in climate, and the spread of so-called killer bees from South America. The decline of domestic bees has worried many farmers, especially those who own orchards. Why should farmers be concerned with bees when they grow apples and peaches?

ANSWERS TO EXERCISES

1. alternation of generation
2. sporophyte
3. gametophyte
4. algae
5. red
6. brown
7. green
8. roots
9. vessels
10. lignin
11. cuticle
12. stomata
13. seeds
14. bryophytes
15. ferns
16. gymnosperms
17. angiosperms
18 false, haploid
19. true

20. true
21. false, temperate
22. false, female
23. true
24. false, nonvascular
25. false, sporophyte
26. true
27. true
28. b
29. a
30. c
31. d
32. b
33. a
34. c
35. a
36. d
37. b
38. a
39. a

40. c
41. a
42. c
43. c
44. b
45. d
46. c
47. a
48. b
49. c
50. e
51. c, d
52. a, b
53. d
54. a
55. c, d
56. a
57. c
58. d

59. a
60. b
61. a
62. a, b, c
63. d
64. a, b
65. b
66. c
67. c
68. b
69. b
70. e
71. sporophyte
72. meiosis
73. gametophytes
74. fertilization
75. diploid
76. haploid

77. Orchard owners are concerned about the decline in native bee populations because bees are the major organisms responsible for cross-pollination of the flowering trees in their orchards, carrying pollen from one tree's flowers to those of another tree. Without adequate numbers of bees, maximum pollination will not occur, leading to reductions in amount of fruit produced.

Chapter 22: The Animal Kingdom

OVERVIEW

This chapter covers the major Phyla of animals including the invertebrates (sponges, cnidarians, flat-, round-, and segmented worms, arthropods, molluscs, and echinoderms) and vertebrates (chordates including fish, amphibians, reptiles, birds, and mammals). **Table 22-1** is an excellent summary of the major characteristics of each phylum of animals discussed in this chapter. Learn as much of it as you can.

1) What Characteristics Define an Animal?

These characteristics collectively define an animal: (1) they have multicellular bodies; (2) They obtain energy by eating the bodies of other organisms; (3) they reproduce sexually; (4) their cells lack cell walls; (5) they are motile; and (6) they respond rapidly to external stimulation.

2) What Are the Major Evolutionary Trends in Animals?

Evolutionarily, animals have increased in complexity in many ways. They have increased cellular organization and specialization: cells are the basic unit of life, **tissues** are groups of similar cells, **organs** are groups of different tissues, and **organ systems** are groups of different organs. Also, there is an increase in **germ (tissue) layers**: sponges have no true tissues; cnidarians have inner **endoderm** lining most hollow organs and outer **ectoderm**, epithelial tissue covering the body and lining the inner cavities, plus the nervous system cells; flatworms and higher animals have a third layer in the middle, the **mesoderm**, which forms muscular, circulatory, and skeletal systems.

Animals have become increasingly symmetrical. Some sponges have no symmetry (asymmetrical) while other sponges, the cnidarians, and some adult echinoderms have wheel-like **radial symmetry**. More complex animals and larval echinoderms have **bilateral symmetry** (right-left or dorsal-ventral mirror images).

Cephalization (concentration of sensory organs and brain in the head) has increased. Radial animals are **sessile** and have no front or back ends. Bilateral animals (flatworms and higher forms) move forward and have concentrations of sensory organs in the **anterior** (front or head) end. The other end is the **posterior** end and typically has a tail

Another evolutionary trend has been in the development of the body cavity between the digestive tube or gut and the outer body wall. Such body cavities free the gut from being attached to the body wall so it can digest food more effectively, and create space for internal organs. Cnidarians and flatworms lack body cavities, roundworms have a **pseudocoelom** (body cavity partially lined with mesoderm tissues), and annelids and higher groups have a true **coelom** (body cavity totally lined with mesoderm tissue). Coeloms develop in one of two general ways. In **protostome** animals (annelids, arthropods, and mollusks), the coelom develops in spaces between the gut and body wall. In **deuterostome** animals (echinoderms and chordates), the coelom develops from outgrowths of the gut.

The structure of the digestive system became much more complex. Sponges have intracellular digestion, while cnidarians and flatworms have an internal digestive sac with one opening serving as mouth and anus and roundworms and higher animals have an efficient, tubular digestive system with separate mouth and anal openings allows for one-way flow of materials. In this system, food is ingested through an anterior mouth, physically pulverized, then enzymatically digested. The nutrients are absorbed and waste is eliminated through the posterior anus.

Segmentation, the presence of similar repeated body parts, first arose in annelid worms.

3) What Are the Major Animal Phyla?

There are two major categories of animals. **Invertebrates**, animals without backbones, are the earliest animals comprising 97% of animals today. **Vertebrates**, animals with backbones (the fish, amphibians, reptiles, birds, and mammals) all are in the phylum Chordata.

4) The Sponges: Phylum Porifera.

Sponges (500 species) lack true tissues and organs. The sponge body is perforated with numerous tiny pores. Three major cell types are present: (1) **epithelial cells** on the body surface, including pore cells; (2) **collar cells** with flagella to control water flow; and (3) **amoeboid cells** to digest and distribute nutrients and make reproductive cells and spines called **spicules**. Sponges may reproduce asexually by **budding** or sexually through fusion of sperm and egg.

5) The Hydra, Anemones, and Jellyfish: Phylum Cnidaria.

Cnidarians (9000 species) are radially symmetrical, have two germ layers (ectoderm and endoderm) with jellylike **mesoglea** between, have true tissues and a **nerve net** to control contractile tissue and feeding, but lack organs and brains. The two body plans are the sessile tubular **polyp**, usually attached to rocks and possessing **tentacles** to attack and seize prey, and the mobile swimming **medusa** ("jellyfish"), with trailing tentacles armed with **cnidocyte** cells that eject poisonous or sticky darts to capture prey. The prey is moved into the digestive sac (**gastrovascular sac**) with one opening, the mouth/anus. Cnidarians can reproduce asexually (budding off replicas) and sexually. One group, the corals, with limestone shells, form reefs that are the basis for a diverse ecosystem.

6) The Flatworms: Phylum Platyhelminthes.

Flatworms have bilateral symmetry, a gastrovascular (GV) digestive cavity, cephalization with **ganglia** (clusters of nerve cells) in the anterior brain, and **nerve cords**. When a free-living planarian encounters food, a muscular tube (**pharynx**) sucks it up into the GV cavity. Flatworms lack circulatory and respiratory systems, relying on diffusion to move molecules. Flatworms reproduce asexually and sexually; most are **hermaphroditic** (possess both male and female sex organs). Some parasitic flatworms are intestinal tapeworms (ingested as encapsulated **cysts**) and liver and blood flukes.

7) The Roundworms: Phylum Nematoda.

Nematodes (10,000 species) have a tubular, one-way digestive tract (mouth ➝ intestine ➝ anus), a fluid-filled pseudocoelom that acts as a **hydrostatic skeleton**, and a head with a brain. They lack circulatory and respiratory systems, relying on diffusion to move molecules. Most reproduce sexually and have separate male and female sexes. Billions thrive in each acre of topsoil. Parasites include trichinella, hookworm, and heartworm.

8) The Segmented Worms: Phylum Annelida.

Annelid (9000 species) bodies consist of a series of repeating segments, each with nerve ganglia excretory structures (**nephridia**), and muscles and a true coelom acting as a hydrostatic skeleton. They have a **closed circulatory system** (blood is confined to the heart and blood vessels). The digestive system of earthworms consists of mouth, pharynx, esophagus, crop, gizzard, intestine, and anus. Segmented worms include earthworms, tube worms, and leeches.

9) The Insects, Arachnids, and Crustaceans: Phylum Arthropoda.

Arthropods (more than 1,000,000 species) are the most successful group of animals on Earth due to these adaptations: **chitin exoskeleton** with jointed appendages; segmentation; efficient gas exchange mechanisms including **gills** (in crustaceans), **tracheae** (insects), or **book lungs** (spiders); **open circulatory systems** (blood flows through vessels and enclosed body cavities called **hemocoels**); and well-developed sensory and nervous systems, including **compound eyes**. The exoskeleton periodically must be shed (**molted**) and replaced with a larger one. Insects are the most diverse and abundant arthropods (850,000 species, class Insecta), with three pairs of legs and two pairs of wings. Some insects undergo complete **metamorphosis**: from the egg to the **larva** (adapted for feeding) to the **pupa** (nonfeeding form in which physical changes occur) to the adult adapted for reproduction. Spiders, scorpions, ticks, and mites are arachnids (50,000 species, class Arachnida) with four pairs of legs and simple eyes. Most are carnivorous predators. Crabs, shrimp, crayfish, lobsters, and barnacles are crustaceans (30,000 species, class Crustacea), the only primarily aquatic arthropods. Crustaceans have two pairs of antennae and many other appendages, compound eyes, and gills.

10) The Snails, Clams, and Squid: Phylum Mollusca.

Molluscs (50,000 species) have a moist, muscular body with a hydrostatic skeleton, an open circulatory system (except for the cephalopods), and a **mantle** (body wall extension that forms a gill chamber). Some have a calcium carbonate shell secreted by the mantle. Snails and slugs are gastropod mollusks (35,000 species, class Gastropoda) with a muscular foot and a rasping **radula** used to scrape algae from rocks for food. Scallops, mussels, clams, and oysters are bivalve mollusks (class Bivalva) with two shells connected by a flexible hinge. Bivalves are filter feeders, using gills for respiration and feeding. Octopuses, squids, nautiluses, and cuttlefish are cephalopod mollusks (class Cephalopoda), mostly marine predators. The foot has evolved into tentacles with suction discs, and they move by jet propulsion caused by forceful expulsion of water from the mantle cavity.

11) The Sea Stars, Sea Urchins, Sand Dollars, and Sea Cucumbers: Phylum Echinodermata.

The echinoderms have free swimming embryos with bilateral symmetry, but the adults have five-parted radial symmetry, lack a head, have an internal **endoskeleton** of calcium carbonate plates, and move slowly by using **tube feet** (rows of suction cups) which are part of the **water vascular system** (water enters through a **sieve plate**, passes through a ring canal and into radial canals, each of which has many tube feet each with a muscular **ampulla** or squeeze bulb). They have no circulatory system. Sea stars can regenerate lost parts.

12) The Tunicates, Lancelets, and Vertebrates: Phylum Chordata.

Chordates share certain characteristics. They have: (1) a **notochord** (stiff flexible anterior-posterior rod made of **cartilage**, for muscle attachment) replaced by the bony backbone (**vertebral column**) in vertebrates; (2) a **dorsal, hollow nerve cord** with an anterior brain; (3) **pharyngeal gill slits** early in development; and (4) a **post-anal tail** early in development. Vertebrates have an endoskeleton of cartilage (sharks) or bone, paired appendages (fins, limbs, wings), and large complex brains.
   There are seven classes of vertebrates:

 Jawless fishes (class Agnatha: marine hagfish and aquatic lampreys) have unpaired fins, cartilage skeletons, no scales, circular gill slits.

Cartilaginous fish (class Chondrichthyes: sharks, rays, skates) have cartilage skeletons, leathery skin with tiny scales, a two-chambered heart, and rows of razor-sharp teeth.

Bony fish (class Osteichytes) have bony skeletons, gills, and swim bladders (for buoyancy) and a two-chambered heart.

Amphibians (class Amphibia: frogs and salamanders) have limbs, primitive lungs, and moist skin (for gas exchange) in adult forms. They have a three-chambered heart, and use external fertilization with juvenile gilled forms developing in water.

Reptiles (class Reptilia: lizards, snakes, turtles, alligators, and crocodiles) have tough, scaly, waterproof skin, and internal fertilization. They have shelled **amniotic eggs** that prevent dessication of the embryos due to the presence of an internal **amnion** membrane that encloses the embryo in a watery environment. Reptiles have efficient lungs, a three-chambered heart, and limbs (except the snakes).

Birds (class Aves) have wings (flight), feathers (heat insulation), hollow bones, internal fertilization, amniotic eggs, warm-bloodedness, a four-chambered heart, and a respiratory system with lungs and air-sacs.

Mammals (class Mammalia) have limbs (modified for running, swimming, flying, grasping), hair (heat insulation), warm-bloodedness, a four-chambered heart, highly developed brains, internal fertilization, **mammary glands** (milk for feeding young offspring), and embryonic development within female uterus (except the egg-laying **monotreme** mammals: platypus, spiny anteater). Most mammals are **placental** (they retain developing young for long periods in a uterine **placenta** where gas, nutrient, and waste exchange occurs). However, the young of **marsupial** mammals (opossums, koalas, kangaroos) leave the uterus and crawl into a protective pouch to continue development.

Major evolutionary advances	
Porifera:	multicellular
Cnidaria:	two tissue layers
	gastrovascular cavity
Platyhelminthes:	true organs
	bilateral symmetry
	three tissue layers
Nematodes:	separate mouth and anal openings to the digestive tract
Annelids:	segmentation
	circulatory system
	internal body cavity
Arthropods:	exoskeleton
	jointed appendages
Mollusca:	enlarged brain
Echinoderms:	internal skeleton
Chordates:	internal cartilaginous or bony skeleton
	complex nervous system

KEY TERMS AND CONCEPTS

Fill-In: Fill in the information requested in the following table.

Phylum	Number of germ (tissue) layers	Type of body symmetry present	Type of body cavity present	Is body Segmentation Present?
Porifera	(1)	(2)	(3)	(4)
Cnidaria	(5)	(6)	(7)	(8)
Platyhelminthes	(9)	(10)	(11)	(12)
Nematoda	(13)	(14)	(15)	(16)
Annelida	(17)	(18)	(19)	(20)
Arthropoda	(21)	(22)	(23)	(24)
Chordata	(25)	(26)	(27)	(28)

Key Terms and Definitions

amnion: one of the embryonic membranes of reptiles, birds, and mammals, enclosing a fluid-filled cavity that envelops the embryo.

amniotic egg: the egg of reptiles and birds. It contains an amnion that encloses the embryo in a watery environment; this allows the egg to be laid on dry land.

amoeboid cell: a cell type found in sponges, which serves to digest and distribute nutrients and make reproductive cells and spines called spicules.

ampulla: part of the water-vascular system in echinoderms; serves as a squeeze-bulb attached to the tube feet to create suction.

anterior: the front, forward, or head end of an animal.

bilateral symmetry: body plan in which only a single plane drawn through the central axis will divide the body into mirror-image halves.

book lungs: thin layers of tissue resembling pages in a book, enclosed in a chamber and used as a respiratory organ by certain types of arachnids.

budding: a form of asexual reproduction in which the adult produces miniature versions of itself that drop off and assume independent existence.

cartilage: flexible, translucent tissue that serves as the forerunner of bone during embryonic development in most vertebrates. In the class Chondrichthyes, cartilage is retained and forms the entire skeleton.

cephalization: the increasing concentration over evolutionary time of sensory structures and nerve ganglia at the anterior end of animals.

chitin: a tough, flexible polysaccharide; an important constituent of the arthropod exoskeleton.

closed circulatory system: a type of circulatory system in which the blood is always enclosed in the heart and vessels.

cnidocyte: a specialized cell found in cnidarians which, when disturbed, ejects a sticky or poisoned thread which traps and stings prey.

coelom: a space or cavity within the body separating the body wall from the inner organs.

collar cells: specialized cells lining the inside channels of sponges. Flagella extend from a sieve-like collar, creating a water current that draws microscopic organisms through the collar to be trapped.

compound eye: an image-forming eye consisting of numerous similar light-gathering and light-detecting elements.

cyst: an encapsulated resting stage in the life cycle of certain invertebrates, such as parasitic flatworms and roundworms.

deuterostome: a group of animals including the echinoderms and chordates in which the coelom develops in spaces as outpockets from the digestive tract.

dorsal: the top, back, or uppermost surface of an animal oriented with its head forward.

ectoderm: the outermost embryonic tissue layer, which gives rise to structures such as hair, the epidermis of the skin, and the nervous system.

endoderm: the innermost embryonic tissue layer, which gives rise to structures such as the lining of the digestive and respiratory tracts.

endoskeleton: a supportive structure within the body; an internal skeleton. It may be nonliving, as in echinoderms and sponges, or living, as in vertebrates.

epithelial cell: cell type that covers the body surfaces and lines body cavities, and gives rise to glands.

exoskeleton: an external, nonliving supporting structure; an external skeleton.

free-living: not parasitic.

ganglion: an aggregation of neurons.

gastrovascular cavity: a saclike opening in the bodies of some invertebrates (such as cnidarians and flatworms) with a single opening serving as both mouth and anus.

germ layer: a tissue layer formed during early embryonic development.

gill: in aquatic animals, a branched tissue richly supplies with capillaries around which water is circulated for gas exchange.

hemocoel: a blood cavity within the bodies of certain invertebrates in which blood bathes tissues directly. A hemocoel is part of an open circulatory system.

hermaphroditic: possessing both male and female sexual organs. Some hermaphroditic animals can fertilize themselves; others must exchange sex cells with a mate.

hydrostatic skeleton: the use of fluid contained in body compartments to provide support for the body and mass against which muscles can contract.

invertebrate: a category of animals that never possess a vertebral column.

larva: an immature form of an organism prior to metamorphosis into its adult form. The caterpillars of moths and butterflies, and the maggots of flies, are larvae.

mammary glands: milk-producing organs used by female mammals to nourish their young.

mantle: an extension of the body wall in certain invertebrates, such as mollusks. It may secrete a shell, protect the gills, and, as in cephalopods, aid in locomotion.

marsupial: a type of mammal whose young are born at an extremely immature stage and undergo further development in a pouch while they remain attached to a mammary gland. Includes kangaroos, opossums, and koalas.

medusa: a bell-shaped, often free-swimming stage in the life cycle of many cnidarians. Jellyfish are one example.

mesoderm: the middle embryonic tissue layer, absent in poriferans and cnidarians, but present in the other groups of animals; gives rise to structures such as the muscular, circulatory, and skeletal systems.

mesoglea: a middle, jellylike layer within the body wall of cnidarians.

metamorphosis: a dramatic change in body form during development, as seen in amphibians (tadpole to frog) and insects (caterpillar to butterfly).

molt: to shed an external body covering, such as an exoskeleton, skin, feathers, or fur.

monotreme: a type of mammal that lays eggs; for example, the platypus.

nephridium: an excretory organ in the annelid worms.

nerve cord: also called the spinal cord of vertebrates, a nervous structure lying along the dorsal side of the body of chordates.

nerve net: a loosely coordinated network of neurons.

notochord: a stiff but somewhat flexible, supportive rod found in all members of the phylum Chordata at some stage of development.

open circulatory system: a type of circulatory system in arthropods and mollusks in which the blood is pumped

through an open space (the hemocoel), where it bathes the internal organs directly.

organ: two or more tissues integrated to perform a specialized function, i.e., the kidney.

organ system: two or more organs that work together to perform a specific function, i.e., the digestive system.

parasite: an organism that lives in or on the body of another organism, causing it harm as a result.

pharyngeal gill slit: openings in the pharynx area of the digestive/respiratory tract through which water flows, allowing gas exchange may occur; develop during early embryonic development in chordates.

pharynx: a portion of the digestive system between the mouth and the esophagus. In flatworms, it is developed as an extensible, muscular organ.

placenta: a tissue rich in blood vessels that develops in the mammalian uterus during pregnancy. Here nutrients and oxygen from maternal blood are exchanged for wastes from the developing embryo.

placental: a characteristic of most mammals in which the developing young spends a long developmental period within the mother's uterus attached to tissues through which gases, nutrients, and wastes are exchanged between mother and offspring.

polyp: the sedentary, vase-shaped stage in the life cycle of many cnidarians. Hydra and sea anemones are examples.

post-anal tail: a characteristic of chordates in which a tail develops posterior to the anal opening during early embryonic development.

posterior: the tail, hindmost, or rear end of an animal.

protostome: a group of animals including the annelid worms, arthropods, and mollusks in which the coelom develops in spaces between the gut and the body wall.

pseudocoelom: "false coelom"; a body cavity with a different embryological origin than a coelom, but serving a similar function; found in roundworms.

pupa: the stage of insect metamorphosis when the larva undergoes physical transformation into an adult.

radial symmetry: a body plan in which any plane drawn along a central axis will divide the body into approximately mirror-image halves. Cnidarians and many adult echinoderms show radial symmetry.

radula: a ribbon of tissue in the mouth of gastropod mollusks that bears numerous teeth on its outer surface and is used to scrape and drag food into the mouth.

segmentation: division of the body into repeated, often similar units.

sessile: not free to move about, usually permanently attached to a surface.

sieve plate: the opening into the echinoderm water-vascular system through which water is taken in.

spicule: subunits of the endoskeleton of sponges, made of protein, silica, or calcium carbonate.

tentacle: an elongate, extensible projection of the body of cnidarians and cephalopod mollusks that may be used for grasping, stinging, and immobilizing prey, and locomotion.

tissue: a group of (usually similar) cells that together carry out a specific function, i.e., muscle.

tracheae: a system of air tubes that branch within the body of insects and some arachnids, carrying air close to every cell.

tube feet: cylindrical extensions of the water-vascular system of echinoderms, used for locomotion, grasping food, and respiration.

ventral: the lower, or underside of an animal whose head is oriented forward.

vertebral column: a backbone; part of the internal skeleton of vertebrates that supports the body and protects the spinal cord.

vertebrate: an animal that possesses a vertebral column.

water-vascular system: a system in echinoderms consisting of a series of canals through which seawater is conducted and used to inflate tube feet for locomotion, grasping food, and respiration.

THINKING THROUGH THE CONCEPTS

True or False: Determine if the statement given is true or false. If it is false, change the underlined word so that the statement reads true.

29._____ Large animals are likely to have less stable internal environments.
30._____ All adult sponges are motile.
31._____ Water leaves a sponge through the osculum.
32._____ Free-swimming cnideria are called medusae.
33._____ Flatworms have radial symmetry.
34._____ Terrestrial arthropods have open circulatory systems.
35._____ Birds have three-chambered hearts.
36._____ Echinoderms have endoskeletons.
37._____ The osteichthyes are cartilaginous fishes.
38._____ Mammals evolved from birds.

Matching: Sponges and jellyfish.

39.____ coral reefs
40.____ some extracellular digestion
41.____ resemble colonies of cells
42.____ have polyp and medusae stages
43.____ hydras
44.____ simplest multicellular animals
45.____ separate mouth and anal openings
46.____ lack true tissues
47.____ have radial symmetry
48.____ predatory
49.____ have nematocysts
50.____ body wall has many pores
51.____ have collar cells
52.____ have bilateral symmetry
53.____ have internal skeletons of spicules

Choices:

a.   poriferans
b.   cnidarians
c.   both of these
d.   neither of these

Matching: Worms.

54.____ have bilateral symmetry
55.____ planarians
56.____ have nerve ganglia
57.____ segmented
58.____ hookworms
59.____ lacks a separate anus
60.____ have coelomic cavities
61.____ have true organs
62.____ heartworms
63.____ have flame cells
64.____ have circulatory systems
65.____ trichinosis
66.____ have nephridia
67.____ tapeworms
68.____ leeches

Choices:

a.   roundworms
b.   flatworms
c.   annelid worms
d.   all worms
e.   no worms

Matching: Arthropods and molluscs.

		Choices:
69.____	octopuses	
70.____	have internal skeletons	a. arthropods
71.____	largest Phylum of animals	b. molluscs
72.____	have limestone shells	c. both of these
73.____	have open circulatory systems	d. neither of these
74.____	have exoskeletons	
75.____	clams	
76.____	have tracheae	
77.____	are segmented	
78.____	evolved from annelid worms	
79.____	metamorphosis during development	
80.____	snails	
81.____	crabs	
82.____	can learn and remember	
83.____	have water-vascular systems	
84.____	barnacles	

Matching: Chordates. (Hint: There may be more than one correct answer for some questions.)

		Choices:
85.____	produce amniotic eggs	
86.____	are segmented	a. cartilage fish
87.____	produce placentas	b. bony fish
88.____	frogs	c. amphibians
89.____	have non-bony skeletons	d. reptiles
90.____	have 2-chambered hearts	e. birds
91.____	dinosaurs	f. mammals
92.____	have notochords at some stage of development	g. all of these
93.____	have four-chambered hearts	h. none of these
94.____	on the boundary between aquatic and terrestrial existence	
95.____	have swim bladders	
96.____	feed milk to their young	
97.____	have open circulatory systems	
98.____	have three-chambered hearts	
99.____	have backbones	
100.____	are warmblooded	
101.____	evolved into amphibians	
102.____	have hollow bones and air sacs	
103.____	whales	
104.____	sharks	

Multiple Choice: Pick the most correct choice for each question.

105.   The simplest multicellular animals are:
a.   hydras
b.   anemones
c.   sponges
d.   jellyfishes
e.   flatworms

106.   Radial symmetry is exhibited by all of the following **except**:
a.   polyp
b.   medusa
c.   sponge
d.   jellyfish

107.   Which characteristic do cnidarians and flatworms have in common?
a.   gastrovascular cavity
b.   bilateral symmetry
c.   flame cells
d.   hydrostatic skeleton
e.   an anus

108.   Oral and anal openings of the digestive tract first developed in
a.   annelids
b.   platyhelminths
c.   nematodes
d.   cnidarians
e.   sponges

109.   Segmentation first appeared in the
a.   segmented worms
b.   insects
c.   roundworms
d.   molluscs
e.   flatworms

110.   The largest and smartest of the invertebrates belong to
a.   phylum Arthropoda
b.   phylum Echinodermata
c.   phylum Mollusca
d.   phylum Chordata
e.   phylum Vertebrata

Short Answer: Based on the following figure, answer the questions below.

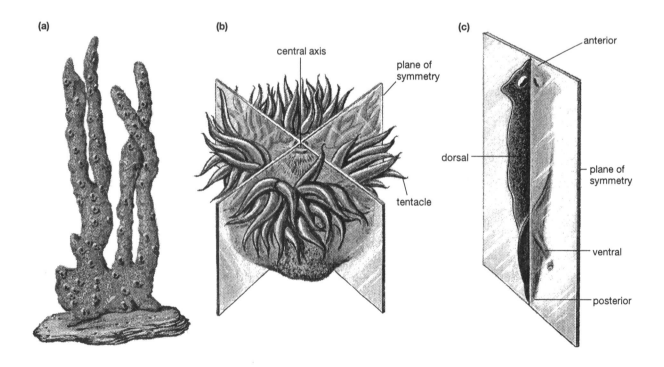

111.   The type of symmetry depicted by (a) is called _____ symmetry.
112.   The type of symmetry depicted by (b) is called _____ symmetry.
113.   The type of symmetry depicted by (c) is called _____ symmetry.

CLUES TO APPLYING THE CONCEPTS

This practice question is intended to sharpen your ability to apply critical thinking and analysis to biological concepts covered in this chapter.

114.   Which groups of land vertebrate animals are found in the Arctic and Antarctic regions, the coldest parts of Earth? Why are they able to live in these regions while other types of animals, especially the arthropods, which are tremendously successful elsewhere, are unable to survive there very well?

ANSWERS TO EXERCISES

1. no true tissues
2. asymmetrical or radial
3. none
4. no
5. two (ectoderm and endoderm)
6. radial
7. none
8. no
9. three (ectoderm, mesoderm, and endoderm)
10. bilateral
11. none
12. no
13. three
14. bilateral
15. pseudocoelom
16. no
17. three
18. bilateral
19. coelom (protostome type)
20. yes
21. three
22. bilateral
23. coelom (protostome type)
24. yes
25. three
26. bilateral

27. coelom (deuterostome type)
28. yes
29. false, more
30. false, sessile
31. true
32. true
33. false, bilateral
34. true
35. false, four-chambered
36. true
37. false, bony
38. false, reptiles
39. b
40. b
41. a
42. b
43. b
44. a
45. d
46. a
47. b
48. b
49. b
50. a
51. a
52. d
53. a
54. d

55. b
56. d
57. c
58. a
59. b
60. c
61. d
62. a
63. b
64. b
65. a
66. c
67. b
68. c
69. b
70. d
71. a
72. b
73. c
74. a
75. b
76. a
77. c
78. c
79. a
80. b
81. a
82. b

83. d
84. a
85. d, e, f
86. g
87. f
88. c
89. a
90. a, b
91. d
92. g
93. e, f
94. c
95. b
96. f
97. h
98. c, d
99. g
100. e, f
101. b
102. e
103. f
104. a
105. c
106. c
107. a
108. c
109. a
110. c

111.   lack of symmetry or asymmetrical

112.   radial symmetry

113.   bilateral symmetry

114.   Just about all the land vertebrates in the Arctic and Antarctic regions are birds (penguins, for instance) and mammals (polar bears, for instance). What sets these vertebrate groups apart from the rest is that they are warm blooded animals. They maintain a high constant body temperature and that enables them to function well even in the constant coldness of the polar regions. Reptiles and Amphibians, the other types of land vertebrates, cannot remain active in the Arctic or Antarctic environments since their body temperature would be as cold as their surroundings. The same is true for all types of terrestrial invertebrates.

Chapter 23: Plant Form and Function

OVERVIEW

This chapter will introduce you to the structure of plants and how the individual structures function so that the plant survives. Plants have evolved specific adaptations such as organ modifications in order to live in particular habitats; these adaptations are presented. Important interactions between the plant and other organisms such as fungi and bacteria are discussed.

1) How Are Plant Bodies Organized?

Plant bodies are organized into the **shoot system**, typically the above-ground portion, and the **root system**, typically the below-ground portion. The shoot system is composed of **stems**, **leaves**, **buds**, and when appropriate, **flowers** and **fruits**. The above-ground portion functions to absorb sunlight energy and conduct photosynthesis, to transport nutrients and other substances between structures, and to produce certain hormones. Flowers and fruits are specialized for reproduction. The root system functions to anchor the plant to a substrate (i.e.: soil), absorb water and **minerals** from the soil, store excess sugars (usually as starch), transport water, minerals, and other substances to the shoot, produce certain hormones, and interact with soil microorganisms.

   The way in which plant structures are organized depends, in part, on their relationship to one of two evolutionary groups. **Monocot** and **dicots** differ in the organization of flowers, leaves, vascular tissue, root systems, and seed structures.

2) How Do Plants Grow?

Most plants grow throughout their life. Growth occurs from undifferentiated **meristem cells**, which divide by mitosis. Some of the daughter cells produced mature into cells specialized in form and function; these are **differentiated cells**. Plants grow taller or longer from the tips of shoots and roots through division of **apical meristems**. Apical meristems are responsible for **primary growth**. Growth that results in increased diameter occurs due to divisions of **lateral meristems**, also called the **cambium**. Growth due to lateral meristems is **secondary growth**.

3) What Are the Tissues and Cell Types of Plants?

The **tissue systems** of plant structures include the **dermal tissue system**, which covers the outer surfaces of the plant body; the **ground tissue system**, which consists of all tissues that are not dermal or vascular; and the **vascular tissue system**, which transports water, nutrients, minerals, and hormones throughout the plant body.

   Dermal tissue may be either **epidermal tissue**, also called the **epidermis**, or peridermal tissue, the **periderm**. The epidermis constitutes the cell layer covering leaves, stems, roots, flowers, seeds, and fruits. The epidermis of above-ground organs, in turn, is covered by a waxy coating, the **cuticle**, which serves to protect and reduce water loss from the plant. Occasional epidermal cells elongate, forming hairs. Hairs located on the root system are called **root hairs** and they serve to greatly increase the absorptive ability of the roots. As woody stems and roots age, the periderm replaces the epidermis. It is made up of specialized waterproof **cork cells**.

Cells that constitute ground tissue include **parenchyma, collenchyma,** and **sclerenchyma**. Parenchyma cells are thin-walled and alive when mature. It is the parenchyma cells that conduct most of the metabolic functions for the plant; therefore, they are quite diverse. Collenchyma are commonly described as having unevenly thickened cell walls. And although they are alive when mature, they do not often divide. Collenchyma help provide support in plants. Sclerenchyma cells have thick cell walls that are hardened with a substance called lignin. These cells are not alive when mature; however, they continue to provide support for the plant.

Vascular tissue is composed of the transport tissues **xylem** and **phloem**. Xylem transports water and minerals from the roots to the shoot. The conducting cells found here are either **tracheids** or **vessel elements**. Tracheids lie end to end, allowing water and minerals to flow through **pits** in the ends of the cells. Vessel elements are larger than tracheids but also lie end to end. The ends may have open pores or they may be completely open, forming a long tube. The phloem, which transports metabolic substances, is made up of **sieve-tube elements** connected end to end forming long, continuous **sieve tubes**. Between each seive-tube element are **sieve plates**, the ends of the seive-tube elements with pores in them allowing substances to flow through the seive tube. Although seive-tube elements are alive when they are mature, they do not contain a nucleus. The functions of seive-tube elements are directed, instead, by **companion cells**.

4) Roots: Anchorage, Absorption, and Storage

Often the first visible sign of seed germination is the appearance of the **primary root**. The primary root of a dicot will grow deep into the soil and mature into a **taproot system**. With taproot systems, the taproot is maintained while smaller, much less developed roots grow out from its sides. The primary root of a monocot will grow down into the soil but soon dies. New roots are produced from the base of the stem, forming a **fibrous root system**. Growth and development of individual roots is similar between taproot and fibrous root systems. A **root cap** is produced at the tip of a root by the apical meristem. The root cap protects the root apical meristem from the soil as the root grows downward. The root apical meristem also produces the water-permeable epidermis, a **cortex**, and a **vascular cylinder**.

Most of the root is made up of parenchyma cells of the cortex. Forming a ring inside the parenchyma cells is the **endodermis**. The cell walls of the endodermis are embedded with wax, forming the **casparian strip**. The endodermal cell surfaces facing the parenchyma cells and the vascular cylinder do not contain the wax. This organization forces water into the cells and to enter the vascular cylinder.

The vascular cylinder of the root contains the xylem and the phloem. Surrounding the xylem and the phloem is the **pericycle**. It is portions of the pericycle that will divide to produce a **branch root** off the existing root.

5) Stems: Reaching for the Light

As the stem grows, it may differentiate into buds, leaves, or flowers. Clusters of meristem cells along the stem form **leaf primordia**, which are the beginnings of new leaves. **Lateral buds** are also produced by the meristem cells. Lateral buds will later develop into branches. Leaves and lateral buds are located at **nodes** on the stem. The areas of the stem between the nodes are called **internodes**.

Stems are formed from dermal tissue, vascular tissue, and two types of ground tissue, cortex and **pith**. Since the waxy cuticle covers the epidermis of stems (and leaves) and retards water loss, gas exchange between the cells and the atmosphere is also compromised. Specialized cells form pores or

stomata in the epidermis allowing CO_2 and O_2 to pass through. As in roots, the cortex is located between the epidermis and the vascular tissue. Additionally, stems have pith which is located inside the vascular tissue. Cells of both the cortex and pith provide support to stems, store starch, and may conduct photosynthesis.

Vascular tissue in stems is similar to that of roots and is continuous from the root system, through the stem structures, to the leaves. It is the apical meristem that produces vascular tissue in young stems. The xylem that is produced is **primary xylem** and the phloem that is produced is **primary phloem**. In woody plants, meristematic tissue called the **vascular cambium** produces **secondary xylem** and **secondary phloem**, giving rise to lateral growth. The young, newly formed secondary xylem is called **sapwood**, while the older xylem is called **heartwood**. In the spring, xylem that is produced tends to contain large cells since water is plentiful; as the season progresses and summer droughts occur, smaller xylem cells are produced. This creates the typical **annual rings** seen in tree trunks. To the outside of the secondary phloem, epidermal cells are stimulated by hormones to produce the **cork cambium**. This meristematic tissue gives rise to **cork cells** that protect the trunk and prevent it from drying out. The **bark** of a tree consists of phloem, cork cambium, and cork cells.

6) Leaves: Nature's Solar Collectors

Most leaves are comprised of a broad, flat **blade** specialized to capture sunlight and a **petiole** that attaches the blade to the stem. Vascular tissue runs through the petiole branching out in the blade, forming **vascular bundles** or **veins**. As in the stem, the cuticle covers the leaf, retarding water loss and compromising gas exchange. Specialized epidermal cells, **guard cells**, form openings called stomata allowing the exchange of CO_2 and O_2. Beneath the epidermis, parenchyma cells specialized for photosynthesis constitute the **mesophyll**. The mesophyll is often organized into two layers of cells, the **palisade cells** and the **spongy cells**.

7) What Are Some Special Adaptations of Roots, Stems, and Leaves?

Plants have evolved to fit their environment and with this has come modifications of their organs. **Roots** may be specialized to store a relatively large amount of food as starch. Other plants, such as the orchid, have roots that are capable of photosynthesis. **Stems** may be used by the plant as a way of starting miniature versions of themselves in a new area. **Runners** serve to propagate plantlets away from the "mother" plant. Stems may also serve as storage receptacles. The stems of cacti store water, while underground stems, **rhizomes**, store carbohydrates. Modified branches such as **thorns** protect the plant from would-be predators, whereas **tendrils** grasp support structures, enabling the plant to climb. **Modified** leaves may store water if they are succulent, or reduce water loss if they are spiny. Other leaves function as tendrils in climbing up a support structure or in nitrogen gathering by capturing and digesting insects.

8) How Do Plants Acquire Nutrients?

Nutrients are taken into root hairs of roots by active transport. The nutrients then diffuse into pericycle cells through **plasmodesmata**, connections between cells. Active transport is again used to get the nutrients out of the pericycle and into the extracellular space around the xylem within the vascular cylinder. From the extracellular space, the nutrients diffuse into the tracheids of the xylem.

Symbiotic relationships between plants and fungi and between plants and soil bacteria increase nutrient acquisition in nutrient-poor soils. Fungi that form a symbiotic relationship with plant roots are called **mycorrhizae**. Mycorrhizae help plants acquire minerals, especially phosphorous. In return, the fungus receives carbohydrates and amino acids from the plant. Nitrogen is also often limited for many plants. **Legumes** have overcome this by forming an association with **nitrogen-fixing bacteria**. These bacteria possess the enzymes needed to convert N_2 from the atmosphere into ammonium or nitrate, the process of **nitrogen fixation**. The bacteria enter the root hairs of a legume and travel to the cortex of the root. The bacteria multiply and stimulate the cortex cells to multiply as well. A root **nodule** is produced. It is within the nodule that nitrogen fixation occurs, supplying the plant with a steady supply of usable nitrogen. In return, the bacteria receive carbohydrates for their energy needs.

9) How Do Plants Acquire Water?

Water flows into root cells by osmosis. The high mineral content in root cells helps to maintain the concentration gradient needed for osmosis to occur. Water also flows freely through epidermal and cortex cell walls. When the water encounters the casparian strip around the endodermal cells, it is goes into the endodermal cells by osmosis. From the endodermal cells, water flows into the vascular cylinder.

10) How Do Plants Transport Water and Minerals?

How does water get to the top of a tree 100 m tall? The pressure created by the active transport, osmosis, and diffusion route into the root xylem is only sufficient to move water up a plant less than a meter tall. According to the **cohesion-tension theory**, the water moving through plants is actually pulled up from the top. The water moves by **bulk flow**, the movement of fluid molecules as one unit rather than as individual molecules. Minerals are carried passively along. The movement of water up a plant is best understood if the cohesion and adhesion properties of water molecules are recalled. The cohesive nature of water molecules holds them together, since they are attracted to one another. The water molecules are also attracted to the walls of the xylem cells; they adhere to them and creep up the walls. These properties account for half of the reason water makes it to the top of a tree. The other half relates to the water that is evaporating from the surface of leaves, **transpiration**. As water molecules leave the plant through transpiration, a tension is created since the water molecules leaving the plant are, in essence, attached to the ones still in the plant. The tension created pulls water molecules up the plant.

Water loss from the leaves is controlled by the epidermal guard cells closing and opening the stomata. The opening and closing of stomata is regulated by potassium concentration in the guard cells. When the potassium concentration is high, water flows into the guard cells by osmosis; they swell, and create the stomatal opening. When the potassium concentration is lowered in the guard cells, water flows out by osmosis, the guard cells shrink, closing the stomatal opening. The movement of potassium is regulated by light reception by the guard cells, CO_2 concentration, and water availability. When water is lost from the plant faster than it can be replaced from the soil, the mesophyll cells release the hormone **abscisic acid**. If abscisic acid has been stimulated, potassium is inhibited from moving into the guard cells. Thus they remain closed, conserving water.

11) How Do Plants Transport Sugars?

The areas of the plant that produce carbohydrates are referred to as a **source**. Any structure that needs carbohydrates produced by a source is referred to as a **sink**. Movement of sugars between sources and sinks must occur. This is believed to occur according to the **pressure flow theory**. Glucose is produced by photosynthesis, usually in leaves. The glucose is converted to sucrose and moved by active transport in to companion cells of the phloem. As the concentration gradient of sucrose increases in companion cells, it is low in sieve-tube elements. Thus, sucrose diffuses into the sieve-tube elements. Water enters the sieve-tube by osmosis from nearby xylem, increasing the hydrostatic pressure within the sieve tube. The reverse sequence occurs unloading sucrose at a sink region. Sucrose is actively transported into the sink, such as a developing fruit, thus lowering the sucrose concentration in the sieve tube. With a lower sucrose concentration, water flows out of the sieve tube by osmosis, lowering the hydrostatic pressure again. The hydrostatic pressure difference between the sieve tube at the source and the sieve tube at the sink causes water to flow down the sieve tube, carrying sucrose with it.

KEY TERMS AND CONCEPTS

Fill-In: From the following list of key terms, fill in the blanks in the following statements.

abscisic acid	bark	apical meristems	annual rings
bulk flow	blade	branch roots	casparian strip
cortex	cambium	differentiated	cohesion-tension theory
endodermis	cuticle	epidermis	cork cambium
internode	lateral buds	ground tissue system	dermal tissue system
minerals	leaves	primary phloem	fibrous root system
node	pericycle	primary growth	heartwood
pit	periderm	primary root	lateral meristem
pith	petiole	root system	leaf primordia
rootcap	primary xylem	secondary xylem	taproot system
stems	sapwood	secondary growth	vascular tissue system
stomata	shoot system	secondary phloem	vascular bundle
theory	sieve plate	vascular cylinder	vascular cambium
transpiration			

The above ground portion of a plant is called the (1) _____ and is composed of (2) _____, (3) _____, buds, flowers, and fruits. The below ground portion of a plant is the (4) _____ which functions to absorb water and (5) _____ from the soil.

The area of the stem where leaves attach is called a (6) _____, while the area of the stem between leaves is an (7) _____. A leaf is composed of the (8) _____, which functions to absorb sunlight, and the (9) _____, which attaches to the stem.

Shoots grow taller and roots grow longer by cell division of the (10) _____ which are responsible for (11) _____. Cells that specialize in form and function have (12) _____. Shoots grow in diameter by division of cells of the (13) _____, also called the (14) _____. The growth in diameter leads to (15) _____.

The three tissue systems of a plant include the (16) _____, (17) _____, and the (18) _____.

Epidermal tissue is also referred to as the (19) _____ or, in older tissue, it is called the (20) _____ made up of cork cells. A waxy layer, the (21) _____ covers the shoot system.

The first plant structure to emerge from a seed is the (22) _____. In dicots, this root matures into a (23) _____, while, in monocots, it matures into a (24) _____. The tip of a growing root is protected by the (25) _____, and water can permeate the cells of the root epidermis, the (26) _____.

Water entering the vascular cylinder is first forced into cells by the presence of the waxy (27) _____ around cells of the (28) _____. Surrounding the vascular cylinder is a layer of meristematic cells, the (29) _____ which may give rise to (30) _____.

New leaves are formed by growth of (31) _____ and branches mature from (32) _____.

In young, developing stems, the vascular tissue is formed by the apical meristem. The xylem that is formed is (33) _____ and the phloem that is formed is (34) _____. As woody plants mature, the (35) _____ produces (36) _____ and (37) _____.

Young, newly formed secondary xylem is called (38) _____, while the older xylem is called (39) _____. Variations in rainfall during the growing season produces alternating spring wood and summer wood, the recognizable (40) _____ of tree trunks.

Outside the secondary phloem, hormones stimulate epidermal cells to produce the (41) _____ which, in turn, produces cork cells. The secondary phloem, cork cambium, and cork cells constitute the (42) _____ of a tree.

Plants move water to the top of plants using the (43) _____ which involves the movement of water as one unit or movement in (44) _____ due to the pull from water leaving the plant through (45) _____. The loss of water from the plant is regulated by the opening and closing of (46) _____ often controlled by the hormone (47) _____.

Key Terms and Definitions

abscisic acid: a plant hormone that strongly inhibits the active transport of potassium into guard cells, shrinking the guard cell and closing stomata.

annual ring: pattern of alternating light (early) and dark (late) xylem of woody stems and roots, formed because of unequal availability of water in different seasons of the year, usually spring and summer.

apical meristem: the cluster of meristematic cells found at the tip of a shoot or root (or one of their branches).

bark: the outer layer of a woody stem, consisting of phloem, cork cambium, and cork cells.

blade: the flat part of a leaf.

branch root: a root that arises as a branch of a preexisting root, through divisions of pericycle cells and subsequent differentiation of the daughter cells.

bulk flow: the movement of many molecules of a gas or fluid in unison from an area of higher pressure to an area of lower pressure.

cambium (pl. cambia): a lateral meristem that runs parallel to the long axis of roots and stems that causes secondary growth of woody plant stems and roots. See cork cambium; vascular cambium.

Casparian strip: a waxy, waterproof band in the cell walls between endodermal cells in a root, which prevents the movement of water and minerals in and out of the vascular cylinder through the extracellular space.

cohesion-tension theory: a model for transport of water in xylem, which states that water is pulled up the xylem tubes, powered by the force of evaporation of water from the leaves (producing *tension*) and held together by hydrogen bonds between nearby water molecules (*cohesion*).

collenchyma: an elongated, polygonal plant cell type with irregularly thickened primary cell walls that is alive at maturity and that provides support to the plant body.

companion cell: a cell adjacent to a sieve-tube element in phloem, involved in control and nutrition of the sieve-tube element.

cork cambium: a lateral meristem in woody roots and stems that gives rise to cork cells.

cork cell: a protective cell of the bark of woody stems and roots; at maturity, cork cells are dead, with thick, waterproofed cell walls.

cortex: the part of a primary root or stem located between the epidermis and the vascular cylinder.

cuticle: a waterproof, waxy layer secreted by epidermal cells onto their outside surfaces to reduce evaporation of water from the plant.

dermal tissue system: a plant tissue system that makes up the outer covering of the plant body.

dicot: short for dicotyledon; a type of flowering plant characterized by embryos with two cotyledons, or seed leaves, modified for food storage.

differentiated cell: a mature cell specialized for a specific function; in plants, differentiated cells usually do not divide.

endodermis: the innermost layer of small, close-fitting cells of the cortex of a root that form a ring around the vascular cylinder.

epidermal tissue: dermal tissue in plants that forms the epidermis, the outermost cell layer that covers young plants.

epidermis: the outermost layer of cells of a leaf, young root, or young stem.

fibrous root system: a root system, commonly found in monocots, characterized by many roots of approximately the same size arising from the base of the stem.

ground tissue system: a plant tissue system consisting of parenchyma, collenchyma, and sclerenchyma cells that makes up the bulk of a leaf or young stem, excluding vascular or dermal tissues. Most ground tissue cells function in photosynthesis, support, or carbohydrate storage.

guard cell: one of a pair of specialized epidermal cells surrounding the central opening of a stoma of a leaf, which regulates the size of the opening.

heartwood: older xylem that contributes to the strength of a tree trunk.

internode: the part of a stem between two nodes.

lateral bud: a cluster of meristematic cells at the node of a stem; under appropriate conditions, it grows into a branch.

lateral meristem: also called cambium; a meristematic tissue that forms cylinders that run parallel to the long axis of roots and stems; usually found between the primary xylem and primary phloem (vascular cambium) and just outside the phloem (cork cambium).

leaf: an outgrowth of a stem, usually flattened and photosynthetic.

leaf primordium (pl. primordia): a cluster of meristem cells located at the node of a stem that develops into a leaf.

legume: a family of flowering plants, including peas, clover, and soybeans, most of which harbor nitrogen-fixing bacteria in nodules on their roots.

meristem cell: an undifferentiated cell that remains capable of cell division throughout the life of a plant.

mesophyll: loosely packed parenchyma cells located beneath the epidermis of a leaf.

mineral: an inorganic substance, especially one found in rocks or soil.

monocot: short for monocotyledon; a type of flowering plant characterized by embryos with one seed leaf, or cotyledon.

mycorrhiza (pl. mycorrhizae): a symbiotic relationship between a fungus and the roots of a land plant that facilitates mineral extraction and absorption.

nitrogen fixation: the process of converting atmospheric nitrogen (N_2) to ammonium (NH_4^+) or nitrates.

nitrogen-fixing bacteria: bacteria that possess the ability to remove nitrogen (N_2) from the atmosphere and combine it with hydrogen to produce ammonium (NH_4^+).

node: a region of a stem at which leaves and lateral buds are located.

nodule: a swelling on the root of a legume or other plant that consists of cortex cells inhabited by nitrogen-fixing bacteria.

palisade cells: columnar mesophyll cells containing chloroplasts just beneath the upper epidermis of a leaf.

parenchyma: a plant cell type that is alive at maturity, usually with thin primary cell walls, that carries out most of the metabolism of a plant. Most dividing meristem cells in a plant are parenchyma.

pericycle: the outermost layer of cells of the vascular cylinder of a root.

periderm: the outer cell layers of roots and a stem that have undergone secondary growth, consisting primarily of

cork cambium and cork cells.

petiole: the stalk that connects the blade of a leaf to the stem.

phloem: a conducting tissue of vascular plants that transports a concentrated sugar solution up and down the plant.

pit: an area in the cell walls between two plant cells in which secondary walls did not form, so that the two cells are separated only by a relatively thin and porous primary cell wall.

pith: cells forming the center of a root or stem.

pressure flow theory: a model for transport of sugars in phloem, which states that movement of sugars into a phloem sieve tube causes water to enter the tube by osmosis, while movement of sugars out of another part of the same sieve tube causes water to leave by osmosis; the resulting pressure gradient causes bulk movement of water and dissolved sugars from the end of the tube into which sugar is transported toward the end of the tube from which sugar is removed.

primary growth: growth in length and development of initial structures of plant roots and shoots, due to cell division of apical meristems and differentiation of the daughter cells.

primary phloem: phloem in young stems produced from an apical meristem.

primary root: the first root that develops from a seed.

primary xylem: xylem in young stems produced from an apical meristem.

rhizome: an underground stem, usually horizontal, that stores food.

root: the part of the plant body, normally underground, that provides anchorage, absorbs water and dissolved nutrients and transports them to the stem, produces some hormones, and in some plants serves as a storage site for carbohydrates.

root cap: a cluster of cells at the tip of a growing root, derived from the apical meristem. The root cap protects the growing tip from damage as it burrows through the soil.

root hair: a fine projection from an epidermal cell of a young root that increases the absorptive surface area of the root.

root system: the part of a plant, usually below ground, that anchors the plant in the soil, absorbs water and minerals, stores food, transports water, minerals, sugars, and hormones, and produces certain hormones.

runner: a horizontally growing stem that may develop new plants at nodes that touch the soil.

sapwood: young xylem that transports water and minerals in a tree trunk.

sclerenchyma: a plant cell type with thick, hardened secondary cell walls, that usually dies as the last stage of differentiation and provides both support and protection for the plant body.

secondary growth: growth in diameter of a stem or root due to cell division in lateral meristems and differentiation of their daughter cells.

secondary phloem: phloem produced from the cells arising toward the outside of the vascular cambium.

secondary xylem: xylem produced from cells arising at the inside of the vascular cambium.

shoot system: all the parts of a vascular plant exclusive of the root. Usually above ground, it consists of stem, leaves, buds, and (in season) flowers and fruits. Its functions include photosynthesis, transport of materials, reproduction, and hormone synthesis.

sieve plate: structures between two adjacent sieve-tube elements in phloem, where holes formed in the primary cell walls interconnect the cytoplasm of the elements.

sieve tube: in phloem, a single strand of sieve-tube elements, which transports sugar solutions.

sieve-tube element: one of the cells of a sieve tube, which form the phloem.

sink: in plants, any structure that uses up sugars or converts sugars to starch and toward which phloem fluids will flow.

source: in plants, any structure that actively synthesizes sugar and away from which phloem fluid will be transported.

spongy cells: irregularly shaped mesophyll cells containing chloroplasts located just above the lower epidermis of a leaf.

stem: the normally vertical, above ground part of a plant body that bears leaves.

stoma (pl. stomata): an adjustable opening in the epidermis

of a leaf, surrounded by a pair of guard cells, that regulates the diffusion of carbon dioxide and water into and out of the leaf.

taproot system: a root system commonly found in dicots, consisting of a long, thick main root that develops from the primary root, and many smaller lateral roots that grow from the primary root.

tendril: a slender outgrowth of a stem that coils about external objects and supports the stem; usually a modified leaf or branch.

terminal bud: the bud at the extreme end of a stem or branch.

thorn: a hard, pointed outgrowth of a stem; usually a modified branch.

tracheid: an elongated xylem cell with tapering ends containing pits in the cell walls; forms tubes that transport water.

transpiration: evaporation of water through the stomata of a leaf.

vascular bundle: a strand of xylem and phloem found in leaves and stems; in leaves, commonly called a vein.

vascular cambium: a lateral meristem located between the xylem and phloem of a woody root or stem and gives rise to secondary xylem and phloem

vascular cylinder: the centrally located conducting tissue of a young root, consisting of primary xylem and phloem.

vascular tissue system: a plant tissue system consisting of xylem (which transports water and minerals from root to shoot) and phloem (which transports water and sugars throughout the plant).

vein: in vertebrates, a large-diameter, thin-walled vessel that carries blood from venules back to the heart; in vascular plants, a vascular bundle, or a strand of xylem and phloem in leaves.

vessel: a tube of xylem composed of vertically stacked vessel elements, with perforated or missing end walls, leaving a continuous, uninterrupted hollow cylinder.

vessel element: one of the cells of a xylem vessel; elongated, dead at maturity, with thick lignified lateral cell walls for support but with end walls either lacking entirely or heavily perforated.

xylem: a conducting tissue of vascular plants that transports water and minerals from root to shoot.

THINKING THROUGH THE CONCEPTS

True or False: Determine if the statement given is true or false. If it is false, change the underlined word so that the statement reads true.

48. _____ The epidermis of the root system is covered by the cuticle.
49. _____ The function of root systems is greatly enhanced by the presence of root hairs.
50. _____ It is the parenchyma cells that store starch in white potatoes.
51. _____ It is the sclerenchyma cells that create the "gritty" texture of pears.
52. _____ Tracheids and vessel elements transport water and minerals through phloem tissue.
53. _____ Sieve-tube elements function independently even though they have no nucleus.
54. _____ A monocot such as a dandelion produces a taproot.
55. _____ The rootcap must be continuously replaced as the root pushes its way through the soil.
56. _____ When regulating stomatal openings, plants must balance the need for CO_2 with the need to conserve water.
57. _____ Some stems are the primary site for the conduction of photosynthesis.
58. _____ Leaves grow from lateral buds.
59. _____ Xylem and phloem cells are differentiated sclerenchyma cells.
60. _____ Removing a ring of bark will ultimately kill a tree.
61. _____ Leaf mesophyll cells are differentiated from parenchyma cells.
62. _____ A tendril may be derived from either a stem or a leaf.
63. _____ The casparian strip plays a vital role in getting water into the vascular cylinder.
64. _____ The cohesion-tension theory explains how plants circulate sugar.
65. _____ Water is pushed through a plant.

66. Complete the following table comparing features of monocots and dicots.

	Monocots	Dicots
flowers		
leaves		
vascular tissue		
root pattern		
embryo in seed		

Matching: Plant structures and their functions.

Choices:

67. _____ leaf   a. absorbs water and minerals
68. _____ stem   b. covers and protects leaves, stems, roots
69. _____ root system   c. waxy layer; prevents water loss
70. _____ petiole   d. transports water and minerals
71. _____ epidermis   e. regulate diffusion of O_2 and CO_2 and water loss
72. _____ internode   f. increases absorptive surface area
73. _____ node   g. transports sugars, amino acids, and hormones
74. _____ root hair   h. absorbs sunlight
75. _____ phloem   I. develops into branches
76. _____ lateral bud   j. stem area between leaves
77. _____ cuticle   k. develop into leaves
78. _____ stomata   l. stalk of leaf, attaches blade to stem
79. _____ xylem   m. supports plant

Matching: Cell types and their functions or characteristics.

80. _____ increase absorption surface area
81. _____ waterproof cells of stems and roots
82. _____ transports water and minerals
83. _____ transports carbohydrates
84. _____ unevenly thickened walls; provides support
85. _____ lie end to end; connect through pits
86. _____ hardened with lignin; not alive when mature
87. _____ conducts photosynthesis
88. _____ controls functions of sieve-tube element
89. _____ lie end to end; open ends
90. _____ controls stomatal openings
91. _____ active cellular division; allows plant growth
92. _____ makes up phloem; open tubes
93. _____ conduct most metabolic functions in a plant; are alive
   when mature

Choices:

a. collenchyma
b. parenchyma
c. sclerenchyma
d. guard cell
e. mesophyll cell
f. companion cell
g. cork cell
h. meristem cell
I. root hair
j. xylem
k. tracheids
l. phloem
m. vessel elements
n. sieve-tube elements

94. Complete the following table of stem and leaf modifications.

Stem modifications	Function	Plant example	Leaf modifications	Function	Plant example
runners			spines		
succulent stems			succulent leaves		
storage			insectivorous		
rhizome			bulb		
thorns			leaf tendrils		
stem tendrils					

CLUES TO APPLYING THE CONCEPTS

These practice questions are intended to sharpen your ability to apply critical thinking and analysis to the biological concepts covered in this chapter.

95. Compare a taproot system and a fibrous root system.

_____
_____
_____
_____
_____
_____
_____
_____

96. Provide two examples of legumes:

a. _____   b. _____

97. Using the following terms discuss how fungi and bacteria aid roots in acquiring nutrients.

ammonium	phosphorous	nodule	nitrogen
carbohydrate	fungus	roots	nitrogen-fixing bacteria
mycorrhizae	legume	N_2	root hairs
nitrogen fixation	mineral	nitrate	

_____

_____

_____

_____

_____

_____

_____

_____

_____

_____

98. Using the following terms discuss how plants transport carbohydrates throughout the plant body.

companion cell	sieve-tube element	osmosis	xylem
hydrostatic pressure	diffusion	phloem	leaf
pressure flow theory	sieve tube	source	sink

_____

_____

_____

_____

_____

_____

_____

_____

_____

_____

_____

ANSWERS TO EXERCISES

1. shoot system
2. leaves
3. stems
4. root system
5. minerals
6. node
7. internode
8. blade
9. petiole
10. apical meristems
11. primary growth
12. differentiated
13. lateral meristem
14. cambium
15. secondary growth
16. dermal tissue system
17. ground tissue system
18. vascular tissue system
19. epidermis
20. periderm
21. cuticle
22. primary root
23. taproot system

24. fibrous root system
25. rootcap
26. cortex
27. casparian strip
28. endodermis
29. pericycle
30. branch roots
31. leaf primordia
32. lateral buds
33. primary xylem
34. primary phloem
35. vascular cambium
36. secondary xylem
37. secondary phloem
38. sapwood
39. heartwood
40. annual rings
41. cork cambium
42. bark
43. cohesion-tension theory
44. bulk flow
45. transpiration

46. stomata
47. abscisic acid
48. false, shoot system
49. true
50. true
51. true
52. false, xylem
53. false, are dependent on companion cells because
54. false, dicot
55. true
56. true
57. true
58. false, branches
59. false, parenchyma
60. true
61. true
62. true
63. true
64. false, transport water and minerals
65. false, pulled

66. Monocots vs Dicots

	Monocots	Dicots
flowers	multiples of 3	multiples of 4 or 5
leaves	smooth, narrow, parallel veins	oval or palmate, net-like veins
vascular tissue	vascular bundles scattered	vascular bundles in ring
root pattern	fibrous root system	taproot system
embryo in seed	one cotyledon	two cotyledons

67. h
68. m
69. a
70. l
71. b
72. j
73. k
74. f
75. g
76. i

77. c
78. e
79. d
80. I
81. g
82. j
83. l
84. a
85. k

86. c
87. e
88. f
89. m
90. d
91. h
92. n
93. b

94. Modification table:

Stem modifications	Function	Plant example	Leaf modifications	Function	Plant example
runners	sprout young plants	strawberry plants	spines	protection	cacti
succulent stems	store water	cactus, baobab	succulent leaves	store water	some desert plants
storage	starch	white potato	insectivorous	nitrogen acquisition	venus fly-traps, sundews
rhizome	horizontal under ground stem	iris	bulb	store nutrients, to over winter	onion, daffodils, tulips
thorns	protection	roses	leaf tendrils	grasping, support	garden peas
stem tendrils	grasping, support	ivy, grapes			

95. Taproot systems are composed of a persistent primary root with many smaller roots growing out from the sides of the taproot. In fibrous root systems, the primary root does not persist. New roots are produced from the base of the stem. These new roots are approximately equal in size.

96. Peas, clover, soybeans, alfalfa

97. Symbiotic relationships between plants and fungi and between plants and soil bacteria increase nutrient acquisition in nutrient poor soils. **Fungi** that form a symbiotic relationship with plant **roots** are called **mycorrhizae**. Mycorrhizae help plants acquire **minerals**, especially **phosphorous**. In return, the fungus receives carbohydrates and amino acids from the plant. **Nitrogen** is also often in limiting quantities to many plants. **Legumes** have overcome this by forming an association with **nitrogen-fixing bacteria**. These bacteria possess the enzymes needed to convert N_2 from the atmosphere into **ammonium** or **nitrate**, the process of **nitrogen fixation**. The bacteria enter the **root hairs** of a legume and travel to the cortex of the root. The bacteria multiply and stimulate the cortex cells to multiply as well. A root **nodule** is produced. It is within the nodule that nitrogen fixation occurs, supplying the plant with a steady supply of usable nitrogen. In return, the bacteria receive **carbohydrates** for their energy needs.

98. An area of the plant that produce carbohydrates such as a **leaf** is referred to as a **source**. Any structure that needs carbohydrates produced by a source is referred to as a **sink**. Movement of sugars between sources and sinks must occur. It is believed that the **pressure flow theory** describes how this happens. Glucose is produced by photosynthesis, usually in leaves. The glucose is converted to sucrose and moved by active transport into **companion cells** of the **phloem**. As the concentration gradient of sucrose increases in companion cells, it is low in sieve-tube elements. Thus, sucrose diffuses into the **sieve-tube elements**. Water enters the **sieve-tube** by osmosis from nearby **xylem**, increasing the hydrostatic pressure within the sieve tube. The reverse sequence occurs, unloading sucrose at a sink region. Sucrose is actively transported into a sink, such as a developing fruit, thus lowering the sucrose concentration in the sieve tube. With a lower sucrose concentration, water flows out of the sieve tube by osmosis, lowering the hydrostatic pressure again. The hydrostatic pressure difference between the sieve tube at the source and the sieve tube at the sink causes water to flow down the sieve tube, carrying sucrose with it.

Chapter 24: Plant Reproduction and Development

OVERVIEW

This chapter presents an overview of how flowering plants reproduce. The organization of the flower and the importance of each part is discussed. The methods of pollination, including some of the animals involved, are identified. These animals have coevolved with plants to develop pollinator-nutrient resource relationships. Once pollination has occurred, fertilization may take place, leading to the development of the embryo plant within a dormant seed. The chapter ends by examining the importance of seed development, dormancy, and germination.

1) What Are the Features of Plant Life Cycles?

The plant life cycle is represented by two distinct generations, the diploid **sporophyte** and the haploid **gametophyte**. Because of this, the plant life cycle is referred to as the **alternation of generations**. Ferns are most commonly used as an example of the alternation of generations. The fern sporophyte produces haploid **spores** through meiosis. The spores are dispersed to the soil by the wind. When conditions are favorable, the spore **germinates** and develops into a mature, haploid gametophyte. The gametophyte produces eggs and sperm without another meiotic event. When egg and sperm fuse, a diploid zygote is formed that will mature into the sporophyte. A similar life cycle occurs in all plants. In primitive land plants, such as mosses and ferns, the gametophyte is an independent plant that requires free water to complete reproduction. More evolved land plants have microscopic gametophytes that are dependent on the sporophyte. In flowering plants the sporophyte produces the **flower**. Within the flower, meiosis produces the spores that develop into gametophytes, still within the flower. A **megaspore** divides by mitosis to produce the female gametophyte, and a **microspore** divides by mitosis to produce the male gametophyte, the **pollen grain**. The female gametophyte produces the egg while the male gametophyte produces the sperm.

When pollen is distributed from one flower to another, the pollen grain germinates, producing a pollen tube that will deliver the sperm to the egg for fertilization. The zygote that is formed develops into an embryo sporophyte, **dormant** within a **seed** until favorable conditions exist.

2) How Did Flowers Evolve?

Flowers evolved as plants acquired characteristics that attracted pollinators. Plants that carried genetic mutations that led to the production of energy-rich nectar or pollen and a way to display its availability to pollinators passed those traits on to their offspring. Ultimately, the diverse assortment of flowers seen today arose. Through the modification of leaves, **sepals**, **petals**, **stamens**, and **carpels** developed. If the flower is a **complete flower**, then sepals, petals, stamens, and carpels are all present. If one or more of the flower parts are not present, then the flower is an **incomplete flower**. The sepals form the outermost whorl of flower parts and serve to protect the bud. The petals are located just inside the sepals. They are often brightly colored as a display. If the stamens are present, they are located just inside the petals. The stamens include a **filament** that supports an **anther** bearing pollen. If the carpel is present, it is the innermost structure. The carpel is composed of one or more **ovaries**, each with a rather long **style** ending in a sticky **stigma**. Within each ovary is an **ovule** containing the female gametophyte. After fertilization of the egg, the ovule will mature into the **seed** and the ovary will develop into the **fruit**.

3) How Have Flowers and Pollinators Adapted to One Another?

As mutations occurred in plants that resulted in the attraction of pollinators, mutations also occurred in the pollinators that resulted in traits that helped them pollinate certain plants. Plants and pollinators have acted as a significant force of natural selection in the evolution of each other. Thus, the two have **coevolved**. For example, bee-pollinated flowers attract the bees by emitting a sweet smell. At the same time, the flowers often have ultraviolet markings that direct the bee to the nectar it desires. When the bee feeds, pollen is brushed onto her back. With a visit to another flower, she deposits some of the pollen from the first flower on the stigma of the second flower. In contrast, flowers that have evolved without a sweet smell and that are red, do not attract bees for pollination. Instead, these flowers attract hummingbirds. The long bill and tongue of the bird are perfect for reaching the copious nectar at the base of the long, tubular flower. Some species have so perfectly coevolved that they rely solely on one another for survival.

4) How Do Gametophytes Develop in Flowering Plants

Within the flower meiosis produces the spores that develop into gametophytes. The anthers of stamens produce diploid **microspore mother cells** within the pollen sacs. Each microspore mother cell will divide by meiosis producing four **microspores**. In turn, each microspore will divide by mitosis producing the male gametophyte, a **pollen grain**. In each pollen grain are two cells, a **tube cell** and, within the tube cell, a **generative cell**.

Within the ovaries of carpels, ovules are produced consisting of layers of cells. The outer layers are called the **integuments**, and they enclose the diploid **megaspore mother cell**. The megaspore mother cell divides by meiosis producing four haploid **megaspores**. Three of the megaspores deteriorate. The surviving megaspore divides by mitosis three times, producing eight haploid nuclei. From these eight nuclei, seven cells are produced. This seven-celled structure is the **embryo sac**, the female gametophyte. The seven cells of the female gametophyte are distributed so that there are three cells at each end of the structure. The remaining cell, the **primary endosperm cell**, contains two nuclei called **polar nuclei**. The egg is located at the bottom of the female gametophyte, near an opening in the integuments.

5) How Does Pollination Lead to Fertilization?

When a compatible pollen grain lands on the stigma of a flower, it absorbs water from the stigma inducing it to germinate. Germination results in elongation of the tube cell down through the style to the ovary and in the production of two sperm from mitotic division of the generative cell. As the tube cell elongates, it carries the two sperm with it. As the tip of the tube cell reaches the ovule, it penetrates the embryo sac and releases both of the sperm. One sperm fertilizes the egg forming the diploid zygote. The second sperm enters the primary endosperm cell, fusing with both polar nuclei. This triploid cell divides mitotically, forming the **endosperm**. Since there are two fertilization events, this process is called **double fertilization**. It is necessary to distinguish between **pollination**, the pollen grain landing on the stigma, and **fertilization**, the fusion of sperm and egg, as two separate events.

·6) How Do Seeds and Fruits Develop?

Once fertilization has occurred, the embryo sac and the integuments develop into the seed. The integuments specifically form the **seed coat**. Fertilization of the primary endosperm cell produces the endosperm, which will provide nutrients for the developing embryo. As the embryo forms from the zygote, it develops the primary root, the shoot, and **cotyledons** or seed leaves. The region of the developing shoot is divided at the attachment site of the cotyledons. The region above the cotyledons is referred to as the **epicotyl** while the region below the cotyledons is the **hypocotyl**.

The mature embryo within the seed enters a state of lowered metabolic activity; it becomes **dormant**. In this dormant state, the embryo can endure adverse environmental conditions, such as a harsh, cold winter, or a period of drought. Often, the dormant state will not cease, and the seed does not germinate, until an adaptive set of requirements has been met. As a protection against early **germination**, many seeds must actually dehydrate. Others must experience a prolonged cold period followed by a warm, moist period. This avoids germination during warm, moist autumn days that will be followed by winter temperatures. It is the wall of the surrounding ovary that will devlop into the fruit. The fruit may be soft and fleshy, firm and rigid, or otherwise modified for a specific type of dispersal. The seed coat, itself, may serve as a deterrent to germination. Cracking or splitting of the seed coat to allow water and or oxygen to enter may be recquired before a seed will germinate.

7) How Do Seeds Germinate and Grow?

When a seed breaks dormancy and begins to germinate, it absorbs a great deal of water, causing it to swell. The seed coat breaks open and the primary root emerges. As the shoot grows, it pushes upward, out of the soil. In order to protect the fragile apical meristem, monocot shoots are protected by the **coleoptile**. Dicots do not produce a coleoptile. Instead, they protect their shoot apical meristem by forming a hypocotyl hook. In other words, the shoot remains bent as it pushes upward. While the young plant is becoming established it receives nourishment from its cotyledons until newly formed leaves develop and begin photosynthesizing.

KEY TERMS AND CONCEPTS

Fill-In: From the following list of key terms, fill in the blanks in the following statements.

alternation of generations	coevolution	dormant	anthers	epicotyl
coleoptile	cotyledons	embryo sac	carpels	fertilize
complete flower	filament	endosperm	flower	fruit
double fertilization	gametophyte	microspore	petals	ovary
generative cell	germination	pollen grain	seed	ovules
incomplete flower	hypocotyl	pollination	sepals	seed coat
megaspore mother cell	integuments	sporophyte	stigma	spores
microspore mother cell	megaspore	tube cell	style	stamens
primary endosperm cell	polar nuclei			

The plant life cycle includes a haploid generation and a diploid generation. The cycling between the two is referred to as (1) _____. During the plant life cycle, the diploid (2) _____ produces (3)_____ and the haploid (4) _____ produces the gametes.

In flowering plants the male gametophyte is produced by a (5)_____ that has divided by mitosis. The male gametophyte is actually the (6)_____. The female gametophyte is produced by mitotic division of the (7)_____. The zygote that is formed after fertilization occurs stays protected in a drought resistant (8)_____

(9) _____ are composed of modified leaves. The (10) _____ function to protect the bud, and the (11) _____ function to attract pollinators.

The male reproductive structures are called the (12) _____ and are composed of a long stalk, the (13)_____ , with the pollen producing (14) _____ at its tip.

The female reproductive structures are called the (15) _____ and are composed of a sticky (16) _____ that is at the tip of an elongated (17)_____. At the base is a bulbous (18) _____ which contains (19) _____. It is the ovary that will develop into the (20) _____ which may be edible.

If a flower contains all flower parts, it is considered to be a (21) _____. If, however, a flower is missing one or more of the structures, it is an (22)_____.

Flowering plants and their animal pollinators have adapted to one another, this is an example of (23) _____.

Development of the male gametophyte begins with division of the diploid (24) _____ producing four haploid microspores. As microspores divide and develop into the pollen grain, two cells are produced with in the pollen grain, the (25) _____ and the (26) _____. When the pollen has matured, it will be distributed to the stigma of another flower. This is referred to as (27) _____.

The female gametophyte is the (28) _____. The process of its development occurs with in an ovule. The outer layers of the ovule are the (29) _____ which surround the diploid cell, the (30) _____. Through meiosis, four haploid megaspores will be produced. The embryo sac is a seven celled structure. One cell, in the center of the embryo sac contains two nuclei. These are the (31)_____ that are contained within the (32) _____.

Once pollination has occurred, the pollen grain germinates producing the pollen tube with two sperm inside. One sperm will (33) _____ the egg, while the second sperm will fuse with the two polar nuclei, producing the triploid (34) _____. The use of two sperm in the fertilization process is termed (35) _____.

After fertilization, the integuments of the ovule develop into the (36) _____, and the endosperm nourishes the developing embryo. Most of the endosperm is absorbed in to the seed leaves or (37) _____. The seed will remain (38) _____ until conditions are favorable for (39) _____.

Once germination is initiated, the primary root will emerge from the seed followed by the young shoot. In monocots, the emerging shoot is protected by the (40) _____. In dicots, the shoot is divided in to regions with respect to the cotyledons. The shoot above the cotyledons is called the (41) _____, and the shoot below them is called the (42) _____.

Key Terms and Definitions

alternation of generations: a life cycle, typical of plants, in which a diploid sporophyte (spore-producing) generation alternates with a haploid gametophyte (gamete-producing) generation.

anther: the uppermost part of the stamen in which pollen develops.

carpel: the female reproductive structure of a flower, composed of stigma, style, and ovary.

coevolution: the evolution of adaptations in two species due to their extensive interactions with one another, so that each species acts as a major force of natural selection upon the other.

coleoptile: a protective sheath surrounding the shoot in monocot seeds, allowing the shoot to push aside soil particles as it grows.

complete flower: a flower that has all four floral parts (sepals, petals, stamens, and carpels).

cotyledon: also called a seed leaf; a leaflike structure within a seed that absorbs food molecules from the endosperm and transfers them to the growing embryo.

dormancy: a state in which an organism does not grow or develop; usually marked by lowered metabolic activity and resistance to adverse environmental conditions.

double fertilization: in flowering plants, a phenomenon in which two sperm nuclei fuse with the nuclei of two cells of the female gametophyte. One sperm fuses with the egg to form the zygote, while the second sperm nucleus fuses with the two haploid nuclei of the primary endosperm cell to form a triploid endosperm cell.

egg: the haploid female gamete, normally large and nonmotile, containing food reserves for the developing embryo.

embryo sac: the haploid female gametophyte of flowering plants.

endosperm: a triploid food storage tissue found in the seeds of flowering plants.

epicotyl: the part of the embryonic shoot located above the cotyledons but below the tip of the shoot.

fertilization: the fusion of male and female haploid gametes, forming a zygote.

filament: in flowers, the stalk of a stamen, which bears an anther at its tip.

flower: the reproductive structure of a flowering plant.

fruit: in flowering plants, the ripened ovary (plus, in some cases, other parts of the flower).

gametophyte: the multicellular haploid stage in the life cycle of plants.

generative cell: in flowering plants, one of the haploid cells of a pollen grain. The generative cell undergoes mitosis to form two sperm cells.

germination: the growth and development of a seed, spore, or pollen grain.

hypocotyl: the part of the embryonic shoot located below the cotyledons but above the root.

incomplete flower: a flower that is missing one of the four floral parts (sepals, petals, stamens, or carpels).

integument: the outer layers of cells of the ovule surrounding the embryo sac; develops into the seed coat.

megaspore: a haploid cell formed by meiosis from a diploid megaspore mother cell. Through mitosis and differentiation, the megaspore develops into the female gametophyte.

megaspore mother cell: a diploid cell contained within the ovule of a flowering plant, which undergoes meiosis to produce four haploid megaspores.

microspore: a haploid cell formed by meiosis from a microspore mother cell. Through mitosis and differentiation, the microspore develops into the male gametophyte.

microspore mother cell: a diploid cell contained within an anther of a flowering plant, which undergoes meiosis to produce four haploid microspores.

ovary: in flowering plants, a structure at the base of the carpel containing one or more ovules; develops into the fruit.

ovule: a structure within the ovary of a flower, inside which the female gametophyte develops. After fertilization, the ovule develops into the seed.

petal: part of a flower, often brightly colored and fragrant, serving to attract potential animal pollinators.

polar nucleus: in flowering plants, one of two nuclei in the primary endosperm cell of the female gametophyte; formed by the mitotic division of a megaspore.

pollen grain: the male gametophyte of a seed plant; also called *pollen*.

pollination: in flowering plants, when pollen lands on the stigma of a flower of the same species; in conifers, when pollen lands within the pollen chamber of a female cone of the same species.

primary endosperm cell: the central cell of the female gametophyte of a flowering plant, containing the polar nuclei (usually two). After fertilization, the primary endosperm cell undergoes repeated mitotic divisions to produce the endosperm of the seed.

seed: the reproductive stage of conifers and flowering plants, usually including an embryonic plant and a food reserve, enclosed within a resistant outer covering.

seed coat: the thin, tough, and waterproof outermost covering of a seed, formed from the integuments of the ovule.

sepal: set of modified leaves that surround and protect a flower bud, often opening into green, leaflike structures when the flower blooms.

spore: in the alternation of generation life cycle of plants, a haploid cell produced through meiosis, that then undergoes repeated mitotic divisions and differentiation of daughter cells to produce a multicellular, haploid organism called the gametophyte.

sporophyte: the multicellular diploid stage of the life cycle of plants.

stamen: the male reproductive structure of a flower, consisting of a filament and an anther in which pollen grains develop.

stigma: the pollen-capturing tip of a carpel.

style: a stalk connecting the stigma of a carpel with the ovary at its base.

tube cell: the outermost cell of a pollen grain; the tube cell digests a tube through the tissues of the carpel, ultimately penetrating into the female gametophyte.

THINKING THROUGH THE CONCEPTS

True or False: Determine if the statement given is true or false. If it is false, change the underlined word so that the statement reads true.

43. _____ The ovule develops into the seed after fertilization occurs

44. _____ During pollination, pollen is attached to the stigma

45. _____ Spores are produced by the gametophyte

46. _____ The pollen grain is the male gametophyte

47. _____ Conifers are wind pollinated, a very efficient process.

48. _____ A complete flower has only stamens

49. _____ Flower evolution occurred separately from the evolution of insects

50. _____ Pollination and fertilization are the same event

51. _____ Double fertilization occurs only in flowering plants

52. _____ The generative cell, from the pollen grain, divides to produce two sperm

53. _____ The megaspore mother cell is haploid

54. _____ The embryo is uniquely triploid

55. _____ The flesh of a peach develops from the ovary

56. _____ The seed coat is mature integuments

57. _____ Flour is made from wheat seed endosperm

58. _____ The purpose of the fruit is to aid in <u>pollen</u> distribution

59. _____ Newly developed seeds often <u>germinate immediately</u>

Matching: Flower characteristics and animal pollinators. There may be more than one answer for a question.

60. _____ odors are sweet, "flowery"   Choices:

61. _____ mimics female   a. bee

62. _____ red flowers   b. moth

63. _____ deep, narrow nectar tubes   c. humming bird

64. _____ bright yellow flowers   d. beetle

65. _____ odors are strong, musky   e. fly

66. _____ produce large amount of nectar   f. wasp

67. _____ no odor

68. _____ odors of rotting flesh

69. _____ tubular flowers

70. _____ white flowers

71. _____ reflect ultraviolet light

Identify the following: Which of these adaptations for seed dispersal are characteristics of dispersal by **animal, edible, wind, water,** or "**shot gun**".

72. _____ seeds with hairy tufts

73. _____ explosive fruits

74. _____ featherweight fruits

75. _____ round and buoyant fruits

76. _____ fruit with hooked spines

77. _____ juicy and tasty fruits

78. _____ winged fruit

79. _____ seeds have supply of fertilizer

Identify: the flower parts on the diagram below.

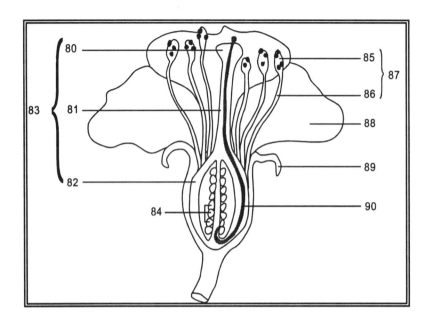

80. _____   86. _____
81. _____   87. _____
82. _____   88. _____
83. _____   89. _____
84. _____   90. _____
85. _____

CLUES TO APPLYING THE CONCEPTS

These practice questions are intended to sharpen your ability to apply critical thinking and analysis to the biological concepts covered in this chapter.

91. What is meant by the term double fertilization?

_____

_____

_____

_____

92 - 97 Identify the seedling structures on the diagram below.

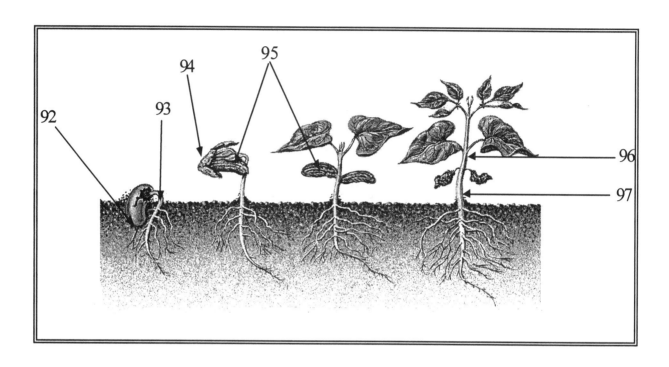

92. _____   95. _____
93. _____   96. _____
94. _____   97. _____

98. Using the labeled diagram above, describe the events that occur as a seed germinates and emerges from the soil. Include the importance of the cotyledons.

_____
_____
_____
_____
_____
_____
_____
_____

ANSWERS TO EXERCISES

1. alternation of generations
2. sporophyte
3. spores
4. gametophyte.
5. microspore
6. pollen grain
7. megaspore
8. seed.
9. flowers
10. sepals
11. petals
12. stamens
13. filament
14. anthers
15. carpels
16. stigma
17. style
18. ovary
19. ovules
20. fruit
21. complete flower
22. incomplete flower
23. coevolution
24. microspore mother cell
25. generative cell
26. vegetative cell
27. pollination
28. embryo sac
29. integuments
30. megaspore mother cell

31. polar nuclei
32. primary endosperm cell
33. fertilize
34. endosperm
35. double fertilization
36. seed coat
37. cotyledons
38. dormant
39. germination
40. coleoptile
41. epicotyl
42. hypocotyl.
43. true
44. true
45. false, sporophyte
46. true
47. false, very inefficient
48. false, all flower parts
49. false, along with
50. false, separate events
51. true
52. true
53. false, diploid
54. false, endosperm
55. true
56. true
57. true
58. false, seed distribution
59. false, are dormant
60. a

61. f
62. c
63. b/c
64. a
65. b
66. c
67. c
68. d/e
69. a/b/c
70. b
71. a
72. wind
73. "shot gun"
74. wind
75. water
76. animal
77. edible
78. wind
79. edible
80. stigma
81. style
82. ovary
83. carpel
84. ouvles
85. anther
86. filament
87. stamen
88. petal
89. sepal
90. pollen tube

91. The term double fertilization is used to describe the fertilization events in flowering plants. One sperm fertilizes the egg, producing the zygote. A second sperm enters the primary endosperm cell and fuses with the two polar nuclei, producing the triploid endosperm.

92. seed
93. primary root
94. seed coat
95. cotyledons
96. epicotyl
97. hypocotyl

98. The seed absorbs water, breaking open the seed coat. The primary root emerges and grows downward in the soil. The hypocotyl elongates, maintaining a hook to protect the apical meristem. The cotyledons emerge from the soil, breaking free of the seed coat. The cotyledons provide nutrients to the developing shoot and new leaves, shriveling as the nutrients become depleted. New leaves expand and begin photosynthesizing.

Chapter 25: Plant Responses to the Environment

OVERVIEW

This chapter presents information on how chemicals produced by plants act as hormones and induce a response to environmental signals. An environmental signal might be gravity or the direction of the sun's rays, which induces a growth response. The signal may be the length of a day, indicating the appropriate time for flowering. Thus, how a plant will grow, flower, or initiate dormancy is due to stimuli from the environment that results in hormone production.

1) How Were Plant Hormones Discovered?

Observations about how plants grow and respond to their environment provided the first clues to the presence of hormones in plants. Darwin reacted to such observations by conducting experiments using grass coleoptiles and their **phototropism** response. It was later determined that the phototropic response was due to chemicals produced by the coleoptile. That chemical was identified as **auxin**.

2) What Are Plant Hormones, and How Do They Act?

Hormones are chemicals that are produced by cells in one area of an organism (the plant) and are transported to another area where they cause a response. Five classes of **plant hormones** have been identified to date. **Auxins** promote elongation of cells in the shoot. They affect roots differently. In low concentrations, they stimulate elongations; however, slightly higher concentrations will inhibit root growth. The distribution of auxin within the plant body is affected by light and gravity. Therefore, the hormone influencing phototropism and **gravitropism** is auxin. **Gibberellins** also stimulate stem elongation. But, they also influence flowering, fruit development, seed germination, and bud sprouting. **Cytokinins** tend to stimulate cell division, thus they influence events relating to cell division, including bud sprouting, fruit development, endosperm production, and embryo growth. Cytokinins also influence plant metabolism. **Ethylene** is a unique hormone because it is a gas at room temperature. Ethylene stimulates fruit ripening and induces leaf drop, as well as flower and fruit drop. **Abscisic acid** controls stomatal opening and closing and acts to inhibit the effects of gibberellins so that seeds and buds remain dormant until conditions are favorable.

3) How Do Hormones Regulate the Plant Life Cycle?

The hormone **abscisic acid** reduces the metabolic rate of the embryo within the seed so that its growth is suspended. The high concentration of abscisic acid must be reduced before germination can occur. This may occur by heavy rains rinsing the abscisic acid out of the seed or by freezing temperatures causing its breakdown. As abscisic acid levels decrease, gibberellin levels increase. Gibberellin triggers the production of enzymes that will digest the endosperm, providing energy and nutrients for the growing embryo. As the embryo grows out of the seed, it must determine which way to grow its shoot system and root system. Organelles in the stem detect gravity and cause auxin to accumulate on the lower side of the stem. This accumulation of auxin causes the lower cells to elongate, curving the stem upward. The effect of auxin on roots may induce opposite results. Again, organelles in the root detect gravity and cause **auxin** to accumulate on the lower side of the root. As **auxin** accumulates on the lower side of a horizontal root, the cells are inhibited from elongating, while the cells on the upper side grow more quickly, causing the root to curve downward. The organelles that detect gravity are starch-filled **statoliths**.

The size and shape of a plant's shoot and root systems are determined by the interactions between auxin and cytokinin. For instance, in some plants the shoot tip produces auxin, inhibiting lateral bud growth. As the auxin concentration decreases down the stem, lateral buds become less inhibited and will grow to produce branches. At the same time, cytokinins produced in the roots move up the stem stimulating lateral bud growth. The interactions between the two hormones give the plant a characteristic shape. The interactions between auxin and cytokinin also influence the size of the shoot system relative to the size of the root system.

Most plants use day length to regulate their production of flowers. However, some, such as snapdragons and tomatoes, flower as soon as the plant has matured. These plants are **day-neutral**. If day length is important, then a critical length of daylight must be reached or not exceeded. If a species is a **long-day plant**, then it must receive daylight for periods longer than a specific critical value. For example, spinach must experience days longer than 13 hours or it will not flower. If a species is a **short-day plant**, then it must not receive daylight for periods longer than a specific critical value. For example, cockleburs will not flower if the day length exceeds 15.5 hours. Day length stimulates the production of hormones called **florigens** to induce flowering. Florigens, however, have yet to be identified. But what is it that detects day length and sets the **biological clock** of a plant? It is the pigment **phytochrome**, which absorbs red light and far-red light. After phytochrome absorbs red light, it is changed to its P_{fr} form. In the P_{fr} form, phytochrome is active and may stimulate or inhibit physiological processes. When phytochrome P_{fr} absorbs far-red light, it is converted to the P_r, or inactive, form. Phytochrome regulates flowering, leaf growth, chlorophyll synthesis, and epicotyl/hypocotyl hook straightening.

Once a flower is mature, pollination followed by fertilization is likely and will lead to fruit and seed development. Hormones are also involved in regulating fruit and seed development. The pollen that lands on the stigma releases auxin or gibberellin, which causes the ovary to begin developing into a fruit. Fertilization further stimulates fruit development as the seeds mature. The ripening of the fruit is caused by the production of ethylene by the fruit cells after they have been exposed to a high concentration of auxin from the mature seeds.   As the plant reaches the end of a growing season, the leaves and fruits under go **senescence**, forming an **abscision layer** at the point of stem attachment. The formation of the abscision layer allows the plant to drop its leaves and ripened fruit when stimulated by ethylene. Senescence and the abscision layer are induced by a decrease in the production of cytokinin by the roots and of auxin by the leaves and fruits. Furthermore, lateral buds become dormant due to the presence of abscisic acid. The plant has ended its growing season and will remain dormant until environmental stimuli activate hormones again in the spring.

4) Do Plants Communicate by Means of Chemicals?

Plants are capable of communicating to one another and to other organisms by releasing chemicals into the air. This occurs when plants are under attack and serves to protect themselves and plants nearby. Virus-infected plants have been found to release salicylic acid and methyl salicylate. Salicylic acid induces an enhanced immune response by the plant to fight off the virus. Methyl salicylate diffuses through the air to nearby plants, which convert it to salicylic acid enhancing their immune defenses to resist the virus. The chemical volicitin is produced by maize plants in response to herbivory by a ·caterpillar. The release of volicitin induces a response in a parasitic wasp that will lay its eggs in the caterpillar eventually killing it.

KEY TERMS AND CONCEPTS

Fill-In: From the following list of key terms, fill in the blanks in the following statements.

abscission layer	cytokinin	biological clock	abscisic acid	apical dominance
auxin	day-neutral plant	gibberellin	ethylene	hormone
plant hormone	gravitropism	long-day plant	florigen	senescence
short-day plant	phytochrome	statolith	phototropism	

(1) _____ can be observed as a plant grows toward the light. This response to an environmental stimulus is induced by the action of a (2) _____ called

(3) _____.

To date, five classes of (4) _____ have been classified. As indicated above, auxins are involved in stem elongation. Auxins also influence root growth downward in response to gravity.

This response is called (5) _____. Organelles called (6) _____ are filled with starch and are believed to be the gravity sensors in a plant.

Fruit development and seed germination are influenced by the hormone (7) _____.

Meanwhile, growth responses involving cell division, such as bud sprouting, are controlled by

(8) _____. One of the more economically important hormones is the gas

(9) _____ since it regulates fruit ripening. Plants growing under drought conditions will be able to regulate water loss by closing their stomata, which are controlled by

(10) _____.

Although a flowering hormone, per say, has not been identified, the term (11) _____ is used to refer to its presence. It is known, however, that plants have a (12) _____ and flowering, in most, is regulated by day length. Plants that need short days (with long nights) are called

(13) _____. In turn, plants that need long days are called

(14) _____. Some species, however, will flower independent of day length if the plant is mature; these plants are (15) _____. There is a pigment

(16) _____ that absorbs red and far-red light, regulating flowering.

At the close of the growing season for a plant, a rapid aging process occurs in leaves and fruit. This event is known as (17) _____ and is followed by leaf or fruit drop. Leaf or fruit drop can only occur after the formation of the (18) _____ at the point of stem attachment .

Key Terms and Definitions

abscisic acid: a plant hormone that generally inhibits the action of other hormones, enforcing dormancy in seeds and buds and causing closing of stomata.

abscission layer: a layer of thin-walled cells at the base of the petiole of a leaf that produces an enzyme that digests the cell wall holding leaf to stem, allowing the leaf to fall off.

apical dominance: the phenomenon whereby a growing shoot tip inhibits the sprouting of lateral buds.

auxin: a plant hormone that influences many plant functions, including phototropism, apical dominance, and root branching. Auxin generally stimulates cell elongation and, in some cases, cell division and differentiation.

biological clock: a metabolic timekeeping mechanism found in most organisms, whereby the organism measures the approximate length of a day (24 hours) even without external environmental cues such as light and dark.

cytokinin: a plant hormone that promotes cell division, fruit growth, sprouting of lateral buds, and prevents aging of plant parts, especially leaves.

day-neutral plant: a plant in which flowering occurs as soon as it has grown and developed, regardless of length of day.

ethylene: a plant hormone that promotes ripening of fruits and leaf and fruit drop.

florigen: one of a group of plant hormones that can both trigger and inhibit flowering; daylength is a stimulus.

gibberellin: a plant hormone that stimulates seed germination, fruit development, and cell division and elongation.

gravitropism: growth with respect to the direction of gravity.

hormone: a chemical produced by one group of cells and transported to other parts of the body, where it exerts specific effects.

long-day plant: a plant that will flower only if the length of daylight is greater than some species-specific duration.

phototropism: growth with respect to the direction of light.

phytochrome: a light-sensitive plant pigment that mediates many plant responses to light, including flowering, stem elongation, and seed germination.

plant hormone: the plant-regulating chemicals auxin, gibberellins, cytokinins, ethylene, and abscisic acid; plant hormones somewhat resemble animal hormones in that they are chemical produced by cells in one location that influence the growth or metabolic activity of other cells, often some distance away in the plant body.

senescence: in plants, a specific aging process, often including deterioration and dropping of leaves and flowers.

short-day plant: a plant that will flower only if the length of daylight is shorter than some species-specific duration.

statoliths: in plants, starch-filled organelles in the stem and root cap that sink to the downward side of cells and act as a sensor for gravity.

THINKING THROUGH THE CONCEPTS

True or False: Determine if the statement given is true or false. If it is false, change the underlined word so that the statement reads true.

19. _____ Charles Darwin conducted hormone experiments using grass seedlings.

20. _____ The high concentration of auxin produced in stems would inhibit the growth of roots.

21. _____ Abscisic acid maintains dormancy in seeds by reducing the metabolic rate of the embryo.

22. _____ Roots "know" to grow downward because of statoliths in cells.

23. _____ The branching pattern of a plant is determined by the interaction of ethylene and water.

24. _____ Pinching off the tip of a *Coleus* plant will create a bushier plant since the dominant influence of auxin has been removed.

25. _____ The environmental cue that induces flowering is temperature.

26. _____ Day-neutral plants flower only under long-day conditions.

27. _____ The hormone phytochrome sets the biological clock of a plant.

28. _____ Phytochrome in the active P_{fr} form absorbs red light and stimulates or inhibits physiological processes.

29. _____ As a shoot emerges from the soil and is exposed to red light, cytokinin changes form, inducing the stem to straighten from its protective bend.

30. _____ When fruits are ripe, they are often green in color.

31. _____ Increased ethylene production by aging leaves causes loosening of cell walls of the abscision layer.

32. _____ Methylsalicylate production decreases the immune defenses of a plant under a virus attack.

33. _____ The chemical volicitin in caterpillar saliva cause maize plants to "signal" a parasitic wasp by releasing a mixture of chemicals.

Identify the Following: Are these growth responses related to **phototropism**, **gravitropism**, or **both**?

34. _____ growth toward light

35. _____ movement induced by statoliths

36. _____ regulated by auxin

37. _____ coleoptile bending

38. _____ root growth into soil

39. _____ bending due to cell elongation

40. _____ growth in response to gravity

41. _____ bending due to stem inhibition

Matching: Responses to inducing hormones.

42. _____ seed germination   Choices:
43. _____ phototropism   a. auxin
44. _____ maintains seed dormancy   b. gibberellins
45. _____ shoot elongation   c. cytokinins
46. _____ leaf drop   d. abscisic acid
47. _____ stimulates lateral bud sprout   e. ethylene
48. _____ influences flowering   f. florigen
49. _____ gravitropism
50. _____ closes stomata
51. _____ prevents lateral bud sprout (apical dominance)
52. _____ fruit development
53. _____ fruit ripening
54. _____ is a gas
55. _____ has not yet been found

CLUES TO APPLYING THE CONCEPTS

These practice questions are intended to sharpen your ability to apply critical thinking and analysis to the biological concepts covered in this chapter.

56. Give a general definition of a hormone.

_____

_____

57. "One bad apple spoils the bunch . . ." is a common phrase often referring to societies ills. It is, however, a true statement. Explain the phenomenon from which this phrase is derived.

_____

_____

_____

_____

_____

_____

58. Briefly explain how the hormones auxin and cytokinin interact to regulate root and stem branching.

_____

_____

_____

_____

_____

_____

_____

_____

_____

59. How does a venus fly trap catch a fly? Briefly explain how the leaves close, trapping the insect inside; and how they open again?

_____

_____

_____

_____

_____

_____

ANSWERS TO EXERCISES

1. phototropism,
2. hormone
3. auxin
4. plant hormones
5. gravitropism
6. statoliths
7. gibberellin
8. cytokinin
9. ethylene
10. abscisic acid
11. florigen
12. biological clock
13. short-day plants
14. long-day plants
15. day-neutral plants
16. phytochrome
17. senescence
18. abscision layer
19. true

20. true
21. true
22. true
23. false, auxin and cytokinin
24. true
25. false, day length
26. false, when mature
27. false, pigment
28. false, far-red
29. false, phytochrome
30. false, red
31. true
32. false, increases
33. true
34. phototropism
35. gravitropism
36. both
37. phototropism

38. gravitropism
39. phototropism
40. gravitropism
41. gravitropism
42. b
43. a
44. d
45. a
46. e
47. c
48. f
49. a
50. d
51. a
52. b
53. e
54. e
55. f

56. Hormones are chemicals that are produced by cells in one area and are transported to another area, where they exert specific effects.

57. "One bad apple spoils the bunch . . ." is derived from the fact that ethylene gas is given off from ripe fruit. Since ethylene influences fruit ripening, any fruit nearby will be affected by the gas being released. If an over-ripe apple is in a barrel, surrounded by less-ripe apples, the ethylene released from the over- ripe apple will cause the surrounding apples to ripen quickly. These apples will, in turn, release more ethylene, affecting the apples around them. In a relatively short time, the entire barrel will be over ripe and spoiled.

58. High concentrations of auxin at the apical meristem suppresses lateral bud growth at the top of the plant. As the concentration of auxin decreases down the stem, the inhibitory affects are lessened. Low concentrations of auxin stimulate root growth and branching. At the same time, cytokinin produced by roots moves up the stem, stimulating lateral bud growth. The two balance the growth of the plant. Large shoot systems will produce auxin in sufficient amounts to stimulate root-system expansion. If the shoot system gets too large for the root system to support nutritionally, cytokinin levels will decrease, reducing the rate of expansion of the shoot system until the root system can catch up.

59. Simply, the triggering of the sensory hairs on the leaves causes rapid growth of an outer layer of cells. This causes the leaves to shut, much like the growth of outer stem cells stimulated by auxin causes the stem to bend toward light. Over time, the cells of the second layer of the leaf grow, balancing the size of the cells of the two layers and slowly opening the trap.

Chapter 26: Homeostasis and the Organization of the Animal Body

OVERVIEW

This chapter presents an overview of the mechanisms that enable animals to regulate and control their bodily functions, even when they inhabit harsh environments or are under harsh conditions. These mechanisms include ways to maintain a stable internal environment as well as organs and organ systems that have specialized functions within that internal environment.

1) How Do Animals Maintain Internal Constancy?

Animals maintain internal consistency (**homeostasis**), in part, through **negative feedback.** Negative feedback systems counteract the effects of the external environment so that the body system is returned to its original condition. An example of a negative feedback response is your body regulating its internal temperature. When specific nerve endings sense a change in body temperature, the message is sent to your hypothalamus. The hypothalamus then activates the mechanisms needed to bring your body temperature back to its set point (98.6 F). When your body temperature has reached its set point, your hypothalamus deactivates the mechanisms used to restore your body temperature to normal. Homeostasis is also maintained through **positive feedback**. Positive feedback systems reinforce changes, thus the change continues in a controlled but self-limiting chain reaction. The initiation of childbirth is such a mechanism. The initial uterine contractions begin a sequence of events that result in stronger and stronger contractions. This self-limiting event ends with the birth of the child and the placenta. Numerous negative and positive feedback systems act in concert to maintain internal consistency. In order to achieve this, the body has evolved complex mechanisms to coordinate the systems and exchange signals.

2) How Is the Animal Body Organized?

The animal body is composed of cells organized into **tissues**. The tissues combine to make up **organs** each with specific functions. The organs, in turn, combine to form **organ systems**.

Epithelial tissues form the **membranes** that cover the body and line organs. They provide a barrier that controls what substances cross into or out of the organ. Epithelial tissues withstand a great deal of wear and tear; thus, they must constantly be replaced through mitotic cell divisions. A second function of epithelial tissue is to serve as **glands**. **Exocrine glands** are connected to the epithelium by a duct. **Endocrine glands** are no longer attached to the epithelium. Most commonly, these glands produce and secrete hormones.

Connective tissues tend to be surrounded by an extracellular substance produced by the tissue and are often complexed with fibers of **collagen**. Immediately beneath the epidermis of the skin is a layer of connective tissue, the **dermis**. **Tendons** function to connect muscles to bones while **ligaments** attach bones to other bones. Covering the ends of bones, forming pads between vertebrae, and structuring respiratory passages, ears, and the nose, is **cartilage**. **Bone**, itself, is a connective tissue that is complexed with deposits of calcium phosphate. **Fat** cells make up the connective tissue called **adipose tissue** that specializes in high energy storage. Also included in connective tissue are **blood** and **lymph** since they are primarily composed of extracellular fluid.

Muscle tissue cells are specialized to contract and relax. **Skeletal muscle** typically contracts in response to a conscious decision to move a portion of the skeleton. **Cardiac muscle**, found only in the heart, contracts and relaxes spontaneously as the heart beats. **Smooth muscle** contractions are slow, rhythmic, and involuntary.

Nerve tissue cells are called **neurons** and are specialized to generate and conduct electrical signals. The **dendrite** of a neuron receives a stimulus from the external environment or from another neuron. The **cell body** controls the functions of the cell. **Axons** of the neuron conduct the signal to the target cell, and **synaptic terminals** transmit the signal to the target cell, either another neuron, a muscle cell, or a gland cell. **Glial cells** allow neurons to function properly by surrounding and protecting them.

When two or more tissue types function together, an organ is formed. The skin, for example, is a typical organ. The **epidermis** is covered and protected by a layer of dead cells containing the protein **keratin**. Beneath the epidermis is the connective tissue layer, the dermis. The dermis is densely imbedded with arterioles and capillaries, lymph vessels, and nerve endings. **Hair follicles** are glands within the dermis that produce hair. Sweat glands cool the skin and remove waste products. **Sebaceous glands** produce oil to moisturize the epithelium.

KEY TERMS AND CONCEPTS

Fill-In: From the following list of key terms fill in the blanks in the following statements.

skeletal muscle	cardiac muscle	hair follicle	dendrite	blood
synaptic terminal	smooth muscle	epidermis	collagen	dermis
connective tissue	adipose tissue	membrane	keratin	tendon
endocrine gland	exocrine gland	glial cells	lymph	glands
negative feedback	organ system	ligament	tissue	axon
sebaceous gland	epithelial tissue	cell body	organ	bone
positive feedback	nerve tissue	cartilage	neuron	fat

(1) _____ refers to the body's ability to maintain a relatively constant internal environment. This consistency is maintained, in part, through (2) _____, which responds to counteract change, and (3) _____, which responds to enhance change.

The cells of multicellular organisms, such as animals, are organized into (4) _____ that have a specific function. These organized cells then form (5) _____, such as the lungs, heart, or stomach.   Several different organs involved in a similar process form an (6) _____.

The body's cavities are lined with (7) _____, creating a (8) _____, which serves to restrict substances moving from inside an organ to the outside as well as the reverse. Epithelial tissue also forms (9) _____ that secrete substances. (10) _____ remain connected to the epithelial tissue by way of a duct. (11) _____ separate from the epithelium and tend to secrete hormones into surrounding extracellular fluid.

The most diverse tissue in structure and function is (12) _____. Most of this tissue is complexed with fibrous strands of the protein (13) _____. The connective tissue located under the skin's epidermis is the (14) _____.

The human body shape is maintained, in part, by the internal skeleton.   Connective tissue plays a part in that (15) _____ connect muscle to bones while (16) _____ connect bones to bones. (17) _____ themselves are connective tissue made rigid by calcium phosphate deposits. Adipose tissue is composed of (18) _____ cells, which specifically store energy. The most unusual connective tissues are the fluids (19) _____ and (20) _____.

Muscle tissue is specialized for contraction and exists as three types. (21) _____ contracts when told to do so by a conscious thought. (22) _____ is found only in the heart and contracts spontaneously. (23) _____ also contracts involuntarily and is found in the gastrointestinal tract.

The brain and spinal cord are composed of (24) _____. Nerve cells, or (25) _____, generate and conduct electrical impulses. The (26) _____ receives signals, the (27) _____ maintains and repairs the neuron, and the (28) _____ conducts the impulse to a target cell. The signal is actually transmitted through (29) _____ , while (30) _____ allow the neuron to function optimally.

The skin as an organ has an outer layer, the (31) _____, that is protected with dead cells containing the protein (32) _____. The connective tissue of the skin, the dermis, contains glands that produce hair. These glands are the (33) _____. Also, the (34) _____ secrete oil to keep the skin lubricated.

Key Terms and Definitions

adipose tissue: tissue composed of fat cells.

axon: a long extension of a nerve cell, extending from the cell body to synaptic endings on other nerve cells or on muscles.

blood: a fluid consisting of plasma in which blood cells are suspended; carried within the circulatory system.

bone: a hard, mineralized connective tissue that is a major component of the vertebrate endoskeleton; provides support and sites for muscle attachment.

cardiac muscle (kar´-de-ak): the specialized muscle of the heart, able to initiate its own contraction, independent of the nervous system.

cartilage (kar´-teh-lij): a form of connective tissue that forms portions of the skeleton; consists of chondrocytes and their extracellular secretion of collagen; resembles flexible bone.

cell body: the part of a nerve cell in which most of the common cellular organelles are located; typically a site of integration of inputs to the nerve cell.

collagen: a fibrous protein found in connectivetissue such as bone and cartilage.

connective tissue: a tissue type consisting of diverse tissues, including bone, fat, and blood, which generally contain large amounts of extracellular material.

dendrite (den´-drit): a branched tendril that extends outward from the cell body of a neuron; specialized to respond to signals from the external environment or from other neurons.

dermis: the layer of skin lying beneath the epidermis, composed of connective tissue and containing blood vessels, muscles, nerve endings, and glands.

epidermis: specialized epithelial tissue that forms the outer layer of skin.

epithelial tissue: a tissue type that forms membranes that cover the body surface and line body cavities, and that also gives rise to glands.

exocrine gland: a gland that releases its secretions into ducts that lead to the outside of the body or into the digestive tract.

fat: fat-storing connective tissue whose cells are packed with triglycerides (fats); also called adipose tissue.

gland: a collection of cells that are specialized to secrete substances.

glial cell: a cell of the nervous system that provides support and insulation for neurons.

hair follicle: a gland in the dermis of mammalian skin, formed from epithelial tissue, that produces a hair.

homeostasis: the maintenance of a relatively constant physiological state within an organism.

keratin: fibrous protein found in hair, nails, and the epidermis of skin.

ligament: a tough connective tissue band connecting two bones.

lymph (limf): a pale fluid, within the lymphatic system, that is composed primarily of interstitial fluid and lymphocytes.

membrane: continuous sheets of epithelial cells that cover the body and line body cavities.

negative feedback: a situation in which a change initiates a series of events that tend to counteract the change and restore the original state. Negative feedback in physiological systems maintains homeostasis.

nerve tissue: the tissue that make up the brain, spinal cord, and nerves; consists of neurons and glial cells.

neuron (nur´-on): a single nerve cell.

organ: a structure (such as the liver, kidney, or skin) composed of two or more distinct tissue types that function together.

organ system: two or more organs that work together to regulate a particular physiological process.

positive feedback: a situation in which a change initiates events that tend to amplify the original change.

sebaceous gland: a gland in the dermis of skin, formed from epithelial tissue, that produces an oily substance called sebum that lubricates the epidermis.

skeletal muscle: the type of muscle that is attached to and moves the skeleton and is under the direct, normally voluntary, control of the nervous system; also called *striated muscle*.

smooth muscle: the type of muscle that surrounds hollow organs, such as the digestive tract, bladder, and blood vessels; normally not under voluntary control.

synaptic terminal: a swelling at the branched ending of an axon; where the axon forms a synapse.

tendon: a tough connective tissue band connecting a muscle to a bone.

tissue: a group of cells similar in structure and/or function. The tissue may include extracellular material produced by its cells.

THINKING THROUGH THE CONCEPTS

True or False: Determine if the statement given is true or false. If it is false, change the underlined wordso that the statement reads true.

35. _____ Cells are the building blocks of organs.
36. _____ The kidney is an example of an organ system.
37. _____ Skin is composed of epithelial tissue.
38. _____ The epidermis is replaced two times a year.
39. _____ Salivary glands are an example of exocrine glands.
40. _____ Endocrine glands are connected to the epithelium by ducts.
41. _____ Blood is a connective tissue.
42. _____ Cartilage forms the internal structure of the ear.
43. _____ Smooth muscle is found only in the heart.
44. _____ Skeletal muscle contractions are voluntary.
45. _____ Blood vessels contain muscle tissue.
46. _____ Dendrites conduct an electrical signal to a target cell.
47. _____ An animal's bladder contracts using smooth muscle.
48. _____ Hair follicles are glands found in the dermis.

Matching: Epithelial tissues and their function or characteristic shape

49. _____ one layer of thin flat cells

50. _____ one layer of cubed cells

51. _____ forms lining of blood vessels

52. _____ forms lining of respiratory tract

53. _____ many layers of thin flat cells

54. _____ one layer of elongated cells

55. _____ forms lining of stomach

56. _____ upper layer of skin

57. _____ one layer of elongated cells, nuclei scattered

58. _____ part of salivary glands

59. _____ forms lining of mouth

60. _____ cells are rapidly replaced

61. _____ cells are ciliated

Choices:

a. simple squamous

b. stratified squamous

c. simple cuboidal

d. simple columnar

e. pseudostratified columnar

Short Answer:

62. Identify the four major categories of animal tissues.

    a. _____   b. _____

    c. _____   d. _____

63. Identify the three types of muscle found in the animal body.

    a. _____   b. _____   c. _____

64. Complete the following table relating the major vertebrate organs systems.

Organ system	Major structures	Physiological role
Respiratory system		
	Lymph, lymph nodes and vessels, white blood cells	
Excretory system		
		Supplies body with nutrients for growth and maintenance
	Smooth muscle, cardiac muscle skeletal muscle	
Endocrine system		
		Provides support, protects organs, muscle attachment sites
Reproductive system		
Circulatory system		
	Brain, spinal cord, peripheral nerves	

65. Provide an example of an organ system listing the organs within the system.

_____

_____

_____

_____

_____

_____

CLUES TO APPLYING THE CONCEPTS

These practice questions are intended to sharpen your ability to apply critical thinking and analysis to the biological concepts covered in this chapter.

66. The term homeostasis actually means unchanging. Is this truly an accurate description of the state of the body's internal environment? Why or why not?

_____

_____

_____

_____

_____

_____

67. Explain why the example of your home thermostat represents a type of negative feedback mechanism.

_____

_____

_____

_____

_____

_____

ANSWERS TO EXERCISES

1. homeostasis
2. negative feedback
3. positive feedback
4. tissues
5. organs
6. organ system
7. epithelial tissue
8. membrane
9. glands
10. Exocrine glands
11. Endocrine glands
12. connective tissue
13. collagen
14. dermis
15. tendons
16. ligaments
17. Bones
18. fat
19. blood
20. lymph
21. Skeletal muscle

22. Cardiac muscle
23. Smooth muscle
24. nerve tissue
25. neurons
26. dendrite
27. cell body
28. axon
29. synaptic terminal
30. glial cells.
31. epidermis
32. keratin
33. hair follicles
34. sebaceous glands.
35. false, tissues
36. false, organ
37. true
38. false, two times per month
39. true
40. false, Exocrine
41. true

42. true
43. false, Cardiac
44. true
45. true
46. false, axon
47. true
48. true
49. a
50. c
51. a
52. e
53. b
54. d
55. d
56. b
57. e
58. c
59. b
60. b
61. e

62. a. epithelial tissue   b. connective tissue   c. muscle tissue   d. nerve tissue
63. a. skeletal   b. cardiac   c. smooth

64. Table relating the major vertebrate organs systems

Organ system	Major structures	Physiological role
Respiratory system	Nose, trachea, lungs, or gills	Gas exchange between blood and environment
Lymphatic/Immune system	Lymph, lymph nodes and vessels, white blood cells	Carries fat, excess fluids to blood, destroys invading microbes
Excretory system	Kidneys, ureters, bladder, urethra	Filters cellular waste, toxins, and excess water and nutrients; stabilizes bloodstream
Digestive system	Mouth, esophagus, stomach, small & large intestines, digestive glands	Supplies body with nutrients for growth and maintenance
Muscular system	Smooth muscle, cardiac muscle skeletal muscle	Movement through digestive tract, large blood vessels; begins and carries out heart contractions; moves skeleton
Endocrine system	Hormone secreting glands: hypothalamus, pituitary, thyroid, pancreas, adrenals	Regulates physiological processes along with nervous system
Skeletal system	Bones, cartilage, tendons, ligaments	Provides support, protects organs, muscle attachment sites
Reproductive system	Testes, seminal vesicles, penis Ovaries, oviducts, uterus, vagina, mammary glands	Produces sperm, inseminates female Produces egg cells, nurtures developing offspring
Circulatory system	Heart, vessels, blood	Transports nutrients, gases, hormones, wastes. Temperature control
Nervous system	Brain, spinal cord, peripheral nerves	Senses environment, directs behavior, controls physiological processes along with endocrine system

65. Respiratory system: nose, trachea, lungs, or gills; Lymphatic/Immune system: lymph, lymph nodes and vessels, white blood cells; Excretory system: kidneys, ureters, bladder, urethra; Digestive system: Mouth, esophagus, stomach, small and large intestines, digestive glands; Muscular system: smooth muscle, cardiac muscle, skeletal muscle; Endocrine system: hormone secreting glands: hypothalamus, pituitary, thyroid, pancreas, adrenals; Skeletal system: bones, cartilage, tendons, ligaments; Reproductive system: testes, seminal vesicles, penis, ovaries, oviducts, uterus, vagina, mammary glands; Circulatory system: heart, vessels, blood; Nervous system: brain, spinal cord, peripheral nerves.

66. The term homeostasis is used to describe the body's ability to maintain an internal constancy even though the animal's body is exposed to an external environment that is continually changing. However, the term suggests a static (unchanging) state. In fact, this is not an accurate description. The body's internal environment is dynamic (always changing). Chemical and physical changes are constantly occurring. At the same time, the changes occur within certain parameters, enabling cells to function. Thus, dynamic equilibrium is a more appropriate term.

67. Negative feedback mechanisms respond to counteract a change. If the temperature in your home cools beyond a certain set point, the sensor on the thermometer will detect the change and signal the heating device to turn on. Once the set-point temperature has been reached, the thermometer again detects the change and signals the heating device to turn off. This system turns on to counteract a drop in temperature. When the temperature has been restored, it turns off.

Chapter 27: Circulation

OVERVIEW

This chapter explores the circulatory system, its components and their functions. The focus of the chapter is on the human system; however, parallels to other vertebrate systems can be made. Since the lymphatic system functions so closely with the circulatory system, it is included here.

1) What Are the Major Features and Functions of Circulatory Systems?

The circulatory system includes **blood**, a fluid that transports nutrients and oxygen to cells and wastes away from cells; **blood vessels**, which carry the fluid to and away from the cells; and the **heart**, which serves as the pump to circulate the fluid. A circulatory system may an **open circulatory system**, such as that found in many invertebrates. In open systems, the organs and tissues are bathed in blood in a **hemocoel**. A circulatory system may a **closed circulatory system**, such as that found in humans and most other vertebrates. In a closed circulatory system, the blood is maintained within vessels and the heart; this is a more efficient way of transporting nutrients and wastes.

    The functions of a circulatory system include the transport of oxygen from the respiratory system to cells and the removal of carbon dioxide from cells, the transport of nutrients to cells and wastes from cells to the liver or kidney, and the distribution of hormones. The circulatory system also helps to regulate body temperature as blood flow is adjusted, and prevents blood loss. Finally, the circulatory system enhances the immune system by circulating antibodies and white blood cells.

2) What Are the Features and Functions of the Vertebrate Heart?

The heart is the pump behind the circulation of blood through the body. Within the heart, the blood collects in **atria** (or one **atrium**). The muscular atria contract, pushing the blood into the **ventricles**. When the ventricles contract, blood is pumped through the body. Oxygen and carbon dioxide are exchanged to and from cells through **capillaries**, thin-walled vessels.

    In the four-chambered vertebrate heart, blood flows from the body through a **vein** into the right atrium. The atrium pumps the blood to the right ventricle. When the right ventricle contracts, it pumps blood through the pulmonary **arteries** to the lungs. This constitutes **pulmonary circulation**. The deoxygenated blood picks up oxygen in the lungs and flows back to the heart through pulmonary veins into the left atrium. The left atrium pumps the oxygen-rich blood into the left ventricle. The left ventricle contracts and pumps the oxygen-rich blood out the aorta and to the rest of the body. This constitutes **systemic circulation**. The alternating contractions of the atria and the ventricles generates the **cardiac cycle**.

    Once blood is pumped from an atrium to a ventricle, it must be kept from flowing back into the atrium. The **atrioventricular valves** open as blood is pumped through to the ventricles and then are kept shut by pressure of the blood in the ventricle. Specifically, the **tricuspid valve** separates the right atrium and the right ventricle, and the **bicuspid valve** separates the left atrium and the left ventricle. In turn, back flow of blood leaving the heart is governed by **semilunar valves** between the right ventricle and the pulmonary artery and the left ventricle and the aorta. The contractions of the chambers are regulated by a cluster of cardiac muscle cells that specialize as a **pacemaker**. The principal pacemaker is found in the wall of the right atrium and is called the **sinoatrial (SA) node**. In order to coordinate atrial and ventricle contractions so that the atria contract before the ventricles, the electrical impulse is delayed at the **atrioventricular (AV) node**. If the pacemaker is unable to regulate contractions, **fibrillation**, or irregular contractions, occurs.

3) What Are the Features and Functions of Blood?

Blood is composed of **plasma** and specialized cells within the fluid plasma. Dissolved within plasma are proteins, such as albumins, globulins, and fibrinogen, nutrients, hormones, gases, wastes, and salts. The specialized cells include **erythrocytes**, the red blood cells. The red color is due to an iron-containing pigment, **hemoglobin**. Hemoglobin functions to carry oxygen to cells and carbon dioxide away from cells. Red blood cells are manufactured in bone marrow and have a life span of only 120 days. The hormone **erythropoietin**, produced in the kidneys, is responsible for maintaining red blood cell production through a negative feedback mechanism. Proteins (A or B) present on the plasma membrane of erythrocytes are used to determine blood type. Another protein, called the **Rh factor**, is also present on erythrocyte membranes. Determining the presence or absence of these proteins is extremely important in the event a blood transfusion is needed or an incompatible pregnancy occurs. An Rh-positive child carried by an Rh-negative mother will induce the mother's body to produce antibodies against the Rh factor. Any subsequent pregnancies involving an Rh-positive child could result in the child being affected by **erythroblastosis fetalis**, which occurs when the mother's antibodies attack the red blood cells of the fetus, inducing severe anemia.

Additional specialized cells within blood are the **leukocytes**, the white blood cells. Leukocytes are involved in the body's immune system. It is the **monocytes** and **neutrophils** that travel to injured sites where bacteria have gained entry. The monocytes differentiate into foreign-cell-"eating" **macrophages** and, together with the neutrophils, feed on the bacteria. **Lymphocytes** produce antibodies in response to disease. If the body is infected by parasites, **eosinophils** are produced to destroy the parasites. **Basophils** are involved in allergic responses and inhibit blood clotting in a wound, allowing blood flow to remove any bacteria and reducing the possibility of infection.

Platelets are membrane-enclosed fragments of **megakaryocytes** that reside in the bone marrow. Platelets are involved in **blood clotting**. Clotting is initiated when platelets encounter an irregular surface, such as a wound. The platelets stick to the irregular surface. In the meantime, production of the enzyme **thrombin** is induced, converting the plasma protein fibrinogen to **fibrin**. Fibrin fibers form a mesh that traps red blood cells and platelets. Eventually, a dense scab is formed over the wound.

4) What Are the Structures and Functions of Blood Vessels?

Blood leaves the heart through arteries. The muscular elastic walls of arteries help maintain a steady flow of blood to smaller vessels. As the diameter of an artery decreases **arterioles** are formed. Arterioles are capable of responding to electrical and hormonal signals when changes in tissue needs are detected. When the diameter of the arterioles diminishes, **capillaries** are formed. Blood flow into the capillaries is controlled by rings of muscle, the **precapillary sphincters**, at the point that arterioles become capillaries. It is the capillaries that function in gas exchange, as well as nutrient and waste exchange, with cells of the body. Pressure within capillaries causes fluid to continuously leak from them into spaces around the capillaries and tissues. This fluid, the **interstitial fluid**, allows the gas, nutrient, and waste exchange to occur. As blood returns to the heart, the capillaries increase in diameter, becoming **venules** and finally veins. Blood always returns to the heart through veins. The walls of veins are thin and expandable. There is little resistance to blood flow through veins. Contractions of skeletal muscle, squeezing veins, moves blood through them. One-way valves prevent backward flow of blood during the muscle contractions.

5) What Are the Structures and Functions of the Lymphatic System?

Within the **lymphatic system** are lymph capillaries and vessels, **lymph nodes**, the **thymus**, and the **spleen**. The lymphatic system functions to remove excess interstitial fluid and molecules that have leaked from capillaries. **Lymph** fluid is the interstitial fluid collected by lymph capillaries and carried back to the circulatory system in lymph vessels. The lymphatic system also functions to carry fat globules from the small intestine to the bloodstream. Additionally, the lymphatic system functions to carry white blood cells throughout the body to fight off bacteria and viruses. Clusters of connective tissue containing large numbers of lymphocytes are found throughout the body. The largest of these are the **tonsils**. Lymphocytes are produced in lymph nodes, in the thymus, and in the spleen. The spleen also acts as a filter of the blood, as blood flows through the spleen, macrophages and lymphocytes remove and destroy foreign material and aged red blood cells.

KEY TERMS AND CONCEPTS

Fill-In: From the following list of key terms, fill in the blanks in the following statements.

arteries	bicuspid valve	blood clotting	atrioventricular valves
arterioles	blood	blood vessels	atrioventricular (AV) node
atria	cardiac cycle	erythropoietin	closed circulatory system
capillaries	erythrocytes	hemocoel	erythroblastosis fetalis
heart	fibrillation	hemoglobin	megakaryocytes
leukocytes	fibrin	interstitial fluid	open circulatory systems
lymph	spleen	lymph nodes	precapillary sphincters
pacemaker	thrombin	lymphatic system	semilunar valves
plasma	tonsils	lymphocytes	sinoatrial (SA) node
platelets	veins	thymus	systemic circulation
Rh factor	ventricles	venules	tricuspid valve

All circulatory systems consist of (1)_____ to transport, (2)_____
to conduct fluid, and (3) a _____ to pump the fluid throughout the body.   A
circulatory system may contain a large open space called the (4) _____, in which
organs and tissues are bathed in circulating blood. This exists in
(5)_____, found in most invertebrates. The (6)_____,
found in humans and other vertebrates, maintains the blood within vessels.

The (7) _____ of the heart collect blood from the body or from the lungs. When these chambers pump, the blood goes into the (8)_____. Unoxygenated blood coming back to the heart from the body and then on to the lungs to release CO_2 and receive O_2 constitutes (9)_____. Oxygenated blood being pumped throughout the body to deliver O_2 to the cells and to receive CO_2 from the cells constitutes (10)_____.

Blood flowing toward the heart is always carried in (11) _____, while blood flowing away from the heart is always carried in (12)_____.
The alternating pumping of the atria and the ventricles is called the (13)_____.

One-way blood flow within and from the heart is controlled by valves. Specifically, one-way blood flow from the atria into the ventricles is controlled by the (14)_____. The valve between the right atrium and the right ventricle is the (15)_____ and the valve between the left atrium and the left ventricle is the (16)_____. Blood flowing to the pulmonary artery or the aorta must pass through the (17)_____ from the respective ventricle.

The rhythm of the beating heart is coordinated by a natural (18)_____. The specialized cluster of cells in the wall of the right atrium that is primarily responsible for the coordination is the (19)_____. In order for the ventricles to pump a fraction of a second after the atria, the electrical signal is delayed at the (20)_____. When the pumping cycle fails to be coordinated, irregular contractions, or (21)_____, occur.

The fluid component of blood is called (22)_____. Within this fluid are specialized cells and (23)_____, the membrane-enclosed fragments of larger cells called (24)_____. The red blood cells of blood are called (25)_____ and are red due to the pigment (26)_____ within them. The number of red blood cells within the body is monitored by a negative feedback system involving the hormone (27)_____ produced in the kidneys.

Proteins embedded in the plasma membrane of red blood cells are used to determine individual blood type as A, B, AB, or O. Another protein, the (28)_____ is a concern when a pregnancy occurs involving an Rh-negative woman carrying and Rh-positive fetus. If the mother's body produces antibodies against the blood of the fetus, (29) _____ could affect the newborn child.

White blood cells within blood are called (30) _____, which function in the body's immune response. Platelets are intimately involved with the process of (31)_____, which reduces the chances of bleeding to death. In response to a wound, platelets adhere to the injured surface of a blood vessel, in part, triggering production of the enzyme (32)_____. This enzyme catalyzes the conversion of fibrinogen to (33)_____.

Arteries branch into smaller vessels called (34)_____, which help regulate the distribution of blood throughout the body. The tiniest of all vessels are the (35)_____ that are responsible for gas and nutrient exchange with the body's cells. There is continuous leakage of fluid from these tiny vessels, forming the (36)_____, which bathes nearly all the body's cells. Blood flow from arterioles into capillaries is regulated by tiny rings of smooth muscle called the (37)_____. Unoxygenated blood in capillaries drains into larger vessels called (38)_____ which empty into large veins.

Closely associated with the circulatory system is a second system of vessels, the (39)_____. This system of vessels receives interstitial fluid leaked from capillaries. The fluid is now referred to as (40)_____ and is carried back to the circulatory system. The function of the system is to help the body defend against foreign invaders. Patches of connective tissue with large numbers of (41)_____ are found in linings of the respiratory, digestive, and urinary tracts. The largest patch of this tissue makes up the (42)_____ behind the mouth. Lymph fluid is passed through kidney-bean-shaped structures called (43)_____. Organs that function as part of the lymphatic system include the (44)_____, which produces lymphocytes, and the (45)_____, which filters blood.

Key Terms and Definitions

angina : chest pain associated with reduced blood flow to the heart muscle caused by obstruction of coronary arteries.

arteriole: a small artery that empties into capillaries. Contraction of the arteriole regulates blood flow to various parts of the body.

artery: a vessel with muscular elastic walls that conducts blood away from the heart.

atherosclerosis: a disease characterized by the obstruction of arteries by cholesterol deposits and a thickening of the arterial walls.

atrioventricular (AV) node: a specialized mass of muscle at the base of the right atrium through which the electrical activity initiated in the sinoatrial node is transmitted to the ventricles.

atrioventricular valve: heart valves between the atria and the ventricles that prevent backflow of blood into the atria during ventricular contraction.

atrium: a chamber of the heart that receives venous blood and passes it to a ventricle.

basophil (bas'-o-fil): a type of white blood cell that releases both substances that inhibit blood clotting and chemicals that participate in allergic reactions and in responses to tissue damage and microbial invasion.

bicuspid valve: the atrioventricular valve between the left atrium and the left ventricle of the heart.

blood: fluid consisting of plasma in which blood cells are suspended, carried within the circulatory system.

blood clotting: a complex process by which platelets, the protein fibrin, and red blood cells block an irregular surface in or on the body, such as a damaged blood vessel, sealing the wound.

blood vessel: a channel that conducts blood throughout the body.

capillary: the smallest type of blood vessel, connecting arterioles with venules. Capillary walls, through which exchange of nutrients and wastes occurs, are only one- cell thick.

cardiac cycle: the alternation of contraction (systole) and relaxation (diastole) of the heart chambers; systole and diastole.

closed circulatory system: the type of circulatory system found in certain worms and vertebrates in which the blood is always confined within the heart and vessels.

eosinophil (e-o-sin'-o-fil): a type of white blood cell that converges on parasitic invaders and releases substances to kill them.

erythroblastosis fetalis: a condition in which the red blood cells of a newborn Rh-positive baby are attacked by antibodies produced by its Rh-negative mother, causing jaundice and anemia. Retardation and death are possible consequences if treatment is inadequate.

erythrocyte: red blood cell active in oxygen transport, containing the red pigment hemoglobin.

erythropoietin: a hormone produced by the kidneys in response to oxygen deficiency that stimulates the production of red blood cells by the bone marrow.

fibrillation: rapid, uncoordinated, and ineffective contractions of heart muscle cells.

fibrin: a clotting protein formed in the blood in response to a wound, it binds with other fibrin molecules and provides a matrix around which a blood clot forms.

heart: a muscular organ within the circulatory system responsible for pumping blood throughout the body.

heart attack: a situation in which blood flow through a coronary artery is severely reduced or blocked, depriving some of the heart muscle of its blood supply.

hemocoel: the blood cavity in an open circulatory system.

hemoglobin: an iron-containing protein that gives red blood cells their color, it binds to oxygen in the lungs and releases oxygen to the tissues.

hypertension: arterial blood pressure that is chronically elevated above the normal level.

interstitial fluid: fluid similar in composition to plasma (except lacking large proteins) that leaks from the capillaries and acts as a medium of exchange between the body cells and the capillaries.

leukocyte: any of the white blood cells circulating in the blood.

lymph: pale fluid within the lymphatic system composed primarily of interstitial fluid and lymphocytes.

lymphatic system: a system consisting of lymph vessels, lymph capillaries, lymph nodes, the thymus and the spleen. It helps protect the body against infection, absorbs fats, and returns excess fluid and small proteins to the blood circulatory system.

lymph nodes: small structures that act as filters for lymph;they contain lymphocytes and macrophages, which inactivate foreign particles such as bacteria.

lymphocyte: a type of white blood cell important in the immune response.

macrophage: a cell derived from white blood cells called monocytes, whose function is to consume foreign particles including bacteria.

megakaryocyte: a large cell type that remains in the bone marrow, pinching off pieces of itself; these fragments enter the circulation as platelets.

monocyte: a type of white blood cell that travels through capillaries to wounds where bacteria have gained entry, leaves the capillaries, and differentiates into a macrophage.

neutrophil: a type of white blood cell that feeds on bacterial invaders or other foreign cells, including cancer cells.

open circulatory system: a type of circulatory system found in some invertebrates, such as arthropods and molluscs, that includes an open space in which blood directly bathes body tissues.

pacemaker: a cluster of specialized muscle cells in the upper right atrium of the heartthat produce spontaneous electrical signals at a regular rate; the sino-atrial node.

plaque: a deposit of cholesterol and other fatty substances within the wall of an artery.

plasma: the fluid, noncellular portion of the blood.

platelet: cell fragment formed from megakaryocytes in bone marrow; it lacks a nucleus, circulates in the blood, and plays a role in blood clotting.

precapillary sphincter: a ring of smooth muscle between an arteriole and a capillary that regulates the flow of blood into the capillary bed.

pulmonary circulation: the circulation of blood from the body, through the right atrium and right ventricle, and to the lungs.

Rh factor: a protein found on the red blood cells of some people (Rh-positive) but not others (Rh-negative). Exposure of Rh-negative individuals to Rh-positive blood triggers antibody production to Rh-positive blood cells.

semilunar valve: the type of valve between the ventricles of the heart and the pulmonary artery and aorta. Semilunar valves prevent backflow of blood into the ventricles when they relax.

sinoatrial (SA) node: a small mass of specialized muscle in the wall of the right atrium; it generates electrical signals rhythmically and spontaneously and serves as the heart's pacemaker.

spleen: an organ of the lymphatic system in which lymphocytes are produce, and blood is filtered past lymphocytes and macrophages, which remove foreign particles and aged red blood cells.

stroke: an interruption of blood flow to part of the brain, caused by the rupture of an artery or the blocking of an artery by a blood clot. Loss of blood supply leads to rapid death of the area of the brain affected.

systemic circulation: the circulation of blood from the left ventricle, through the rest of the body (except the lungs), and back to the heart.

thrombin: an enzyme produced in the blood as a result of injury to a blood vessel; it catalyzes the production of fibrin, a protein that assists in blood-clot formation.

thymus: an organ of the lymphatic system in which lymphocytes are produced, particularly in young children.

tonsils: patches of lymphatic tissue consisting of connective tissue containing many lymphocytes, located in the pharynx and throat.

tricuspid valve: the valve between the right ventricle and the right atrium of the heart.

vein: a large-diameter, thin-walled vessel that carries blood from venules back to the heart.

ventricle: the lower muscular chamber on each side of the heart, which pumps blood out through the arteries. The right ventricle sends blood to the lungs, and the left to the rest of the body.

venule: a narrow vessel with thin walls that carries blood from capillaries to veins.

THINKING THROUGH THE CONCEPTS

True or False: Determine if the statement given is true or false. If it is false, change the underlined word so that the statement reads true.

46. _____ An <u>open</u> circulatory system is an efficient means of transporting nutrients and wastes.

47. _____ A function of the circulatory system is to distribute <u>hormones</u> from their point of production to their target tissue.

48. _____ When <u>atria</u> contract, blood is sent throughout the body.

49. _____ Blood flows away from the heart in <u>veins</u>.

50. _____ At a resting heart rate, the cardiac cycle is complete in under <u>one second</u>.

51. _____ The right atrium is separated from the right ventricle by the <u>semilunar valves</u>.

52. _____ The delay between the pumping of the atria and the ventricles is maintained by the <u>atrioventricular node</u>.

53. _____ The <u>atrioventricular node</u> serves as the "pacemaker" of the heart.

54. _____ <u>Fibrillation</u> of the ventricles can be fatal.

55. _____ Heart rate is influenced by the hormone <u>epinephrine</u>.

56. _____ The average human has <u>five to six liters</u> of blood.

57. _____ The protein <u>fibrinogen</u> in erythrocytes carries oxygen to the body's cells.

58. _____ Red blood cells <u>are capable of mitotic division</u> after living 120 days.

59. _____ Both red and white blood cells are produced by cells from <u>bone marrow</u>.

60. _____ White blood cells are part of the <u>circulatory system and the lymphatic system</u>.

61. _____ Platelets are <u>cells</u> derived from megakaryocytes.

62. _____ <u>Interstitial fluid</u> contains water and dissolved nutrients, gases, wastes, and blood proteins.

63. _____ Red blood cells pass through capillaries in <u>pairs</u>.

64. _____ Veins contain <u>one-way valves</u> to prevent backflow of blood.

Matching: Leukocytes and their functions.

65. _____ involved in parasitic infections

66. _____ become macrophages

67. _____ inhibit blood clotting

68. _____ produce antibodies

69. _____ feed on foreign invaders along with macrophages

70. _____ involved in allergic reactions

71. _____ amoeba-like cells that engulf bacteria

Choices:

a. basophils
b. lymphocytes
c. neutrophils
d. macrophages
e. monocytes
f. eosinophils

CLUES TO APPLYING THE CONCEPTS

These practice questions are intended to sharpen your ability to apply critical thinking and analysis to biological concepts covered in this chapter.

72. Identify the three major plasma proteins and their functions

Plasma protein	Function

73. Discuss the differences between blood flow in a three chambered heart and a four chambered heart. Trace the flow of blood through the heart, including its path to and from the lungs.

_____

_____

_____

_____

_____

_____

_____

_____

_____

_____

74. Explain how arterioles function to help regulate body temperature on extremely cold or extremely hot days.

_____

_____

_____

_____

_____

_____

_____

_____

_____

_____

75. Outline differences between capillaries of the circulatory system and those of the lymphatic system.

_____

_____

_____

_____

_____

_____

_____

_____

_____

ANSWERS TO EXERCISES

1. blood
2. blood vessels
3. heart
4. hemocoel
5. open circulatory systems
6. closed circulatory system
7. atria
8. ventricles
9. pulmonary circulation
10. Systemic circulation
11. veins
12. arteries
13. cardiac cycle
14. atrioventricular valves
15. tricuspid valve
16. bicuspid valve
17. semilunar valves
18. pacemaker
19. sinoatrial (SA) node
20. atrioventricular (AV) node
21. fibrillation
22. plasma
23. platelets
24. megakaryocytes
25. erythrocytes

26. hemoglobin
27. erythropoietin
28. Rh factor
29. erythroblastosis fetalis
30. leukocytes
31. blood clotting
32. thrombin
33. fibrin
34. arterioles
35. capillaries
36. interstitial fluid
37. precapillary sphincters
38. venules
39. lymphatic system
40. lymph
41. lymphocytes
42. tonsils
43. lymph nodes
44. thymus
45. spleen
46. false, closed
47. true
48. false, ventricles
49. false, arteries

50. true
51. false, tricuspid valve
52. true
53. false, sinoatrial node
54. true
55. true
56. true
57. false, hemoglobin
58. false, are not capable of division
59. true
60. true
61. false, membrane enclosed cell fragments
62. true
63. false, single file
64. true
65. f
66. e
67. a
68. b
69. c
70. a
71. d

72. 3 major plasma proteins and their functions

Plasma protein	Function
albumins	help maintain osmotic pressure of blood which controls water flow across plasma membranes
globulins	transport nutrients and play a role in the immune system
fibrinogen	important in blood clotting: is converted to fibrin by the enzyme thrombin in response to an injury

73. Three-chambered hearts have two atria and a single ventricle. Deoxygenated blood returns from the body to the right atrium while oxygenated blood returns from the lungs into the left atrium. Both atria empty into the one ventricle. There is some mixing of oxygenated and unoxygenated blood within the ventricle, but for the most part, the two types of blood remain on their respective sides of the ventricle. Mostly oxygenated blood gets pumped throughout the body and unoxygenated blood gets pumped to the lungs. This type of heart works fine for cold-blooded animals; however, warm-blooded birds and mammals require a more efficient method of oxygen delivery to their tissues. A four-chambered heart provides this. In a four-chambered heart, blood returns to the heart to the right atrium through the tricuspid valve and into the right ventricle. Blood leaving the right ventricle passes through the semilunar valve as it enters the pulmonary artery into the lungs. After being reoxygenated, the blood leaves the lungs through the pulmonary vein and enters the left atrium. Blood is pumped from the left atrium through the bicuspid valve and into the left ventricle. Oxygenated blood is pumped through the semilunar valve into the aorta for distribution to the body.

74. Muscles in the arteriole walls are directly controlled by nerves, hormones, and chemicals produced by nearby tissues. Therefore, as the needs of nearby tissues change, arterioles contract and relax in response. The circulatory system helps to regulate body temperature. By relaxing and expanding, the arterioles bring blood flow closer to the surface of the skin; thus, the heat dissipates, cooling the body. During cold weather, arteriole walls constrict, pulling the blood away from the skin's surface to maintain body warmth. Under extreme conditions, the blood is shunted to the body's internal organs, specifically, to the heart and brain.

75. The capillaries of the circulatory system are composed of cells that have plasma membranes with pores that allow only water, dissolved nutrients, hormones, wastes, and small blood proteins to leak out. White blood cells ooze through openings between capillary cells. The capillaries of the circulatory system form a continuous network between the arterial system and the venous system. The capillaries of the lymphatic system contain cells with one-way openings between them. Thus, larger molecules can be carried into the. These capillaries deadend in tissues, serving to collect extra interstitial fluid and its contents.

Chapter 28: Respiration

OVERVIEW

This chapter covers the structures of the respiratory system and their specialized functions. Not all organisms use lungs for gas exchange; these adaptations are reviewed. The focus of the chapter, however, is on vertebrate respiratory systems, specifically of humans.

1) What Are Some Evolutionary Adaptations for Gas Exchange?

All respiratory systems use diffusion for gas exchange. The surface across which diffusion will occur must be moist and must have a large surface area that is in contact with the environment. Some aquatic animals do not have specialized respiratory structures. Oxygen and carbon dioxide simply diffuse across their body. This works well for extremely small animals, thin flat animals, or for those that do not move quickly. Other aquatic animals circulate water from their environment through their bodies. Other animals use a large surface area of moist skin and their circulatory system to distribute gases. Most animals, however, have specialized respiratory structures that work closely with the circulatory system to exchange gases between cells and the environment. Gas exchange occurs by diffusion and **bulk flow**. Bulk flow occurs when fluids or gases move across relatively large spaces from an area of high pressure to an area of low pressure.

Many larger aquatic animals utilize **gills** to exchange O_2 and CO_2 with the environment. Gills have a dense network of capillaries just beneath their outer membrane, allowing gas exchange to occur. Fish have an **operculum**, or protective flap, covering their gills. The operculum serves to streamline the body and deter predators from eating the gills. Terrestrial animals evolved respiratory structures that are protected from drying out and have internal support. The respiratory structures of terrestrial animals are internal to reduce the amount of water lost in keeping the respiratory surface moist. The complex system of internal tubes called **tracheae** are the respiratory structures found in insects. (The tubes also support the exoskeleton.) Air, which may be actively pumped, enters the tracheae at valved openings along the abdomen called **spiracles**. Most terrestrial vertebrates primarily respire using saclike **lungs**. Amphibians, however, may also use their skin for supplemental gas exchange, as long as it remains moist.

2) What Are the Features and Functions of the Human Respiratory System?

The respiratory system of humans consists of the **conducting portion**, a series of passageways that carry air, and the **gas-exchange portion**, where O_2 and CO_2 are exchanged with the blood in sacs within the lung. Air flowing through the conducting portion passageways first enters the nose or mouth and passes through the **pharynx** to the **larynx**. The **vocal cords** are located within the larynx. When the vocal cords partially block the larynx, exhaled air causes them to vibrate, producing sounds. From the larynx, air passes through the **trachea**. Within the chest, the trachea divides into branches, forming two **bronchi**, one going into each lung. Within the lungs each **bronchus** branches many times forming smaller tubes called bronchioles. The bronchioles terminate in microscopic **alveoli**, where gas exchange will occur. The alveoli function to increase the gas-exchange surface area of lung tissue, and are infused with capillaries. Gases dissolve in the water covering each alveolus and diffuse through alveolar and capillary membranes. Oxygen in the inhaled air diffuses into the oxygen-poor blood returning from the body. In turn, carbon dioxide in the blood diffuses into the alveoli. Oxygen-rich blood is then transported to the cells and tissues.

Within the blood cells, the iron-containing protein **hemoglobin** binds loosely with up to four oxygen molecules. This process maintains a low oxygen gradient within the blood to facilitate diffusion from inhaled air. Oxygenated blood appears bright red due to the shape of hemoglobin when it is bound to oxygen. Roughly 20% of CO_2 binds to hemoglobin, while 70% reacts with water to form bicarbonate ions that diffuse in the plasma along with the remaining CO_2 molecules. This process provides the mechanism needed to maintain a low CO_2 gradient in the blood to facilitate CO_2 diffusion into the bloodstream from cells.

Air enters the respiratory system actively through **inhalation**. During inhalation, the chest cavity is enlarged by the contraction of **diaphragm** muscles drawing the diaphragm downward. Muscles within the rib cage also contract moving the ribs up and out. Since the lungs are held in a vacuum, the expansion of the chest causes the lungs to expand as well, drawing in air. **Exhalation** of air occurs passively as the muscles relax and the chest cavity decreases in size again. The lungs are protected within the chest cavity by the rib cage, the diaphragm, and the neck muscles and connective tissues. An airtight seal between the chest wall and the lungs is enhanced by the **pleural membranes**. The inhalation-exhalation process occurs without conscious thought through muscle contractions stimulated by the **respiratory center** of the brainstem. The breathing rate is determined by a combination of sensors detecting high CO_2 levels, low O_2 levels, and an increase in activity level.

KEY TERMS AND CONCEPTS

Fill in: From the following list of key terms fill in the blanks in the following statements.

alveoli	inhalation	bronchiole	bulk flow
bronchi	larynx	conducting portion	hemoglobin
diaphragm	lungs	exhalation	pleural membranes
gills	trachea	gas-exchange portion	respiratory center
spiracles	tracheae	operculum	vocal cords
		pharynx	

Fluids or gases moving through large spaces move by (1) _____ from areas of high pressure to areas of low pressure. Diffusion, in contrast, is the movement of individual molecules.

The respiratory structures of aquatic animals are the (2) _____. In fish, these structures are covered by a flap called the (3) _____. The respiratory structures of terrestrial animals are internal. Insects respire using (4) _____. Air enters through openings called (5) _____ along the sides of the abdomen of the insect. Other terrestrial animals use (6) _____, which are usually moist sacs.

The respiratory system is divided into two parts. The (7)_____ carries air to the
(8)_____ , where gases are exchanged.

After air enters the nose or mouth, it passes through the (9)_____ behind the mouth.
The air then passes through the (10)_____ which is protected by the epiglottis. The
(11)_____ are found here, allowing sounds to be made during exhalation. Semicircular
bands of cartilage form the (12)_____ . Within the chest, this tube splits into two tubes
called (13)_____. Inside the lungs, each tube repeatedly splits into smaller
(14)_____. Microscopic chambers within the lung, called (15)_____,
increase the surface area of lung tissue immensely.

The iron-containing protein (16)_____ binds oxygen within red blood cells.
The process of breathing occurs with the help of the muscular (17)_____. As
these muscles contract and the rib cage expands, air is drawn into the lungs. This is called
(18)_____. Air is passively pushed out during (19)_____ when the
muscles relax and the chest decreases in size again. The rhythm of breathing is maintained in an area
of the brain stem called the (20)_____.

A double layer of membranes, the (21)_____, help to provide an airtight seal between
the lungs and the chest wall.

Key Terms and Definitions

alveolus: (pl. alveoli): a tiny air sac within the lungssurrounded by capillaries where gas exchange with the blood occurs.

bronchiole: a narrow tube formed by repeated branching of the bronch, that conducts air into the alveoli.

bronchus: a tube that conducts air from the trachea to each lung.

bulk flow: the movement of many molecules of a gas or fluid in unison from an area of higher pressure to an area of lower pressure.

chronic bronchitis: a persistent lung infection characterized by coughing, swelling of the lining of the respiratory tract, an increase in mucus production, and a decrease in the number and activity of cilia.

conducting portion: the portion of the respiratory system in lung-breathing vertebrates that carries air to the lungs.

diaphragm: a dome-shaped muscle forming the floor of the chest cavity; contraction of this muscle pulls it downward, enlarging the cavity and causing air to be drawn into the lungs.

emphysema: a condition in which the alveoli become brittle and rupture, causing decreased area for gas exchange.

exhalation: the act of releasing air from the lungs that results from relaxation of the respiratory muscles.

gas-exchange portion: the portion of the respiratory system in lung-breathing vertebrates where gas is exchanged in the alveoli of the lungs.

gill: in aquatic animals, a branched tissue richly supplied with capillaries around which water is circulated for gas exchange.

hemoglobin: an iron-containing protein that gives red blood cells their color; it binds to oxygen in the lungs and releases it to the tissues.

inhalation: the act of drawing air into the lungs by enlarging the chest cavity.

larynx: the portion of the air passage between the pharynx and the trachea; it contains the vocal cords.

lung: respiratory organ consisting of inflatable chambers within the chest cavity in which gas exchange occurs.

operculum: an external flap, supported by bone, that covers and protects the gills of most fish.

pharynx: a chamber at the back of the mouth shared by the digestive and respiratory systems.

pleural membrane: membrane that lines the chest and surrounds the lungs.

respiratory center: a cluster of neurons in the brainstem that sends rhythmic bursts of nerve impulses to the respiratory muscles, resulting in breathing.

spiracle: opening in the abdominal segments of insects, through which air enters the tracheae.

trachea: (in birds and mammals) a rigid but flexible tube supported by rings of cartilage, that conducts air between the larynx and the bronchi.

tracheae: (in insects) elaborately branching tubes that ramify through the bodies of insects and carry air close to each body cell; air enters the tracheae through openings called spiracles.

vocal cord: one of two folds of tissue that extend across the opening of the larynx and produce sound when air is forced between them; muscles alter the tension on the vocal cords and control the size and shape of the opening, which in turn determines whether sound is produced and at what pitch.

THINKING THROUGH THE CONCEPTS

True or False: Determine if the statement given is true or false. If it is false, change the <u>underlined</u> word so that the statement reads true.

22. _____ Very small animals may exchange O_2 and CO_2 through their <u>skin</u>.

23. _____ Gases must be <u>dissolved in water</u> when they diffuse in or out of cells.

24. _____ <u>Diffusion</u> is the movement of gases through relatively large spaces from an area of high pressure to an area of low pressure.

25. _____ Gills are extensively branched or folded to <u>increase</u> surface area.

26. _____ The first lungs may have arisen as extensions of the <u>digestive tract</u> in freshwater fish.

27. _____ As some animals develop, <u>both gills and lungs</u> are produced as they go through their life stages.

28. _____ Birds acquire <u>CO_2</u> as they inhale and exhale.

29. _____ The <u>tongue</u> functions to guard the opening of the larynx.

30. _____ The respiratory tract is lined with <u>mucus</u>, which serves to trap bacteria and debris from inhaled air.

31. _____ The internal lung surface area equals about 75 square meters due to the structure of <u>alveoli</u>.

32. _____ Each hemoglobin molecule can bind with <u>eight oxygen molecules</u>.

33. _____ Because of the function of hemoglobin, blood carries <u>less</u> oxygen compared with oxygen simply dissolved in plasma.

34. _____ Hemoglobin binds to CO <u>more strongly</u> than it binds to O_2^-.

35. _____ The enzyme carbonic anhydrase converts <u>CO_2 and water</u> to the bicarbonate ion in plasma.

36. _____ Inhalation is <u>a passive</u> process.

37. _____ Breathing is controlled by the respiratory center in the <u>cerebellum</u>.

Identify the Following: Do these statements about the results of smoking describe **atherosclerosis**, **emphysema**, or **chronic bronchitis**?

38. _____ causes heart attacks

39. _____ mucus is produced in large quantities

40. _____ breathing is labored

41. _____ action of cilia decreases

42. _____ air flow to alveloi decreases

43. _____ arterial walls are thick with fatty deposits

44. _____ alveoli are destroyed

45. _____ lungs resemble blackened Swiss cheese

46. _____ number of cilia decreases

47. _____ lung tissue becomes brittle

48. _____ respiratory tract lining is swollen

APPLYING THE CONCEPTS:

These practice questions are intended to sharpen your ability to apply critical thinking and analysis to biological concepts covered in this chapter.

49. Outline the general path of gas exchange through the body. Identify whether movement occurs through bulk flow or diffusion in each step.

_____

_____

_____

_____

_____

_____

_____

_____

50. Explain how the respiratory center regulates breathing rate.

_____

_____

_____

_____

_____

_____

_____

51. Why is the respiratory center less sensitive to O_2 concentrations than CO_2 concentrations in the blood?

_____

_____

_____

_____

_____

_____

_____

ANSWERS TO EXERCISES

1. bulk flow
2. gills
3. operculum
4. tracheae
5. spiracles
6. lungs
7. conducting portion
8. gas-exchange portion
9. pharynx
10. larynx
11. vocal cords
12. trachea
13. bronchi
14. bronchioles
15. alveoli
16. hemoglobin
17. diaphragm

18. inhalation
19. exhalation
20. respiratory center
21. pleural membranes
22. true
23. true
24. false, bulk flow
25. true
26. true
27. true
28. false, O_2
29. false, epiglottis
30. true
31. true
32. false, 4 oxygen molecules
    or 8 oxygen atoms

33. false, 70 times more
34. true
35. true
36. false, active
37. false, brain stem
38. atherosclerosis
39. chronic bronchitis
40. emphysema
41. chronic bronchitis
42. chronic bronchitis
43. atherosclerosis
44. emphysema
45. emphysema
46. chronic bronchitis
47. emphysema
48. chronic bronchitis

49. Air or water containing O_2 is moved past a respiratory surface by <u>bulk flow</u>. The flow may be facilitated by muscular breathing movements. O_2 and CO_2 are exchanged through the respiratory surface by <u>diffusion</u>; O_2 diffuses into the capillaries of the circulatory system and CO_2 diffuses out. The gases are transported in the circulatory system from the respiratory system to the tissues by <u>bulk flow</u> of the blood as it is pumped throughout the body by the heart. The gases are exchanged between the tissues and the circulatory system by <u>diffusion</u>; O_2 diffuses out of the capillaries and CO_2 diffuses in. The gases are transported back to the respiratory system by <u>bulk flow</u> of the blood. At the respiratory surface, O_2 <u>diffuses</u> into the capillaries of the circulatory system and CO_2 <u>diffuses</u> out again.

50. The respiratory center, located in the brainstem, receives signals from CO_2 receptor neurons. The CO_2 receptors are sensitive to CO_2 levels in the bloodstream. If CO_2 levels rise only 0.3% above the body's acceptable level, the receptors signal the center to increase breathing rate and the depth of breathes.

51. The respiratory center is less sensitive to the O_2 concentration in the blood stream because under normal conditions, there is an overabundance of O_2. A relatively small drop in O_2 levels will not present a problem to tissues in need of O_2. If, however, a sudden drop in O_2 levels occurs, receptors in the aorta and carotid arteries will stimulate the respiratory center.

Chapter 29: Nutrition and Digestion

OVERVIEW

This chapter provides an overview of the nutrients animals need and how they get those nutrients from their food. **Digestion** is the physical grinding and chemical breakdown of food. Since the energy an animal needs for body maintenance, growth, and reproduction amounts to the energy that remains after eating and digestion is complete, animals that have evolved the most efficient means of digesting their food will survive. Many different mechanisms accomplishing this have arisen; however, this chapter will focus on the relatively unspecialized, but well-orchestrated, digestive system of humans that allows us to ingest and digest a wide variety of food items.

1) What Nutrients Do Animals Need?

Animals need lipids, carbohydrates, proteins, minerals, and vitamins to provide energy for cellular metabolism, the basic molecules for building complex molecules of body function, and vitamins and minerals for metabolic reactions. Fats, carbohydrates, and proteins are used for energy. The energy available in food is measured in **calories**. In order for an animal to synthesize the fats it needs, it may need to acquire **essential fatty acids** from its food. Carbohydrates are synthesized from fats, amino acids, or other carbohydrates in the diet and are stored as **glycogen** in the liver and muscles. The human body also uses a small amount of its own protein for energy. When this protein is broken down, the waste product **urea** is produced. The protein is replaced by protein eaten in the diet. Of the 20 amino acids that the body needs to make proteins, the liver can synthesize 9. The other 11 amino acids, called **essential amino acids**, must be acquired through the diet. **Minerals**, elements, and small inorganic molecules, must be obtained from food items or dissolved in drinking water. **Vitamins** are essential organic compounds required in small amounts for the proper functioning of the human body.

2) How Is Digestion Accomplished?

Before digestion can begin, food items must be ingested. That is, food is put into the digestive tract, usually through an opening called the **mouth**. The food is then physically broken down into smaller pieces and exposed to chemical breakdown by digestive enzymes. The resulting small molecules are then absorbed by cells. What could not be used or broken down is eliminated from the body. The way in which ingestion, digestion, absorption, and elimination occurs differs, depending on the animal.

Digestion may occur, as in sponges, within cells. **Intracellular digestion** occurs after microscopic food particles have been engulfed by cells. The food particles are enclosed in a **food vacuole**. The vacuole fuses with **lysosomes** containing digestive enzymes, and the food is broken down and transported to the cell cytoplasm. **Extracellular digestion** occurs in larger animals that have evolved specialized digestive chambers, such as a **gastrovascular cavity**. Food is ingested and eliminated through the single opening to the cavity; thus, one meal must be eaten, completely digested, and eliminated before another can begin. Evolution of a digestive tube allows animals to eat more frequently. In the specialized digestive tube of earthworms, a **pharynx** connects the mouth with the esophagus. Beyond the esophagus is the **crop**, which stores food particles. Slowly the food is passed on to the muscular **gizzard**, where the food is ground using sand particles. The food is passed on to the intestines for chemical digestion and absorbed by the cells of the intestinal lining. Organisms that ingest plant material as their primary energy source have evolved digestive systems specialized to digest cellulose. Within the first of several digestive chambers, the rumen, microorganisms produce **cellulase** to break cellulose into its sugar components. After exposure to one round of cellulase, the cud, as the plant material is now called, is regurgitated and reswallowed to the rumen for further exposure to cellulase. Eventually, the cud is passed on to the three other chambers of the system before entering the intestines.

3) How Do Humans Digest Food?

Within the mouth, food is ground by the teeth and mixed with saliva by the tongue. Within saliva is the enzyme **amylase**, which acts to break down starch to sugar. As the food is swallowed, the tongue pushes it into the **pharynx**. The **epiglottis** covers the opening of the trachea as the food is swallowed into the esophagus. The muscles of the **esophagus** contract, pushing the food toward the **stomach**, in a process called **peristalsis**. Once in the stomach, food is mixed and churned by peristalsis and other muscular contractions before it is slowly released into the small intestine through the **pyloric sphincter**. While in the stomach, however, the food is mixed with chemicals secreted by cells of the stomach. These chemicals include **gastrin**, a hormone that stimulates the production and secretion of hydrochloric acid by stomach cells, and pepsin, a **protease** enzyme that digests proteins into shorter peptides. The churning and mixing that occurs in the stomach converts the food into **chyme**. Chyme is slowly released into the **small intestine**, where it will be digested into small molecules to be absorbed into the bloodstream. Digestion in the small intestine is accomplished by secretions from cells of the small intestine, the liver, and the pancreas. The **liver** functions in digestion by producing **bile**, which is stored in the **gall bladder**. Bile is a mixture of **bile salts**, water, cholesterol, and other salts. Bile salts help digest lipids by acting as detergents, breaking fats into microscopic particles. The microscopic lipid particles are then digested by **lipase** enzymes. The **pancreas** secretes **pancreatic juice** into the small intestine, which neutralizes the acid in the chyme and digests carbohydrates, lipids, and proteins. The small intestine is the primary site of chemical digestion and the primary site of nutrient **absorption**. In order to increase the absorptive surface area of the small intestine, it is extensively folded and is covered internally by **villi**. Each **villus**, in turn, is covered by **microvilli**. Each villus contains numerous blood capillaries and one **lacteal**, a lymph capillary. **Segmentation movements** move the chyme around in the small intestine, so that nutrients contact the absorptive surface. Peristalsis moves what is left after absorption into the **large intestine**, where water is absorbed to form feces. The large intestine consists of the **colon** and the **rectum**.

The digestion process is regulated by both the nervous system and hormones. The nervous system is stimulated by the sight, smell, taste, or even the thought of food, as well as by the process of chewing it. The nervous system readies the mouth and stomach for the arrival of food. The stomach begins to secrete acid, stimulating its cells to produce the hormone gastrin. As chyme enters the small intestine, its acidity stimulates the production of **secretin**. Secretin is released into the bloodstream to stimulate the pancreas to dump sodium bicarbonate into the small intestine, neutralizing the acidity. When chyme enters the stomach, the hormone **cholestokinin** is produced, stimulating the pancreas to release additional digestive enzymes into the small intestine and stimulating the gall bladder to contract and dump bile into the small intestine. When sugars and fatty acids are detected in chyme, the small intestine secretes **gastric inhibitory peptide** to reduce the rate at which the stomach releases chyme. This results in an increase in the length of time chyme spends in the small intestine for absorption.

KEY TERMS AND CONCEPTS

Fill in: From the following list of key terms, fill in the blanks of the following statements.

absorption	Calorie	amylase	essential amino acids
bile salts	colon	crop	essential fatty acids
calorie	esophagus	digestion	gastrovascular cavity
cellulase	extracellular	gastrin	intracellular digestion
chyme	food vacuole	gizzard	large intestine
epiglottis	gallbladder	microvilli	pancreatic juice
minerals	glycogen	mouth	pyloric sphincter
pancreas	lacteal	peristalsis	rectum
stomach	lipase	pharynx	segmentation movement
urea	liver	protease	small intestine
vitamins	lysosome	villi	

(1)_____ is the physical grinding and chemical breakdown of food.

A (2) _____ is the amount of energy needed to raise the temperature of gram of water by one degree Celsius. Whereas, a (3)_____ is 1000 of these units and is the term most commonly used to express the calorie content in foods. Animals store energy as the carbohydrate (4)_____ in the liver and muscles. A small amount of protein is metabolized for energy creating the waste product (5)_____.

When digestion occurs within a cell it is termed (6)_____. This type of digestion encloses food items in a membrane bound (7)_____. The food item is digested when enzymes from a (8)_____ are released into the vacuole.

When digestion occurs by enzymes outside the cell, it is termed (9)_____, which typically occurs in a digestive sac with a single opening. This sac is referred to as the (10)_____.

The earthworm's tubular digestive system includes a thin-walled storage organ called a (11)_____ and a muscular (12)_____ where food is ground.

The human diet requires (13)_____ to build lipids, (14)_____ to build proteins, carbohydrates, (15)_____ and (16)_____.

The enzyme (17)_____ breaks starch into simple sugars, while the enzymes

(18)_____ breaks fats into glycerol and fatty acid molecules and

(19)_____ breaks proteins into peptides.

Food enters the digestive tract through an opening called the (20)_____. After the food is chewed, it passes through the (21)_____ on its way to the esophagus. The (22)_____ prevents the food from entering the trachea while it is being swallowed. As food goes down the (23)_____, the smooth muscles contract rhythmically in a process called (24)_____. Food moving from the (25)_____ to the small intestine is called (26)_____ and its entrance into the small intestine is controlled by a ring of muscle, the (27)_____. (28)_____ and the microscopic (29)_____ greatly increase the absorptive surface area of the (30) _____. Each villus contains a lymph capillary called a (31)_____. Once absorption is complete, the contents remaining in the small intestine move into the (32)_____, which is divided into two portions; the first part is the (33)_____, and the last 15 cm is called the (34)_____.

Secretion of acid by the stomach is controlled by the hormone (35)_____, while many of the digestive enzymes that are secreted into the small intestine are produced in the (36)_____ and make up the substance called (37)_____. The (38)_____ produces bile, which is stored in the (39)_____. Bile is a complex mixture of (40)_____, water, cholesterol, and other salts.

The primary functions of the small intestine are the chemical breakdown of food and (41)_____ of small molecules into the bloodstream. (42)_____ increase the nutrients and other molecules absorbed from chyme while it is in the small intestine.

Key Terms and Definitions

absorption: the process by which nutrients are taken into cells.

amylase: an enzyme found in saliva and pancreatic secretions that catalyzes the breakdown of starch.

bile: a liquid secretion of the liver stored in the gallbladder and released into the small intestine during digestion. Bile is a complex mixture of bile salts, water, other salts, and cholesterol.

bile salts: substances synthesized in the liver from cholesterol and amino acids that assist in the breakdown of lipids by dispersing them into small particles on which enzymes may act.

calorie: a measure of the energy derived from food. When capitalized (that is, Calorie), this unit is the amount of energy required to raise the temperature of 1 liter of water 1 degree Celsius. It represents 1000 calories (with a lowercase "c"). The energy content of foods is measured in Calories.

Calorie: a unit of energy, in which the energy content of foods is measured; the amount of energy required to raise the temperature of 1 liter of water 1 degree Celsius; also called a kilocalorie, equal to 1000 calories.

cellulase: an enzyme that catalyzes the breakdown of the carbohydrate cellulose into its component glucose molecules. Cellulase is almost entirely restricted to microorganisms.

cholecystokinin: a digestive hormone produced by the small intestine that stimulates release of pancreatic enzymes.

chyme: an acidic, souplike mixture of partially digested food, water, and digestive secretions that is released from the stomach into the small intestine.

colon: the longest part of the large intestine, exclusive of the rectum.

crop: an organ found in both earthworms and birds in which ingested food is temporarily stored before being passed to the gizzard, where it is pulverized.

digestion: the process by which food is physically and chemically broken down into molecules that can be absorbed by cells.

epiglottis: a flap of cartilage in the lower pharynx that covers the opening to the larynx during swallowing. In doing so, the epiglottis directs the food down the esophagus.

esophagus: a muscular passageway connecting the pharynx to the next chamber of the digestive tract, the stomach in humans and other mammals.

essential amino acids: amino acids that are required nutrients. The body is unable to manufacture them, thus they must be supplied in the diet.

essential fatty acids: fatty acids that are required nutrients. The body is unable to manufacture them,thus they must be supplied in the diet.

extracellular digestion: the physical and chemical breakdown of food that occurs outside of a cell, usually in a digestive cavity.

food vacuole: a membrane-bound space within a single cell in which food is enclosed. Digestive enzymes are released into the vacuole, where intracellular digestion occurs.

gallbladder: a small sac next to the liver in which the bile secreted by the liver is stored and concentrated. Bile is

released from the gallbladder to the small intestine through the bile duct.

gastric inhibitory peptide: a hormone produced by the small intestine that inhibits the activity of the stomach.

gastrin: a hormone produced by the stomach that stimulates acid secretion in response to food.

gastrovascular cavity: a chamber with digestive functions, found in simple invertebrates. A single opening serves as both mouth and anus, while the chamber provides direct access of nutrients to the cells.

gizzard: a muscular organ found in earthworms and birds in which food is mechanically broken down prior to chemical digestion.

glycogen: a long, branched polymer of glucose that is stored by animals in the muscles and liver and metabolized as a source of energy.

intracellular digestion: the chemical breakdown of food that occurs within single cells.

lacteal: a single lymph capillary that penetrates each villus of the small intestine.

large intestine: the final section of the digestive tract consisting of the colon and the rectum, where feces are formed and stored.

lipase: an enzyme that catalyzes the breakdown of lipids such as fats.

liver: an organ with varied functions including bile production, glycogen storage, and detoxification of poisons.

lysosome (li´-so-som): a membrane-bound organelle containing intracellular digestive enzymes.

microvilli: a fringe of microscopic projections of the cell membrane of each villus that serves to increase its surface area.

mineral: an inorganic substance, especially one in rocks or soil.

mouth: the opening of a tubular digestive system into which food is first introduced.

pancreas: an organ lying next to the stomach that secretes enzymes for fat, carbohydrate, and protein digestion into the small intestine and neutralizes the acidic chyme.

pancreatic juice: a mixture of water, sodium bicarbonate, and enzymes released by the pancreas into the small intestine.

peristalsis: rhythmic coordinated contractions of the smooth muscles of the digestive tract that move substances through the digestive tract.

pharynx: a chamber located behind the mouth. In vertebrates, it is common to both the respiratory and digestive systems.

protease: an enzyme that digests proteins.

pyloric sphincter: a circular muscle at the base of the stomach that regulates the passage of chyme into the small intestine.

rectum: the terminal portion of the vertebrate digestive tube, where feces are stored until they can be eliminated.

secretin: hormone produced by the small intestine that stimulates production and release of digestive secretions by the pancreas and liver.

segmentation movements: unsynchro-nized contractions of the small intestine that result in mixing of partially digested food and digestive enzymes. The move-ments also bring nutrients into contact with the absorptive intestinal wall.

stomach: muscular sac between the esophagus and small intestine where food is stored and mechanically broken down, and in which protein digestion begins.

small intestine: portion of the digestive tract between the stomach and large intestine in which most digestion and absorption of nutrients occurs.

urea (u-re´-uh): a water-soluble, nitrogen-containing waste product of amino acid breakdown; one of the principal components of mammalian urine.

villus: fingerlike projections of the wall of the small intestine that increase its absorptive surface area.

vitamin: any one of a group of diverse chemicals that must be present in trace amounts in the diet to maintain health. Vitamins are used by the body in conjunction with enzymes in a variety of metabolic reactions.

THINKING THROUGH THE CONCEPTS

True or False: Determine if the statement given is true or false. If it is false, change the underlined word so that the statement reads true.

43. _____ An advantage of the evolution of a digestive tract is that it enables an organism to eat frequently.

44. _____ The nematode has a complex digestion tube that is unspecialized along its length.

45. _____ Animals with tubular digestive systems use intracellular digestion to break down their food.

46. _____ The primary function of the stomach is water reabsorption.

47. _____ Bacteria in the large intestine serve no purpose to the human body.

48. _____ The secretion of saliva, stimulated by the sight of food, the thought of food, or the presence of food in the mouth, is controlled by the nervous system.

49. _____ Mucus produced by the stomach serves to neutralize the acid conditions.

50. _____ Within the esophagus, the enzyme pepsin breaks proteins into shorter polypeptides.

51. _____ As food moves into the small intestine, the pancreas is stimulated to produce sodium hydroxide to neutralize the acidic chyme.

52. _____ Bile salts emulsify fats.

53. _____ Absorption of nutrients and small food molecules occurs in the large intestine.

54. _____ The purpose of villi and microvilli is to increase the surface area of the <u>stomach</u>.

55. _____ The <u>vitamins of the B-complex and vitamin C</u> are used to aid enzymes in chemical reactions.

56. _____ Fat soluble vitamins may be <u>toxic</u> if taken in high doses.

Matching: Match the following enzymes with their substrate:

57. _____ cellulase	Choices:	
58. _____ pepsin	a. proteins	
59. _____ lipase	b. disaccharides	
60. _____ lactase	c. individual amino acids	
61. _____ chymotrypsin	d. peptides	
62. _____ amylase	e. lipids	
63. _____ peptidases	f. starches	
64. _____ sucrase	g. cellulose	
65. _____ carboxypeptidase		
66. _____ maltase		

67. Complete the table below by identifying the functions of the structures listed.

Area of the digestive tract	Function
mouth	
esophagus	
stomach	
small intestine	
liver	
pancreas	
large intestine	

Identify the Following: Are the following vitamins soluble in **fat** or **water**?

68. _____ vitamin A		71. _____ vitamin D	
69. _____ vitamin B-complex		72. _____ vitamin E	
70. _____ vitamin C		73. _____ vitamin K	

Identify the Following: To which hormone are the following statements are related?

74. _____ stimulates hydrochloric acid production by stomach cells

75. _____ stimulates release of pancreatic digestive enzymes

76. _____ stimulates bicarbonate release into small intestine

77. _____ inhibits movement of chyme into the small intestine

78. _____ stimulates contraction of the gall bladder

79. _____ is regulated by acid production

80. _____ release is stimulated by the presence of fatty acids in chyme

81. _____ release is stimulated by acidic chyme in the small intestine

Choices:

a. gastrin

b. secretin

c. cholecystokinin

d. gastric inhibitory peptide

Short Answer:

82. List the five tasks a digestive system must accomplish, regardless of its level of complexity.

_____

_____

_____

_____

_____

83. What are the two primary functions of a digestive tract?

_____

_____

_____

_____

_____

84. Where are trypsin, chymotrypsin, and carboxypeptidase found? What is the function of each?

_____

_____

_____

_____

_____

Matching: Match the following statements to the related hormone.

74. _____ stimulates hydrochloric acid production by stomach cells

75. _____ stimulates release of pancreatic digestive enzymes

76. _____ stimulates bicarbonate release into small intestine

77. _____ inhibits movement of chyme into the small intestine

78. _____ stimulates contraction of the gall bladder

79. _____ is regulated by acid production

80. _____ release is stimulated by the presence of fatty acids in chyme

81. _____ release is stimulated by acidic chyme in the small intestine

Choices:

a. gastrin

b. secretin

c. cholecystokinin

d. gastric inhibitory peptide

Short Answer:

82. List the five tasks a digestive system must accomplish, regardless of its level of complexity.

_____

_____

_____

_____

_____

83. What are the two primary functions of a digestive tract?

_____

_____

_____

_____

_____

84. Where are trypsin, chymotrypsin, and carboxypeptidase found? What is the function of each?

_____

_____

_____

_____

_____

ANSWERS TO EXERCISES

1. digestion
2. calorie
3. Calorie
4. glycogen
5. urea
6. intracellular digestion
7. food vacuole
8. lysosome
9. extracellular digestion
10. gastrovascular cavity
11. crop
12. gizzard
13. essential fatty acids
14. essential amino acids
15. vitamins
16. minerals
17. amylase
18. lipase
19. protease
20. mouth
21. pharynx
22. epiglottis
23. esophagus

24. peristalsis
25. stomach
26. chyme
27. pyloric sphincter
28. Villi
29. microvilli
30. small intestine
31. lacteal
32. large intestine
33. colon
34. rectum
35. gastrin
36. pancreas
37. pancreatic juice
38. liver
39. gallbladder
40. bile salts
41. absorption
42. segmentation movements
43. true
44. false, simple
45. false, extracellular digestion

46. false, store food
47. false, synthesize vitamins B$_{12}$, thiamin, riboflavin, K
48. true
49. false, protects stomach cells
50. false, stomach
51. false, sodium bicarbonate
52. true
53. false, small intestine
54. false, small intestine
55. true
56. true
57. g
58. a
59. e
60. b
61. a/d
62. f
63. d
64. b
65. c
66. b

67.

Area of the digestive tract	Function
mouth	mechanical (chewing) and chemical (amylase) breakdown of food
esophagus	transfers food from mouth to stomach
stomach	stores food, mechanical breakdown of food, secretes enzymes
small intestine	digests food into small molecules, absorbs food molecules
liver	produces bile to emulsify fats in small intestine
pancreas	produces pancreatic juice to digest carbohydrates, fats, and protein, and to neutralize chyme
large intestine	absorbs water and salts from chyme, absorbs vitamins made by bacteria

68. fat
69. water
70. water
71. fat
72. fat

73. fat
74. a
75. c
76. b
77. d

78. c
79. a
80. d
81. b

82. Ingestion of food item; mechanical breakdown (food is physically broken into smaller pieces); chemical breakdown (food is exposed to digestive enzymes; absorption of small molecules); elimination of wastes.

83. Digestion of food; absorption of nutrients

84. All three enzymes are in pancreatic juice secreted by the pancreas into the small intestine. Trypsin and chymotrypsin digest proteins and large peptides into small peptide chains. Carboxypeptidase removes individual amino acids from the ends of peptide chains.

85. Thick mucus secreted by cells of the stomach lining protects the stomach cells from the acid environment. The enzyme pepsin is produced in an inactive form, pepsinogen, that is converted to pepsin in the acid environment of the stomach. Again, the mucus lining the stomach would protect the cells from being digested by pepsin. However, cells lining the stomach are damaged or partially digested and must be replaced every few days. In extreme cases, an ulcer forms where the lining and deeper tissues of the stomach have been digested.

86. The bacteria living in the large intestine synthesize the vitamins B_{12}, thiamin, riboflavin, and K, which are absorbed by the large intestine. The normal diet does not provide these vitamins in sufficient amounts. You could include in your argument that without the production of these vitamins a person could be affected by the following vitamin deficiency symptoms:

　　B_{12} -- pernicious anemia, neurological disorders

　　Thiamin -- muscle weakness, peripheral nerve changes, edema, heart failure

　　Riboflavin -- red, cracked lips, eye lesions

　　K -- internal hemorrhages, failure of blood to clot.

87. An antibiotic that you would be taking for a general bacterial infection would most likely be a broad-spectrum antibiotic. That is, it acts against a wide range of bacteria. This would include the bacteria active in your large intestine. The antibiotic does not distinguish between "good" and "bad" bacteria. With the environment of your large intestine disrupted, diarrhea may result. Eating yogurt containing "active cultures" (live acidophilus bacteria) while taking the antibiotic regimen may help restore the bacterial population in your large intestine, reducing the side effects of the antibiotic.

Chapter 30: The Urinary System

OVERVIEW

This chapter presents the role of the urinary system in simple animals, then focuses on the form and function of the urinary system in vertebrates, humans in particular. The urinary system of animals plays a very important role in the maintenance of homeostasis. It removes excess water, salts, nutrients, and minerals in a precise and regulated manner. From simple animals to the more complex vertebrates, the structures responsible function in rather similar ways.

1) What Are the Functions of Vertebrate Urinary Systems?

Urinary systems of animals play a very important role in the maintenance of a stable internal environment, or maintaining **homeostasis**. The **kidneys** collect the fluid plasma from blood, and reabsorb nutrients and water while removing toxins, waste products, and excess water, salts, nutrients, and minerals. What is removed is channeled and stored until it is **excreted** from the body. **Urea** is produced when the body digests protein and removes the amino group from an amino acid; the amino group is released as **ammonia**. Since ammonia is highly toxic to mammals, it is converted to urea in the liver. Urea leaves the body as **urine**.

2) How Do Simpler Animals Regulate Excretion?

Flatworms possess **protonephridia**, the simplest structure specialized for excretion. This system uses **flame cells**, which are single-celled bulbs that filter water and dissolved wastes. More advanced excretory structures are found in some invertebrates. These kidney-like **nephridia** consist of a funnel-shaped **nephrostome** with cilia that conduct body fluids along the long, twisted tube. As the fluid passes through the tube, nutrients and salts are reabsorbed while water and wastes form urine. The urine is released from the body through the **excretory pore**.

3) How Does the Human Urinary System Function?

Blood to be filtered by the human urinary system enters the pair of kidneys through the **renal artery**. Once filtration is complete, the blood leaves the kidneys through the **renal vein**. Urine is transported from each kidney to the **bladder** through a tube called the **ureter**. The ureter is muscular and uses peristaltic contractions to move the urine to the bladder. Urine leaves the body through the **urethra**.

The kidneys, themselves, are complex. A hollow inner chamber, the **renal pelvis**, funnels urine to the ureter. Outside the renal pelvis is the **renal medulla** covered by the **renal cortex**. **Nephrons** fill the cortex and may extend into the medulla. Each nephron is composed of a dense collection of capillaries that filter the blood, a **glomerulus** surrounded by a **Bowman's capsule**, and a long, twisted **tubule**. The tubule, in turn, is divided into the **proximal tubule**, the **loop of Henle**, and the **distal tubule**. The distal tubule leads to the **collecting duct**.

Blood flowing through the glomerulus undergoes **filtration**. The fluid **filtrate**, containing water and dissolved substances, is collected in the Bowman's capsule and transported through the tubule. The filtered blood, now very thick and concentrated, moves through highly porous capillaries winding around the tubule. When the capillaries come in contact with the tubule, water and nutrients in the filtrate are reabsorbed by the capillaries. This is called **tubular reabsorption** and occurs in the proximal tubule using active transport, osmosis, and diffusion. By **tubular secretion**, any wastes still in the blood move into the filtrate in the distal tubule for excretion. The loop of Henle enables the filtrate (urine) to become concentrated. There is a greater osmotic concentration gradient in the interstitial fluid surrounding the loop. As the collecting duct passes through the concentrated interstitial fluid, additional water moves out of the duct by osmosis. The amount of water that is reabsorbed into the blood is controlled by the level of **antidiuretic hormone (ADH)** in the blood. ADH makes the cells of the distal tubule more permeable to water and is regulated according to the volume of blood in the body.

The kidneys are also involved in the regulation of blood pressure. If blood pressure is too low, the kidneys will secrete **renin** into the bloodstream. Renin stimulates the production of **angiotensin**, which causes arterioles to constrict, elevating blood pressure again. The rate of filtration by the kidneys is decreased by the constriction of the arterioles, reducing the amount of water removed from the blood, increasing blood volume, thus increasing blood pressure. If the blood is low in oxygen, the kidneys release the hormone **erythropoietin** to stimulate blood cell production by the bone marrow.

KEY TERMS AND CONCEPTS

Fill-In: From the following list of key terms, fill in the blanks in the following statements.

antidiuretic hormone (ADH)	ammonia	angiotensin	bladder	filtrate
excretory pore	Bowman's capsule	distal tubule	dialysis	kidney
hemodialysis	collecting duct	excretion	filtration	renin
loop of Henle	erythropoietin	glomerulus	flame cell	urea
protonephridia	homeostasis	renal arteries	nephridia	ureter
renal medulla	nephrostome	renal cortex	nephrons	urethra
tubular secretion	proximal tubule	renal pelvis	renal veins	urine
tubular reabsorption				

(1) _____ is defined as the maintenance of a stable environment. The

(2) _____ play a role in maintaining this stability. The wastes that the urinary system

filters from the blood are removed from the body through a process called (3) _____.

During amino acid digestion, the amine group may be removed forming the highly toxic chemical

(4) _____. This toxic chemical is taken to the liver where it is converted to

(5) _____. Eventually, the waste is excreted from the body as (6) _____.

The simplest structures specialized for excretion are the (7) _____ found in flatworms.

This structure uses single celled bulbs with cilia called (8) _____. Some invertebrates

use more advanced, kidney-like structures called (9)_____. Coelomic fluid is

conducted into the funnel-shaped (10) _____. After collecting in a bladderlike

sac, the urine leaves the body through the (11) _____.

In the human, a pair of kidneys receive blood to be filtered through the (12) _____. After it has been filtered, the blood leaves the kidneys through the (13) _____. Urine leaves each kidney through a long tube called a (14)_____ and is stored in the (15)_____. Urine leaves the body by way of the (16) _____.

Within the human kidney is a hollow collecting chamber called the (17) _____. Outside this chamber is the fan-shaped (18) _____ covered by the (19)_____. Microscopic structures, the (20) _____ act as filters.

Each nephron contains a mass of capillaries called the (21) _____ which is surrounded by a cuplike structure, the (22) _____. A long twisted tubule is subdivided into the (23)_____, the (24) _____, and the (25) _____, which leads to the (26) _____. Within the glomerulus, water and dissolved substances are removed from the blood in a process called (27)_____. The fluid that is removed is called the (28) _____.

When the proximal tubule reabsorbs water and nutrients from the filtrate, the process of (29)_____ is occurring. When the distal tubule removes wastes that were not filtered out initially, the process of (30) _____ is occurring. The amount of water reabsorbed by the proximal tubule is regulated by the hormone (31) _____.

The kidneys help to regulate blood pressure by releasing the hormone (32) _____ into the bloodstream if blood pressure falls. This stimulates the release of an arteriole constricting hormone (33)_____. The kidneys also aid in increasing blood oxygen levels by releasing the hormone (34) _____ when low oxygen levels are detected.

Individuals with kidney damage are treated with an artificial kidney. The process, called (35) _____ uses an artificial semipermeable membrane to passively filter substances from the blood. Since it is blood that is being filtered, the process is specifically called (36)_____.

Key Terms and Definitions

ammonia: NH₃; a highly toxic nitrogen-containing waste product of amino acid breakdown. In the mammalian liver, it is converted to urea.

angiotensin (an-je-o-ten′-sun): a hormone that functions in water regulation in mammals by stimulating physiological changes that increase blood volume and blood pressure.

antidiuretic hormone: also called ADH; a hormone produced by the hypothalamus and released into the bloodstream by the posterior pituitary gland when blood volume is low. It increases the permeability of the distal tubule and the collecting duct to water, allowing more water to be reabsorbed into the bloodstream.

bladder: a hollow muscular storage organ for urine.

Bowman's capsule: the cup-shaped portion of the nephron in which blood filtrate is collected from the glomerulus.

collecting ducts: conducting tubes within the kidney that collect urine from many nephrons and conduct it through the medulla into the renal pelvis. Urine may become concentrated in the collecting ducts if ADH is present.

dialysis (di-al′-i-sis): the passive diffusion of substances across an artificial semipermeable membrane.

distal tubule: in the nephrons of the mammalian kidney, the last segment of the renal tubule through which the filtrate passes just before it empties into the collecting duct; a site of selective secretion and reabsorption as water and ions pass between the blood and the filtrate across the tubule membrane.

erythropoietin (e-rith′-ro-po-e′-tin): a hormone produced by the kidneys in response to oxygen deficiency that stimulates the production of red blood cells by the bone marrow.

excretion: elimination of waste substances from the body. Excretion can occur from the digestive system, skin glands, urinary system, or lungs.

excretory pore: opening in the body wall of certain invertebrates, such as the earthworm, through which urine is excreted.

filtrate: fluid produced by filtration; in the kidneys, fluid produced by filtration of the blood through the glomerular capillaries.

filtration: within Bowman's capsule in each nephron of a kidney, the process by which blood is pumped under pressure through permeable capillaries of the glomerulus, forcing out water, dissolved wastes, and nutrients.

flame cell: in flatworms, a specialized cell containing beating cilia that conducts water and wastes through the branching tubes that serve as an excretory system.

glomerulus: a dense network of thin-walled capillaries located within the Bowman's capsule of each nephron. Here, blood pressure forces water and dissolved nutrients through capillary walls for filtration by the nephron.

hemodialysis (he-mo-di-al′-luh-sis): a procedure that simulates kidney function in individuals with damaged or ineffective kidneys; blood is diverted from the body, artificially filtered, and returned to the body.

homeostasis: the relatively constant environment required for optimal functioning of cells, maintained by the coordinated activity of numerous regulatory mechanisms, including the respiratory, endocrine, circulatory, and excretory systems. The precise regulation of the composition of fluid bathing the body cells is a homeostatic function performed largely by the urinary system.

kidney: one of a pair of organs of the excretory system located on either side of the spinal column. Kidneys filter blood, removing wastes and regulating the composition and water content of the blood.

loop of Henle: a specialized portion of the tubule of the nephron in birds and mammals that creates an osmotic concentration gradient in the fluid immediately surrounding it. This gradient in turn makes possible the production of urine more osmotically concentrated than blood plasma.

nephridium: a type of excretory organ found in earthworms, mollusks, and certain other invertebrates. It somewhat resembles a single vertebrate nephron.

nephron: the functional unit of the kidney, where blood is filtered and urine formed.

nephrostome: the funnel shaped opening of the nephridium of some invertebrates such as earthworms, into which coelomic fluid is drawn for filtration.

protonephridium (pro-to-nef-rid′-e-um; pl., protonephridia): an excretory system consisting of tubules that have external opening but lack internal openings; for example, the flame-cell system of flatworms.

proximal tubule: in nephrons of the mammalian kidney, the portion of the renal tubule just after the Bowman's capsule; receives filtrate form the capsule and is the site where selective secretion and reabsorption between the filtrate and the blood begins.

renal artery: the artery carrying blood to each kidney.

renal cortex: the outer layer of the kidney where nephrons are located.

renal medulla: the layer of the kidney just inside the renal cortex where loops of Henle produce a highly concentrated interstitial fluid that is important for producing concentrated urine.

renal pelvis: the inner chamber of the kidney where urine from the collecting ducts accumulates before entering the ureters.

renal vein: the vein carrying cleansed blood away from each kidney

renin: an enzyme that is released (in mammals) when blood pressure and/or sodium concentration in the blood drops below a set point; initiates a cascade of events that restores blood pressure and sodium concentration.

tubular reabsorption: the process by which cells of the tubule of the nephron remove water and nutrients from the filtrate within the tubule and return them to the blood.

tubular secretion: the process by which cells of the tubule of the nephron remove additional wastes from the blood, actively secreting them into the tubule.

tubule: the tubular portion of the nephron. It includes a proximal portion, the loop of Henle, and a distal portion. Urine is formed from the blood filtrate as it passes through the tubule.

urea: water-soluble, nitrogen-containing waste product of amino acid breakdown that is one of the principal components of mammalian urine.

ureter: a tube that conducts urine from each kidney to the bladder.

urethra: a tube that conducts urine from the bladder to the outside of the body.

uric acid: a nitrogen-containing waste product of amino acid breakdown, which is a relatively insoluble white crystal. Uric acid is excreted by birds, reptiles, and insects.

urine: fluid produced and excreted by the urinary system containing water and dissolved wastes, such as urea.

THINKING THROUGH THE CONCEPTS

True or False: Determine if the statement given is true or false. If it is false, change the underlined word so that the statement reads true.

37. _____   Urea is formed when carbohydrates are digested.

38. _____   Flame cells function to heat up body fluids.

39. _____   In the earthworm body, each segment has its own pair of nephridia.

40. _____   Urine is carried to the bladder in the ureter.

41. _____   Urination is under voluntary control of the external sphincter.

42. _____   An adult bladder can hold 0.5 L of urine.

43. _____   Within the kidney, nephrons may extend into the renal medulla.

44. _____   Filtration in the glomerulus occurs because of a difference in diameter between the arterioles bringing blood in and the arterioles taking blood out.

45. _____   Tubular reabsorption occurs in the distal tubule.

46. _____   During tubular secretion, wastes that are actively secreted into the tubule may include drugs.

47. _____   Each drop of blood passes through a kidney 10 times a day.

48. _____   ADH is released when receptors in the hypothalamus detects an inappropriate osmotic level in the blood.

49. _____   A long loop of Henle is found in the kidneys of aquatic animals.

CLUES TO APPLYING THE CONCEPTS

These practice questions are intended to sharpen your ability to apply critical thinking and analysis to the biological concepts covered in this chapter.

50. Using the diagram of the nephron, answer the questions below.

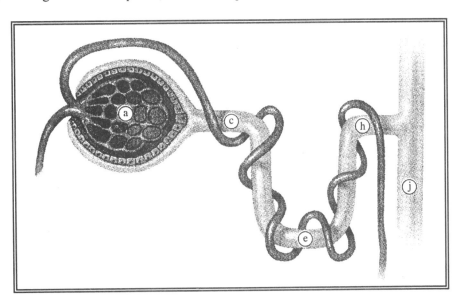

a. _____ identify this structure   h._____ identify this structure

b._____ what is the function of (a)?   I._____ what is the function of (h)?

c. _____ identify this structure   j. _____ identify this structure

d._____ what is the function of (c)?   k._____ what is the function of (j)?

e. _____ identify this structure

f. _____ if this structure was extremely long, in what sort of habitat would this animal live?

g._____ if this structure was rather short, in what sort of habitat would this animal live?

51. Identify the 6 ways the mammalian urinary system helps maintain homeostasis.

_____

_____

_____

_____

_____

52. Trace the path of urine from its production in the kidneys to its point of excretion from the body.

Kidneys ----> _____ ----> _____ ----> _____ ----> excretion

ANSWERS TO EXERCISES

1. homeostasis
2. kidneys
3. excretion
4. ammonia
5. urea
6. urine
7. protonephridia
8. flame cells
9. nephridia
10. nephrostome
11. excretory pore
12. renal arteries
13. renal veins
14. ureter
15. bladder
16. urethra
17. renal pelvis
18. renal medulla
19. renal cortex
20. nephrons
21. glomerulus

22. Bowman's capsule
23. proximal tubule
24. loop of Henle
25. distal tubule
26. collecting duct
27. filtration
28. filtrate
29. tubular reabsorption
30. tubular secretion
31. ADH, antidiuretic hormone
32. renin
33. angiotensin
34. erythropoietin
35. dialysis
36. hemodialysis
37. false, protein
38. false, filter
39. true
40. true

41. true
42. true
43. true
44. true
45. false, proximal tubule
46. true
47. false, about 350 times
48. true
49. false, desert
50. a. glomerulus
    b. filtration
    c. proximal tubule
    d. reabsorption
    e. loop of Henle
    f. desert
    g. aquatic
    h. distal tubule
    i. tubular secretion
    j. collecting duct
    h. concentrate urine

51. Regulation of blood ion levels. Regulation of blood water content. Maintenance of blood pH. Retention of important nutrients. Secretion of hormones. Elimination of cellular waste products.
52. Kidneys ----> ureter ----> bladder ---->urethra ----> excretion.

Chapter 31: Defenses Against Disease: The Immune Response

OVERVIEW

This chapter will present the ways in which the body defends itself against bacteria, fungi, protists, and environmental toxins. The body is designed so that it is difficult for foreign substances to enter it. But, when they do, the body has a well-developed system to detect and destroy invaders. It is able to determine foreign cells from body cells and it is able to remember invaders that have been present before. Occasionally, the system fails, however. Examples of what occurs when the immune system breaks down are discussed.

1) How Does the Body Defend Against Invasion?

The human body is designed to keep invaders from entering. The **skin** provides a barrier to microbes. The outer layer of skin is dry and devoid of nutrients, and the dead cells slough off. Sweat and sebaceous glands produce natural antibiotics. Microbes have a difficult time actually getting into the body through the skin. **Mucous membranes** lining the digestive and respiratory tracts trap microbes; and they contain antibiotic substances, as well. Cilia of cells lining these tracts move the mucous and trapped particles up and out of their respective areas. The mucous is then either coughed or sneezed out, or swallowed.

When microbes do gain entry, the body has a line of defense to remove them. **Phagocytic cells** such as the white blood cells called **macrophages**, ingest microbes by phagocytosis. **Natural killer cells** destroy the body's cells that have been attacked by viruses. When viruses enter a cell they usually leave viral proteins on the outside of the cell. Natural killer cells respond to those proteins by killing the cell. When the skin is cut, an **inflammatory response** is elicited. The damaged cells release **histamine**, relaxing smooth muscle of arterioles and making capillaries leak. This increases blood flow to the area causing it to become red, swollen, and warm. Phagocytes brought to the wound by the increased blood flow ingest microbes infecting the injury. If the phagocytes cannot remove all of the microbes, the microbes may infect larger areas of the body initiating a **fever**. A fever is part of the body's natural defense against infection. It increases phagocyte activity while reducing microbial reproduction. Fever also induces the production of virus-fighting **interferon**. Macrophages responding to an infection release **endogenous pyrogens** that stimulate the hypothalamus to reset the body's thermostat and to set in motion the responses that lead to an increase in body temperature.

2) What Are the Key Characteristics of the Immune Response?

The **immune response** is specifically directed toward a particular invading organism. Lymphocytes called **B cells** and **T cells** recognize foreign organisms, initiate an attack against them, and remember the organisms in order to combat future attacks.

3) How Are Threats Recognized?

B cells produce proteins called **antibodies**. Antibodies may remain attached to the surface of the B cell or they may be secreted into the bloodstream. In the bloodstream, antibodies are called immunoglobulins. T cells produce proteins called **T-cell receptors** that remain attached to them.

Antibodies consist of two heavy peptide chains and two light peptide chains. Both the heavy and the light chains have a **constant region** and a **variable region**. Antibodies on the surface of a B cell act as receptors, projecting outward, ready to bind to appropriate **anti**body **gen**erating molecules (**antigens**). Molecules acting as antigens include proteins, polysaccharides, and glycoproteins. When an antigen binds to a receptor, a response is stimulated in that cell. Antibodies circulating in the bloodstream neutralize or destroy antigens in the blood. There are five classes of antibodies. IgM molecules join together forming a five-molecule immune complex. IgG antibodies coat microbes such that macrophages recognize and engulf them. IgA is active in the digestive and respiratory tracts, binding to microbes so they do not enter the body. IgE is also active in the digestive and respiratory tracts and is responsible for allergic responses. IgD molecules are found on the surface of B cells. Antibodies are capable of binding to specific antigens because their variable regions are highly specific. The constant regions determine how an antibody will act against invaders. It may determine that the antibody will serve as a receptor on the surface of a cell, or it may bind to **complement** proteins in the blood promoting microbe destruction.

T-cell receptors function only as receptors on the surface of T cells. When they bind to an antigen, they elicit a response in the T cell.

The plasma membranes of your body's cells contain proteins and polysaccharides that indicate to your antibodies that these cells belong to your body. These proteins form the **major histocompatability complex (MHC)** and are unique to you (and perhaps to your identical twin). During embryological development immune cells that react to "self" cells are destroyed to avoid destroying cells with the body's own antigens.

4) How Are Threats Overcome?

When the body is threatened by an infection, two types of immunity are engaged. **Humoral immunity** involves the B cells and the antibodies secreted into the bloodstream. When antibodies on B cells bind to antigens from an invading organism, the B cell is induced to divide rapidly, producing genetically identical cells. This is referred to as **clonal selection**. The daughter clones differentiate into **plasma cells** or **memory cells**. Plasma cells produce large quantities of their specific antibodies and release them into the bloodstream. Memory cells are involved in future immunity if reinfection with the same organism occurs.

Antibodies carry out their effect using four mechanisms. Antibodies may **neutralize** a toxic antigen preventing damage to the body. They may coat the surface of a microorganism promoting **phagocytosis** by white blood cells. An antibody may bind to two microbes at a time. Because additional antibodies also bind with more than one organism, **agglutination** occurs, enhancing phagocytosis. Antibodies circulating in the blood attach to microbe antigens with their variable regions while their stems attach to proteins of the **complement system** enhancing phagocytosis.

Cell-mediated immunity involves three types of T cells that attack organisms within body cells. When receptors on **cytotoxic T cells** bind to antigens on an infected cell, the cytotoxic T cells produce and release proteins that will disrupt the infected cell's plasma membrane. When antigens bind to receptors on the surface of **helper T cells**, the cells produce and release chemicals that act as hormones, stimulating both cytotoxic T cells and B cells to divide and differentiate. Once an infection has been successfully fought, **suppressor T cells** shut off the immune response of both cytotoxic T cells and B cells.

5) How Are Invaders "Remembered"?

When the body is infected with an organism that has antigens the immune system has responded to before, memory cells specific to that antigen will recognize the organism. The memory cells multiply rapidly and stimulate the production of huge populations of plasma cells and cytotoxic T cells. This second round of defense occurs much more quickly than the first.

6) How Does Medical Care Augment the Immune Response?

The use of antibiotics serves to compromise the ability of microbes to grow and reproduce. With their population growth in check, the immune system can fight the infection and stands a better chance of winning. Antibiotics are effective against bacteria, fungi, and protists; but *not* viruses. The overuse or improper use of antibiotics augments their already strong qualities of natural selection. This has given rise to many antibiotic resistant strains of microbes.

By injecting dead or weakened disease agents into a healthy individual, the individual does not acquire the disease however; the immune system responds to the presence of the foreign antigens. The response includes the production of a large number of memory cells. Thus, the practice of giving **vaccinations** confers immunity against potentially deadly diseases.

7) Can the Immune System Malfunction?

Yes, unfortunately, the immune system can malfunction. **Allergies** are a common example of such a malfunction. When pollen or another foreign substance such as poison ivy toxin enters the bloodstream and is identified by a B cell as an antigen, the B cell multiplies. The proliferation of B cells produce plasma cells that generate IgE antibodies against the antigen. When the pollen antigen binds to IgE molecules on **mast cells** in the respiratory or digestive tract, the mast cells are stimulated to produce and release histamine. The release of histamine induces the typical "hay fever" symptoms. **Autoimmune diseases** occur when the immune system does not recognize its body's own cells. Thus, a portion of the body is attacked. This is the cause of some forms of anemia and diabetes.

Children born with **severe combined immune deficiency (SCID)** are unable to produce sufficient numbers of immune cells, if any. For the first few months of life, the children are protected from disease by immunity conferred by the mother during pregnancy. Bone marrow transplants, to supply the child with healthy white blood cell producing marrow, is one form of therapy offered to these children.

Other individuals lose the effectiveness of their immune system. Individuals with **acquired immune deficiency syndrome (AIDS)** are infected with the **human immunodeficiency virus** either **type 1** or **2 (HIV-1 or HIV-2)**. These viruses attack helper T cells. Thus, both the humoral and cell-mediated responses are severely hindered, and these individuals are susceptible to a number of otherwise easily combated diseases. The HIV-1 and HIV-2 viruses are **retroviruses**. That is, they are RNA viruses that transcribe their RNA into DNA using its own enzyme **reverse transcriptase**. Their DNA is then inserted into the DNA of the helper T cell. When the gene is activated in the helper T cell, HIV particles are manufactured and released into the bloodstream, killing the helper T cell in the process.

When the immune system is overwhelmed, its effectiveness is severely reduced. This is, in part, what happens when the body develops cancer. A **tumor** occurs when a population of cells grow at an abnormal rate. If the tumor is benign, the growth remains local. If the tumor is malignant, the growth is uncontrolled and increasingly uses the body's nutrient and energy supplies. **Cancer** is the disease resulting from uncontrolled growth of a malignant tumor. When a cell becomes cancerous, one of two types of genes are probably at fault. **Proto-oncogenes** and **tumor-suppressor genes**, in a normally functioning cell, are involved in controlling cell reproduction. Cancerous cells fail to respond to the control signals. When a mutation occurs in a proto-oncogene, an **oncogene** is formed. Oncogenes continuously produce growth-stimulating proteins. When a mutation occurs in a tumor-suppressor gene, the gene fails to produce the proteins necessary to stop cell division. It actually requires several mutations in the same cell before a cancerous cell is formed. When cancerous cells are formed, changes occur in their plasma membrane proteins. This usually enables natural killer cells and cytotoxic T cells to destroy them before they can multiply. However, the cancerous cells may multiply faster than the immune system can destroy them, they may suppress the immune system, or they may be resistant to immune attack. Any or all result in a cancerous growth.

KEY TERMS AND CONCEPTS

Fill in: From the following list of key terms, fill in the blanks in the following statements:

acquired immune deficiency syndrome (AIDS)	constant region	agglutinate	allergy
autoimmune disease	cytotoxic T cells	antibodies	antigens
cell-mediated immunity	helper T cells	complement	B cells
clonal selection	humoral immunity	histamine	cancer
complement system	immune response	interferon	fever
endogenous pyrogens	mucous membranes	macrophages	mast cells
human immunodeficiency virus (HIV)	neutralization	memory cells	retrovirus
inflammatory response	phagocytic cells	oncogene	skin
major histocompatibility complex (MHC)	proto-oncogene	plasma cells	T cells
natural killer cells	reverse transcriptase	vaccination	tumor
severe combined immune deficiency (SCID)	suppressor T cells	variable region	
tumor-suppressor gene	T-cell receptors		

The (1) _____ and (2) _____ provide the body barriers to invasion by foreign substances.

The body's nonspecific defenses use (3) _____, specifically (4) _____ to destroy microbes by phagocytosis and (5) _____ to destroy virus infected cells. An injury provokes the (6) _____ . Injured cells release (7) _____ making capillaries leaky and smooth muscles relax. If a major infection becomes established, the body produces a (8) _____ to slow microbial growth and reproduction and to help fight viral infections by increasing the production of (9) _____ . Macrophages, producing the hormones (10) _____ signal the hypothalamus to "turn up" the body's thermostat.

When nonspecific defenses fail, the body initiates the highly specific (11) _____.
This response involves two specific lymphocytes. The (12) _____ produce protein
(13) _____ on their plasma membranes or release them into the bloodstream. The
(14)_____ produce (15) _____ on their cell surfaces.

Antibodies consist of a (16) _____ that is similar within an antibody class, and a
(17)_____that differs even within an antibody class. When antibodies are attached to
the plasma membrane of a B cell, they serve as receptors. When they circulate in the bloodstream, they
serve as effectors. In both roles, the variable regions respond and bind to (18) _____
produced by a foreign substance. The constant region may bind to the B cell plasma membrane or it may
bind to (19) _____ proteins in the blood.

The immune system is able to differentiate "self" cells from "nonself" cells because of plasma membrane
proteins making up the (20) _____.

In order to attack microbes before they enter cells, the body uses B cells and antibodies to provide
(21) _____. When antibodies bind to antigens, B cells are stimulated to divide
rapidly. Since the cells are identical, this is called (22) _____. The daughter cells
will differentiate into (23) _____ and (24) _____.

Antibodies in the blood may defend the body against foreign substances by (25) _____
of a toxic antigen, by identifying a microbe to be engulfed, thereby promoting phagocytosis, by causing
microbes to clump together or (26) _____, or by triggering blood proteins of the
(27) _____ which induces a series of reactions, the complement reactions.

T cells are used in (28) _____ to attack cells infected by an invader. This type of
response involves (29) _____ that disrupt the plasma membrane of infected cells,
(30) _____ that assist other immune cells, and (31) _____ that
shut off the immune system when an infection has been fought.

A (32) _____ is an injection of weakened or killed microbes causing an immune
response. This decreases the likelihood of an individual contracting a severe case of the disease.

When the immune system recognizes pollen or mold spores as antigens, an (33) _____
has developed to those substances. When the antigen binds to IgE antibodies on (34) _____
in the respiratory tract, histamine is released.

When the immune system does not recognize cells as belonging to "self," and "anti-self" antibodies are produced, the result is an (35) _____. If a child is born with little or no ability to produce immune cells, that child has (36) _____. However, due to the (37) _____, many individuals in the world are now combating the effects of (38)_____. This virus is a (39) _____, which uses its RNA and the enzyme (40) _____ to make DNA and insert it into the genome of helper T cells.

A population of cells that are no longer regulated and are growing at an abnormal rate is called a (41) _____. This growth is benign if it is localized or malignant if it grows uncontrollably. (42) _____ is the disease caused by a malignant growth. This disease occurs when two classes of genes mutate. (43) _____ produce proteins used in growth stimulating pathways. When these genes mutate, forming (44) _____, a continuous supply of growth-stimulating proteins are produced. The second class of genes includes the (45) _____. When they mutate, proteins needed to inhibit cell division are not produced.

Key Terms and Definitions

Acquired immune deficiency syndrome: (AIDS). The fatal disease caused by human immunodeficiency virus (HIV) infection, in which helper T cells are attacked and reduced in number, leading to the body's inability to fight off other infectious diseases.

agglutination: clumping of foreign substances or microbes, caused by binding with antibodies.

allergy: an inflammatory response produced by the body in response to invasion by foreign materials, such as pollen, which are themselves harmless.

antibody: a protein produced by cells of the immune system that combines with a specific antigen and usually facilitates its destruction.

antigen: a complex molecule, usually protein or polysaccharide, that stimulates the production of a specific antibody.

autoimmune disease: a disorder in which the immune system produces antibodies against the body's own cells.

B cell: a type of lymphocyte that participates in humoral immunity; gives rise to plasma cells that secrete antibodies into the circulatory system and to memory cells.

cancer: a disease in which some of the body's cells escape from normal regulatory processes and divide without control.

cell-mediated immunity: an immune response in which foreign cells or substances are destroyed by contact with T cells.

clonal selection: the mechanism by which the immune response gains specificity; an invading antigen elicits a response from only a few lymphocytes, which proliferate to form a clone of cells that attack only the specific antigen that stimulated their production.

complement: a group of blood-borne proteins that participates in the destruction of foreign cells to which antibodies have bound.

complement reactions: interactions among foreign cells, antibodies, and complement proteins, resulting in the destruction of the foreign cells.

complement system: a series of reactions in which complement proteins bind to antibody stems, attracting to the site phagocytic white blood cells that destroy the invading cell that triggers the reactions.

constant region: part of an antibody molecule that is similar in all antibodies.

cytotoxic T cell: a type of T cell that directly destroys foreign cells upon contacting them.

endogenous pyrogen: a chemical produced by the body that stimulates the production of a fever (elevated body temperature).

fever: an elevation in body temperature caused by chemicals (pyrogens) released by white blood cells in response to infection.

helper T cell: a type of T cell that aids other immune cells to recognize and act against antigens.

histamine: a substance released by certain cells in response to tissue damage and invasion of the body by foreign substances. Histamine promotes dilation of arterioles, leakiness of capillaries and triggers some of the events of the inflammatory response.

human immunodeficiency virus (HIV): a pathogenic retrovirus that causes acquired immune deficiency syndrome (AIDS) by attacking and destroying the immune system's T cells.

humoral immunity: an immune response in which foreign substances are inactivated or destroyed by antibodies circulating in the blood.

hybridoma: a cell produced by fusing an antibody-producing cell with a myeloma cell; used to produce monoclonal antibodies.

immune response: a specific response by the immune system to invasion of the body by a particular foreign substance or microorganism, characterized by recognition of the foreign material by immune cells and its subsequent destruction by antibodies or cellular attack.

inflammatory response: a nonspecific, local response to injury to the body, characterized by phagocytosis of foreign substances and tissue debris by white blood cells, and "walling off" of the injury site by clotting of fluids escaping from nearby blood vessels.

interferon: a protein released by certain virus-infected cells that increases the resistance of other uninfected cells to viral attack.

macrophage: a type of white blood cell that engulfs microbes. Macrophages destroy microbes by phagocytosis and also present microbial antigens to T cells, helping to stimulate the immune response.

major histocompatibility complex (MHC): proteins, usually located on the surfaces of body cells, that identify the cell as "self"; MHC proteins also are important in stimulating and regulating the immune response.

mast cell: a cell of the immune system that synthesizes histamine and other molecules that are used in the body's response to trauma and are a factor in allergic reactions.

memory cell: a long-lived descendant of a B or T cell that has been activated by contact with antigen. Memory cells are a reservoir of cells that rapidly respond to reexposure to the antigen.

monoclonal antibody: an antibody produced in the laboratory by the cloning of hybridoma cells; each clone of cells produces a single antibody.

mucous membrane: the lining of the inside of respiratory and digestive tracts.

natural killer cell: a type of white blood cell that destroys some virus-infected cells and cancerous cells on contact. Part of the nonspecific internal defense against disease.

neutralization: the process of covering up or inactivating a toxic substance with antibody.

oncogene: a gene that, when transcribed, causes a cell to become cancerous.

phagocytic cell (fa-go-sit´-ik): a type of immune system cell that destroys invading microbes by using phagocytosis to engulf and digest the microbes.

plasma cell: an antibody-secreting descendant of a B cell.

proto-oncogene (pro-to-onk´-o-jen): a normal cellular gene that can cause cancer if a mutation transforms it into an oncogene.

retrovirus: a virus that uses RNA as its genetic material. When a retrovirus invades a eukaryotic cell, it "reverse transcribes" its RNA into DNA, which then directs the synthesis of more viruses, using the transcription and translation machinery of the cell.

reverse transcriptase: an enzyme found in retroviruses that catalyzes the synthesis of DNA from an RNA template.

severe combined immune deficiency (SCID): a disorder in which no immune cells, or very few, are formed; the immune system is incapable of responding properly to invading disease organisms, and the victim is very vulnerable common infections.

skin: the tissue that makes up the outer surface of an animal body.

suppressor T cell: a type of T cell that depresses the response of other immune cells to foreign antigens.

T cell: a type of lymphocyte that recognizes and destroys specific foreign cells or substances or that regulates other cells of the immune system.

T-cell receptor: a protein receptor, located on the surface of a T cell, that binds a specific antigen and triggers the immune response of the T cell.

tumor: a mass that forms in otherwise normal tissue; caused by the uncontrolled growth of cells.

tumor-suppressor gene: a gene that encodes information for a protein that inhibits cancer formation, probably by regulating cell division in some way.

vaccination: an injection into the body that contains antigens characteristic of a particular disease organism and that stimulates an immune response.

variable region: the part of an antibody molecule that differs among antibodies; the ends of the variable regions of the light and heavy chains form the specific binding site for antigens.

THINKING THROUGH THE CONCEPTS

True or False: Determine if the statement given is true or false. If it is false, change the underlined word so that the statement reads true.

46. _____ The skin provides an <u>excellent</u> breeding ground for microbes.

47. _____ Mucous membranes <u>secrete enzymes</u> to destroy microbes.

48. _____ A <u>fever</u> is actually a defense mechanism against disease.

49. _____ <u>Natural killer cells</u> recognize and destroy cancerous cells.

50. _____ Chemicals produced by wounded cells serve to <u>attract</u> phagocytic cells to an injury.

51. _____ <u>Pus formation</u> is an indication that your immune system is working at a wound site.

52. _____ Interferon <u>decreases</u> a cell's resistance to viral attack.

53. _____ <u>Antibodies</u> are formed from a wide array of genes, some of which mutate easily

54. _____ Antibodies <u>provide no protection against</u> poisons such as snake venom.

55. _____ Antibodies <u>are custom made to fit a specific antigen</u>.

56. _____ B cells, differentiated into <u>plasma cells</u>, release antibodies into the bloodstream.

57. _____ <u>Helper T cells</u> turn the immune system off after an infection has been eradicated from the body.

58. _____ Humans continue to suffer from the common cold because <u>our immune system is unable to produce memory cells for the cold virus</u>.

59. _____ Antibiotics help fight diseases caused by <u>viruses</u>.

60. _____ An anthrax vaccine has been produced using <u>synthetic antigens</u>.

61. _____ <u>Mast cells</u> produce histamine in response to an allergin antigen.

62. _____ Autoimmune diseases can be <u>cured</u> with today's medical technology.

Identify the Following: Do the following statements relate to **B cells**, **T cells**, or **both**.

63. _____ precursor cells produced in bone marrow

64. _____ differentiate in bone marrow

65. _____ differentiate in thymus

66. _____ long lived, can provide future immunity

67. _____ secrete antibodies into the bloodstream

68. _____ destroy cells infected with viruses

69. _____ shut down the immune response when appropriate

70. _____ activate both B and T cells

71. _____ provide humoral immunity

72. _____ provide cell-mediated immunity

73. Label the antibody diagram below.

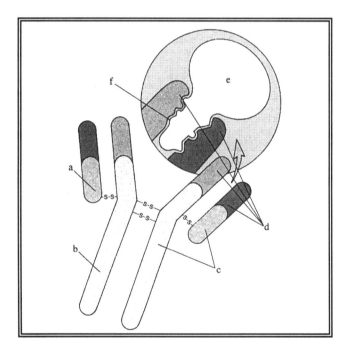

a. _____

b. _____

c. _____

d. _____

e. _____

f. _____

Short Answer:

74. Identify the body's nonspecific defenses against microbial invasion.

_____

_____

_____

_____

75. Explain the four ways antibodies destroy extracellular microbes and molecules.

_____

_____

_____

_____

Complete the following questions specifically about AIDS.

76. How do HIV-1 and HIV-2 affect the immune system? Once they are in a cell, how are more viruses made?

_____

_____

_____

_____

_____

77. If AIDS does not directly kill its victims, what does cause them to die?

_____

_____

_____

_____

_____

_____

78. How do drugs like AZT and protease inhibitors function against HIV? Are they a cure?

_____

_____

_____

_____

79. Why is an effective vaccine against HIV so difficult to produce?

_____

_____

_____

_____

_____

_____

80. Where are the most AIDS cases found in the world and who is most at risk of being infected with HIV?

_____

_____

_____

_____

Complete the following questions specifically about cancer.

81. How are proto-oncogenes and tumor suppressor genes involved in cancer development?

_____

_____

_____

_____

_____

_____

82. Since humans are constantly exposed to substances that cause mutations in our cells, why does cancer not affect many of us until later in life?

_____

_____

_____

_____

_____

_____

83. How does the immune system combat the cells that do become cancerous?

_____

_____

_____

_____

CLUES TO APPLYING THE CONCEPTS

These practice questions are intended to sharpen your ability to apply critical thinking and analysis to biological concepts covered in this chapter.

84. Antihistamines are often taken by individuals suffering from allergies. Using what you now know about the role of histamine in the immune response, how would antihistamines decrease your allergy symptoms? Would taking antihistamines lessen your symptoms from a cold virus?

_____

_____

_____

_____

_____

_____

85. Why will your body reject an organ received during a transplant?

_____

_____

_____

_____

_____

_____

86. Why is it common for cancer patients to experience hair loss, nausea, and severe dry mouth during chemotherapy treatments?

_____

_____

_____

_____

_____

_____

ANSWERS TO EXERCISES

1. skin
2. mucous membranes
3. phagocytic cells
4. macrophages
5. natural killer cells
6. inflammatory response
7. histamine
8. fever
9. interferon
10. endogenous pyrogens
11. immune response
12. B cells
13. antibodies
14. T cells
15. T-cell receptors
16. constant region
17. variable region
18. antigens
19. complement
20. major histocompatibility complex (MHC)
21. humoral immunity
22. clonal selection
23. plasma cells
24. memory cells
25. neutralization
26. agglutinate
27. complement system
28. cell-mediated immunity
29. cytotoxic T cells

30. helper T cells
31. suppressor T cells
32. vaccination
33. allergy
34. mast cells
35. autoimmune disease
36. severe combined immune deficiency (SCID)
37. human immunodeficiency virus (HIV)
38. acquired immune deficiency syndrome (AIDS)
39. retrovirus
40. reverse transcriptase
41. tumor
42. Cancer
43. Proto-oncogenes
44. oncogenes
45. tumor suppressor genes
46. false, poor
47. true
48. true
49. true
50. true
51. true
52. false, increases
53. true

54. false, neutralize
55. false, fit reasonably well to specific antigens
56. true
57. false, suppressor T cells
58. false, the cold virus mutates frequently
59. false, bacteria, fungi, and protists
60. true
61. true
62. false, suppressed
63. both
64. B cells
65. T cells
66. both
67. B cells
68. T cells
69. T cells
70. T cells
71. B cells
72. T cells
73. a. light chain
    b. heavy chain
    c. constant regions
    d. variable regions
    e. antigen
    f. antigen binding site

74. phagocytic cells and natural killer cells, the inflammatory response, and fever
75. Antibodies destroy extracellular microbes and molecules by neutralization of a toxic antigen; coating a microbe, thus promoting phagocytosis; agglutination of microbes, and by inducing complement reactions with the proteins of the complement system, promoting phagocytosis.
76. HIV-1 and HIV-2 infect and destroy helper T cells. Once in a helper T cell, the RNA uses its reverse transcriptase enzyme to create a DNA fragment molecule. The DNA fragment is inserted into the helper T cell's genome, where it will eventually be used to make more HIV viruses.
77. Because the helper T cell population declines in number as the cells are destroyed, there are fewer and fewer cells to assist other immune cells in their defenses. The hormone-like chemicals that helper T cells release to trigger immune cell division and differentiation, antibody production, and cytotoxic cell development are not circulated. Thus, the person with an active HIV infection becomes increasingly susceptible to other diseases. It is one of the "opportunistic" diseases that will cause the patient's death.

78. AZT inhibits reverse transcriptase slowing the production of DNA from the viral RNA. The hope is that DNA will never be properly formed from the RNA, thus new virus will not be made. Unfortunately, AZT does not successfully block the reverse transcription. Protease inhibitors inactivate an enzyme responsible for assembling new viral particles. Protease inhibitors are most effective when used in combination with reverse transcriptase inhibitors. Thus the term "cocktail" of drugs has come into common use. No, these drugs are not a cure. That search continues.

79. An effective vaccine against HIV has been difficult to develop for several reasons. One reason is that antibodies naturally produced against HIV seem to offer little protection against infection. This indicates that a vaccine would need to function in a way that produced more effective immune responses than normal HIV does. Another reason is that HIV mutates at a phenomenal rate. Therefore, a vaccine developed for one strain may have absolutely no effect on the person vaccinated. To further complicate the situation, HIV can exist in different forms within the same person.

80. Most individuals infected with HIV live in developing countries. Sub-Saharan Africa has the highest rate of infection with approximately 20.8 million people infected, while India and Southeast Asia have 6 million people infected. Transmission through heterosexual sex is the most common avenue of infection. Individuals most at risk of being infected are those individuals not practicing "safe sex," regardless of the gender of the individuals involved.

81. Proto-oncogenes contain the genetic codes for many of the proteins that stimulate cell growth. When proto-oncogenes mutate, becoming oncogenes, growth stimulating proteins are continuously produced. Thus, cell growth continues. Tumor-suppressor genes code for the proteins responsible for inhibiting cell growth. When tumor-suppressor genes have mutated, they fail to produce the "stop" signal proteins. Thus cell division continues unabated.

82. Several mutations of proto-oncogenes and/or tumor-suppressor genes must occur in the same cell before the cell would become cancerous. It may take years for the necessary number of mutations to occur. Individuals who inherit already mutated genes may develop cancer earlier in life.

83. Natural killer cells and cytotoxic cells detect the protein changes that occur on the plasma membranes of cancerous cells. They destroy nearly all cancerous cells that occur in our bodies before they have a chance to proliferate.

84. Histamines are produced when an allergin antigen binds to IgE antibodies on mast cells. The result of histamine circulating in the bloodstream is increased mucus production and the inflammatory response increasing discomfort. Antihistamines would inhibit the production of histamine or block its affect, thus reducing the production of mucus and inflammation. Since the cold virus does not typically stimulate mast cells to produce histamine, congestion symptoms due to a cold will not be alleviated by taking antihistamines.

85. The cell plasma membranes contain large proteins and polysaccharides that indicate to the immune system that they belong to "self." These proteins make up the major histocompatibility complex. They are unique to each individual (and an identical twin). If you receive an organ transplant, a "match" will be made in an attempt to reduce the number of different proteins. Since no match will be perfect (except from an identical twin), the body will naturally mount an immunological attack against the foreign cells, rejecting the transplanted organ.

86. Chemotherapy agents target cells that are rapidly growing and frequently dividing. Therefore, any cell in the body that tends to be replaced frequently would be affected by the chemotherapy. Hair follicles and cells lining the stomach and the mouth are body cells that divide and grow rapidly. They are,therefore, damaged along with any cancerous cells.

Chapter 32: Chemical Control of the Animal Body: The Endocrine System

OVERVIEW

This chapter covers how the body's internal chemistry is controlled by hormones. The authors discuss the four classes of animal hormones and the endocrine glands that produce them. Hormone regulation by negative feedback is outlined and the major mammalian endocrine glands are presented in detail.

1) What Are Animal Hormones?

A **hormone** is a chemical secreted by cells in one part of the body that is transported in the bloodstream to other parts of the body, where it affects particular **target cells**. Hormones are released by the cells of major endocrine glands and endocrine organs located throughout the body. Hormones can be grouped into four general classes: (1) **peptide hormones**, made of chains of amino acids; (2) amino acid derivatives, formed from single amino acids (such as epinephrine derived from tyrosine); (3) **steroid hormones**, most of which are derived from cholesterol, secreted by the ovaries, testes, and adrenal cortex; and (4) **prostglandins**, composed of two fatty acid carbon chains attached to a five-carbon ring. Prostaglandins are produced by nearly all cell types and act mainly on nearby cells.

Hormones function by binding to specific receptors on target cells that respond to particular hormone molecules. Receptors for hormones are found either on the plasma membrane or inside the cell, normally within the nucleus. Most peptide and amino acid–based hormones cannot penetrate the plasma membrane and must react with protein receptors that protrude from its outer surface. Most hormones binding to a receptor on the outside cell surface trigger the release inside the cell of a chemical (the **second messenger**) that initiates a cascade of biochemical reactions. **Cyclic AMP** (cAMP) is a common second messenger.

Steroid hormones and thyroid-produced hormones are lipid soluble and pass through the plasma membrane, binding to intracellular receptors, typically to protein receptors in the nucleus. The receptor-hormone complex binds to DNA and stimulates particular genes to become active.

Hormones are regulated by feedback mechanisms. In animals, the secretion of a hormone stimulates a response in target cells that inhibits further secretion of the hormone, a mechanism called negative feedback. For example, loss of water through perspiration triggers the pituitary gland to produce *antidiuretic hormone* (ADH), which causes kidneys to reabsorb water and to produce very concentrated urine. Drinking water replaces what has been lost, triggering negative feedback to turn off ADH secretion when blood water content returns to normal. In a few cases, positive feedback controls hormone release. For example, contractions of the uterus early in childbirth cause the release of the hormone *oxytocin* that stimulates stronger contractions of the uterus.

2) What Are the Structures and Functions of the Mammalian Endocrine Hormones?

Mammals have both **exocrine glands** (which produce secretions released outside the body or into the digestive tract through tubes or openings called **ducts**) and **endocrine glands** (which are ductless glands secreting hormones into capillaries of the bloodstream). Exocrine glands include the sweat and sebaceous glands in the skin, lacrimal glands of the eyes, mammary glands, and glands that produce digestive secretions.

The **hypothalamus** controls the secretions of the **pituitary gland**. The hypothalamus is part of the brain containing **neurosensory cells** that make and store peptide hormones and release them when stimulated. The pituitary gland hangs from the hypothalamus and has two lobes: the **anterior pituitary** and the **posterior pituitary**. The anterior pituitary produces and releases a variety of peptide hormones, four of which regulate hormone release in other endocrine glands. **Follicle-stimulating hormone (FSH)** and **luteinizing hormone (LH)** stimulate production of sperm and testosterone in males and eggs, estrogen, and progesterone in females. **Thyroid-stimulating hormone (TSH)** stimulates release of thyroid gland hormones, and **adrenocorticotropic hormone (ACTH)** stimulates release of hormones from the adrenal cortex. Other hormones of the anterior pituitary do not act on other endocrine glands. **Prolactin** helps stimulate development of the mammary glands during pregnancy. **Endorphins** inhibit the perception of pain by binding to brain receptors. **Melanocyte-stimulating hormones (MSH)** stimulate production of the skin pigment melanin. **Growth hormone** regulates body growth (too little causes *dwarfism*, too much can cause *gigantism*). Growth hormone can be made by genetic engineering. At least nine hypothalamic peptide hormones exert control over release of hormones from the anterior pituitary. Some are **releasing hormones** and others are **inhibiting hormones**, depending on whether they stimulate or prevent release of pituitary hormones.

The posterior pituitary releases two types of peptide hormones produced by cells in the hypothalamus: **antidiuretic hormone (AH)** and **oxytocin**. AH helps prevent dehydration by increasing the permeability of the collecting ducts of kidney nephrons, causing water to be reabsorbed from the urine and retained in the body. Alcohol inhibits the release of ADH. Oxytocin triggers the "milk letdown reflex" in nursing mothers by causing breast muscle tissue to contract during breastfeeding. Oxytocin also has behavioral effects in rats, increasing maternal activities. Oxytocin also aids in male ejaculation.

The thyroid gland produces two major hormones: thyroxin and calcitonin. **Thyroxin** (a iodine-containing modified amino acid) raises the metabolic rate of most cells by stimulating the synthesis of enzymes that break down glucose and provide energy. Thyroxin helps regulate body temperature and helps the body respond to stress. In juvenile animals, thyroxin helps regulate growth by stimulating both metabolic rate and development of the nervous system. Too little thyroxin in early life causes *cretinism*. Thyroxin levels in the bloodstream are controlled by negative feedback loops. Thyroxin release is stimulated by TSH, which is stimulated by a releasing hormone from the hypothalamus. High thyroxin

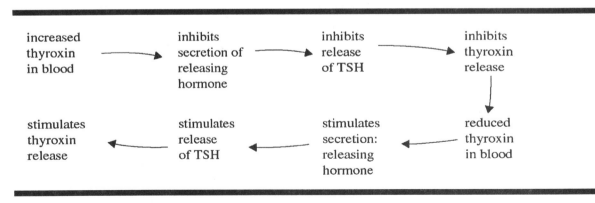

blood levels inhibit secretion of both TSH and releasing hormone, thus inhibiting further release of thyroxin from the thyroid. A diet deficient in iodine causes thyroid enlargement, which bulges from the neck producing a condition called **goiter**. Using iodized salt prevents this condition.

The four small discs of the **parathyroid gland** are embedded in the back of the thyroid. The parathyroids secrete **parathormone**, which along with **calcitonin** from the thyroid, controls the concentration of calcium in the blood by regulating calcium absorption and release by the bones. If blood calcium levels are too low, parathormone is released to stimulate release of calcium from bones; if blood calcium levels are too high, calcitonin is released to inhibit release of calcium from bones.

The **pancreas** is both an exocrine gland (digestive secretions released from the pancreatic duct into the small intestine) and an endocrine gland (clusters of **islet cells** that produce the peptide hormones **insulin** and **glucagon**). Insulin and glucagon work in opposition to regulate carbohydrate and fat metabolism: insulin reduces blood glucose levels and glucagon increases it by activating a liver enzyme that breaks down glycogen. Glucagon also promotes lipid breakdown, releasing fatty acids for metabolism. Defects in insulin production, release, or reception by target cells result in **diabetes mellitus**, a condition in which blood glucose levels are high and fluctuate wildly with sugar intake. Human insulin can be made by genetic engineering.

The sex organs secrete steroid hormones. The **testes** and **ovaries** are endocrine glands producing steroid hormones. Male testes make a group of hormones collectively called **androgens**, and female ovaries make **estrogen** and **progesterone**. Sex hormones play key roles in development, pregnancy, the menstrual cycle, and puberty (the physiological changes that lead to reproductive capacity and secondary sexual characteristics). Puberty occurs when the hypothalamus secretes increasing amounts of releasing hormones that stimulate the anterior pituitary to secrete LH and FSH, which stimulate cells in the gonads to make increasing amounts of sex hormones.

The **adrenal glands** have two parts that secrete different hormones.   The interior **adrenal medulla** contains secretory cells of nervous system origin and control and produces two amino acid–derived hormones: **epinephrine** (*adrenaline*) and **norepinephrine** (*noradrenaline*). These hormones prepare the body for emergency action; they increase heart and respiratory rates, increase blood glucose levels, and direct blood flow away from digestion and toward muscles and the brain. The outer adrenal layer forms the **adrenal cortex**, which secretes three types of steroid hormones called **glucocorticoids**. These hormones help control glucose metabolism. Stressful stimuli (trauma, infection, hot or cold temperatures) stimulate the hypothalamus to secrete releasing hormones, which then stimulate the anterior pituitary to release ACTH, which in turn stimulates the adrenal cortex to release glucocorticoid hormones. The adrenal cortex also secretes **aldosterone**, which regulates blood sodium content. If blood sodium levels fall, the adrenal cortex releases aldosterone, which causes kidneys and sweat glands to retain sodium. When blood sodium levels rise to normal, aldosterone secretion ceases. The adrenal cortex also secretes small amounts of testosterone. Adrenal tumors in females can lead to excessive testosterone release, causing masculinization of women.

Many types of cells produce **prostaglandins**, which are modified fatty acids derived from plasma membranes. Many types of prostaglandins are known. One type causes arteries to constrict to stop bleeding from umbilical cords; another type stimulates uterine contractions during labor. Overproduction of uterine prostaglandins causes menstrual cramping, some cause inflammation and stimulate pain receptors, others expand the air passages of the lungs, and still others stimulate the production of the protective mucus that lines the stomach.

The **pineal gland** lies near the brain and produces the amino acid–derived hormone **melatonin**, which is secreted in a daily rhythm regulated by the eyes of mammals. Melatonin appears to regulate the seasonal reproductive cycles of many mammals. Melatonin may influence sleep-wake cycles in humans.

The **thymus** is in the chest cavity and produces white blood cells and the hormone **thymosin**, which stimulates production of T-cells that play a role in the immune system. The kidneys produce the hormone **angiotensin** from blood proteins; it raises blood pressure by constricting arterioles and stimulates aldosterone release by the adrenal cortex, causing the kidneys to retain sodium. The heart produces a peptide hormone called **atrial natriuretic peptide (ANP),** which is released by cells of the atria when blood volume increases. The ANP causes a reduction in blood presure by decreasing the release of ADH and aldosterone. The stomach and small intestine produce a variety of peptide hormones that help regulate digestion (**gastrin, secretin,** and **cholecystokinin**).

KEY TERMS AND CONCEPTS

Fill-In: From the following list of key terms, fill in the following statements.

adrenocorticotropic	growth	pineal
amino acid derivatives	hormone	prolactin
antidiuretic	insulin	prostglandins
cyclic AMP	luteinizing	second messenger
endocrine	melanocyte-stimulating	steroid
endorphins	melatonin	target
exocrine	oxytocin	thymosin
follicle-stimulating	pancreas	thymus
glucagon	peptide	thyroid-stimulating

A (1)_____ is a chemical secreted by cells in one part of the body that is transported in the bloodstream to other parts of the body, where it affects particular (2)_____ cells.

Hormones can be grouped into four general classes. (3)_____ hormones are made of chains of amino acids. (4)_____ are formed from single amino acids (for example, epinephrine is derived from tyrosine). Most (5)_____ hormones are derived from cholesterol, and are secreted by the ovaries, testes, and adrenal cortex. (6)_____ are composed of two fatty acid carbon chains attached to a five-carbon ring.

Most hormones binding to a receptor on the outside cell surface trigger the release inside the cell of a chemical called the (7)_____ that initiates a cascade of biochemical reactions. (8)_____ is a common second messenger.

Mammals have both (9)_____ glands, which produce secretions released outside the body or into the digestive tract through tubes or openings called ducts, and (10)_____ glands, which are ductless glands secreting hormones into capillaries of the bloodstream.

The anterior pituitary produces and releases a variety of peptide hormones, four of which regulate hormone release in other endocrine glands. (11)_____ hormone and (12)_____ hormone stimulate production of sperm and testosterone in males and eggs, estrogen, and progesterone in females. (13)_____ hormone stimulates release of thyroid gland hormones. (14)_____ hormone stimulates release of hormones from the adrenal cortex.

Other hormones of the anterior pituitary do not act on other endocrine glands. (15)_____ helps stimulate development of the mammary glands during pregnancy. (16)_____ inhibit the perception of pain by binding to brain receptors. (17)_____ hormone stimulates production of the skin pigment melanin. (18)_____ hormone regulates body growth: too little causes dwarfism, too much can cause gigantism).

The posterior pituitary releases two types of peptide hormones produced by cells in the hypothalamus: (19)_____ hormone, which helps prevent dehydration by increasing the permeability of the collecting ducts of kidney nephrons (causing water to be reabsorbed from the urine and retained in the body), and (20)_____, which triggers the "milk letdown reflex" in nursing mothers by causing breast muscle tissue to contract during breast feeding.

In the (21)_____, clusters of islet cells produce the peptide hormones (22)_____, which reduces blood glucose levels, and (23)_____, which increases blood glucose levels by activating a liver enzyme that breaks down glycogen.

The (24)_____ gland lies near the brain and produces the amino acid-derived hormone (25)_____, which is secreted in a daily rhythm regulated by the eyes of mammals.

The (26)_____ is in the chest cavity and produces white blood cells and the hormone (27)_____, which stimulates production of T-cells that play a role in the immune system.

Key Terms and Definitions

adrenal cortex: the outer part of the adrenal gland, which secretes steroid hormones that regulate metabolism and salt balance.

adrenal gland: a mammalian endocrine gland, adjacent to the kidney; secretes hormones that function in water regulation and in the stress response

adrenal medulla: the inner part of the adrenal gland, which secretes epinephrine (adrenalin) and norepinephrine (noradrenaline).

adrenocorticotropic hormone: a hormone, secreted by the anterior pituitary, that stimulates the release of hormones by the adrenal glands, especially in response to stress.

aldosterone: a hormone, secreted by the adrenal cortex, that helps regulate ion concentration in the blood by stimulating the reabsorption of sodium.

androgen: the collective term for male sex hormones.

angiotensin: a hormone that functions in water regulation in mammals by stimulating physiological changes that increase blood volume and blood pressure.

anterior pituitary: a lobe of the pituitary gland that produces prolactin and growth hormone as well as hormones that regulate hormone production in other glands.

antidiuretic hormone (ADH): a hormone produced by the hypothalamus and released into the bloodstream by the posterior pituitary when blood volume is low; increases the permeability of the distal tubule and the collecting duct to water, allowing more water to be reabsorbed into the bloodstream.

atrial natriuretic peptide (ANP) : a hormone, secreted by cells in the mammalian heart, that reduces blood volume by inhibiting the release of ADH and aldosterone.

calcitonin: a hormone, secreted by the thyroid gland, that inhibits the release of calcium from bone.

cholecystokinin: a digestive hormone, produced by the small intestine, that stimulates the release of pancreatic enzymes.

cyclic AMP: a cyclic nucleotide formed within many target cells as a result of the reception of modified amino acid or protein hormones, which causes metabolic changes in the cell; often called a second messenger.

diabetes mellitus: a disease characterized by defects in the production, release, or reception of insulin, characterized by high blood glucose levels that fluctuate with sugar intake.

duct: a tube or opening through which exocrine secretions are released.

endocrine gland: a ductless, hormone-producing gland consisting of cells that release their secretions into the extracellular fluid from which the secretions diffuse into nearby capillaries.

endocrine system: an animal's organ system for cell-to-cell communication, composed of hormones and the cells that secrete them and receive them.

endorphin: one of a group of peptide neuromodulators in the vertebrate brain that, by reducing the sensation of pain, mimics some of the actions of opiates.

epinephrine: a hormone, secreted by the adrenal medulla, that is released in response to stress and that stimulates a variety of responses, including the release of glucose from skeletal muscle and an increase in heart rate.

erythropoietin: a hormone produced by the kidneys in response to oxygen deficiency that stimulates the production of red blood cells by the bone marrow.

estrogen: in vertebrates, a female sex hormone, produced by follicle cells of the ovary, that stimulates follicle development, oogenesis, the development of secondary sex characteristics, and growth of the uterine lining.

exocrine gland: a gland that releases its secretions into ducts that lead to the outside of the body or into the digestive tract.

follicle-stimulating hormone (FSH): a hormone, produced by the anterior pituitary, that stimulates spermatogenesis in males and the development of the follicle in females.

gastrin: a hormone, produced by the stomach, that stimulates acid secretion in response to the presence of food.

glucagon: a hormone, secreted by the pancreas, that increases blood sugar by stimulating the breakdown of glycogen (to glucose) in the liver.

glucocorticoid: a class of hormones, released by the adrenal cortex in response to the presence of ACTH, that make additional energy available to the body by stimulating the synthesis of glucose.

goiter: a swelling of the neck caused by iodine deficiency, which affects the functioning of the thyroid gland and its hormones.

growth hormone: a hormone, released by the anterior pituitary, that stimulates growth, especially of the skeleton.

hormone: a chemical synthesized by one group of cells, secreted, and carried in the bloodstream to other cells, whose activity is influenced by reception of the hormone.

hypothalamus: a region of the brain that controls the secretory activity of the pituitary gland, and also synthesizes, stores, and releases certain peptide hormones.

inhibiting hormone: a hormone secreted by the neurosecretory cells of the hypothalamus that inhibits the release of specific hormones from the anterior pituitary gland.

insulin: a hormone, secreted by the pancreas, that lowers blood sugar by stimulating the conversion of glucose to glycogen in the liver.

islet cells: clusters of cells in the endocrine portion of the pancreas that produce insulin and glucagon.

luteinizing hormone (LH): a hormone, produced by the anterior pituitary, that stimulates testosterone production in males and the development of the follicle, ovulation, and the production of the corpus luteum in females.

melanocyte-stimulating hormone (MSH): a hormone, released by the anterior pituitary, that regulates the activity of skin pigments in some vertebrates.

melatonin: a hormone, secreted by the pineal gland, that is involved in the regulation of circadian cycles.

neurosecretory cell: a specialized nerve cell that synthesizes and releases hormones.

norepinephrine: a neurotransmitter, released by neurons of the parasympathetic nervous system, that prepares the body to respond to stressful situations; also called noradrenaline.

ovary: in animals, the gonad of females.

oxytocin: a hormone, released by the posterior pituitary, that stimulates the contraction of uterine and mammary gland muscles.

pancreas: a combined exocrine and endocrine gland located in the abdominal cavity. Its endocrine portion secretes the hormones insulin and glucagon, which regulate glucose concentrations in the blood.

parathormone: a hormone, secreted by the parathyroid gland, that stimulates the release of calcium from bones.

parathyroid glands: a set of four small endocrine glands embedded in the surface of the thyroid gland that produce parathormone, which (with calcitonin from the thyroid) regulates calcium ion concentration in the blood.

peptide hormone: a hormone consisting of a chain of amino acids. Small proteins that function as hormones are included in this category.

pineal gland: a small gland within the brain that secretes melatonin. The pineal controls the seasonal reproductive cycles of some mammals.

pituitary gland: an endocrine gland located at the base of the brain that produces several hormones, many of which influence the activity of other glands.

posterior pituitary: a lobe of the pituitary gland that is an outgrowth of the hypothalamus and releases antidiuretic hormone and oxytocin.

progesterone: a hormone, produced by the corpus luteum, that promotes the development of the uterine lining in females.

prolactin: a hormone, released by the anterior pituitary, that stimulates milk production in human females.

prostaglandin: a family of modified fatty acid hormones manufactured by many cells of the body.

releasing hormone: a hormone secreted by the hypothalamus that causes the release of specific hormones by the anterior pituitary gland.

renin: an enzyme that is released (in mammals) when blood pressure and/or sodium concentration in the blood drops below a set point; initiates a cascade of events that restores blood pressure and sodium concentration.

second messenger: a term applied to intracellular chemicals, such as cyclic AMP, that are synthesized or released within a cell in response to the binding of a hormone or neurotransmitter (the first messenger) to receptors on the cell surface. Second messengers bring about specific changes in the metabolism of the cell.

secretin: a hormone, produced by the small intestine, that stimulates the production and release of digestive secretions by the pancreas and liver.

steroid hormone: a class of hormone whose chemical structure resembles cholesterol that is secreted by the ovaries and placenta, the testes, and the adrenal cortex.

target cell: a cell upon which a particular hormone exerts its effect.

testis (pl., testes): the gonad of male mammals.

testosterone: in vertebrates, a hormone produced by the interstitial cells of the testis; stimulates spermatogenesis and the development of male secondary sex characteristics.

thymosin: a hormone, secreted by the thymus, that stimulates the maturation of cells of the immune system

thymus: a gland located in the upper chest in front of the heart. The thymus functions in the immune system by secreting thymosin, which stimulates lymphocyte maturation. It begins to degenerate at puberty and has little function in the adult.

thyroid gland: an endocrine gland, located in front of the larynx in the neck, that secretes the hormones thyroxine (affecting metabolic rate) and calcitonin (regulating calcium ion concentration in the blood).

thyroid-stimulating hormone (TSH): a hormone, released by the anterior pituitary, that stimulates the thyroid gland to release hormones.

thyroxine: a hormone, secreted by the thyroid gland, that stimulates and regulates metabolism.

THINKING THROUGH THE CONCEPTS

True or False: Determine if the statement given is true or false. If it is false, change the underlined word so that the statement reads true.

28. _____   Endocrine glands usually have ducts.

29. _____   The action of hormones is dependent upon gland cells.

30. _____   Endocrine glands produce chemicals that exert their effects outside the body of the animal producing them.

31. _____   Neurohormones are hormones that affect nerve cells.

32. _____   Peptide hormones may enter cells more easily.

33. _____   Animals regulate hormone release through positive feedback.

34. _____   An increase in antidiuretic hormone will decrease blood pressure.

35. _____   The anterior pituitary is more like part of the brain than an endocrine gland.

36. _____   The posterior pituitary produces the most types of hormones.

37. _____   The breakdown of glycogen into glucose is favored by the presence of insulin.

38. _____   Insulin increases blood glucose.

39. _____   Adrenalin and noradrenalin stimulate the parasympathetic nervous system.

40. _____   Adrenalin causes blood to flow toward the stomach.

41. _____   The adrenal medulla secretes a female hormone.

42. _____   Aspirin enhances the effects of prostaglandins.

Matching: Types of glands.

43. _____   clusters of hormone-producing cells embedded in capillaries

44. _____   some release hormones into ducts leading to the outside of the body

45. _____   sweat glands

46. _____   release hormones into extracellular spaces surrounded by capillaries

47. _____   mammary glands

Choices:

a. endocrine glands

b. exocrine glands

Matching: Pituitary hormones. There may be more than one correct answer.

48.____ affect sex cell release in both sexes

49.____ produced by the posterior pituitary

50.____ regulates growth of bones

51.____ increases water permeability by the kidney nephrons, reducing dehydration

52.____ stimulates development of the mammary glands during pregnancy

53.____ causes uterine contraction during childbirth

54.____ stimulates release of hormone by the thyroid gland

55.____ allows milk to flow from the breasts during lactation

56.____ causes release of hormones from the cortex of the adrenal glands

57.____ allows ejaculation to occurs in males

Choices:

a. ACTH

b. ADH

c. FSH

d. LH

e. growth hormone

f. oxytocin

g. prolactin

h. TSH

Matching: Adrenal gland hormones. There may be more than one correct answer.

58.____ produced by the adrenal medulla

59.____ release is stimulated by ACTH

60.____ help control glucose metabolism

61.____ male sex hormone

62.____ acts like glucagon

63.____ secretion regulated by sodium levels in the blood

64.____ over production, caused by certain tumors, can result in "bearded ladies"

Choices:

a. adrenalin

b. aldosterone

c. glucocorticoids

d. noradrenalin

e. testosterone

Multiple Choice: Pick the most correct choice for each question.

65. All the following are chemical types of animal hormones **except**
   a. steroids
   b. neurotransmitters
   c. prostaglandins
   d. proteins

66. An example of a gland with exocrine as well as endocrine function is the
   a. stomach
   b. pituitary
   c. pancreas
   d. thyroid
   e. kidney

67. The pituitary gland is located
    a.   in the side of the neck
    b.   near the kidneys
    c.   in the upper thoracic cavity
    d.   at the base of the brain
    e.   near the sex organs

68. During childbirth, contraction of the uterus
    is stimulated by
    a.   oxytocin
    b.   prolactin
    c.   progesterone
    d.   estrogen

69. Growth of the body is regulated partially by
    a.   secretin
    b.   insulin
    c.   parathormone
    d.   somatotrophin
    e.   thyroxine

CLUES TO APPLYING THE CONCEPTS

This practice question is intended to sharpen your ability to apply critical thinking and analysis to biological concepts covered in this chapter.

70. Antidiuretic hormone (ADH) acts on very specific target cells in the kidney, while the insulin hormone affects every cell in the body. Why is it that some hormones are very specific as to which cells they affect and other hormones have a very general effect on many target organs?

    _____
    _____
    _____
    _____
    _____
    _____
    _____
    _____
    _____

ANSWERS TO EXERCISES

1. hormone
2. target
3. Peptide
4. Amino acid derivatives
5. steroid
6. Prostglandins
7. second messenger
8. Cyclic AMP
9. exocrine
10. endocrine
11. Follicle-stimulating
12. Luteinizing
13. Thyroid-stimulating
14. Adrenocorticotropic
15. Prolactin
16. Endorphins
17. Melanocyte-stimulating
18. Growth
19. antidiuretic
20. oxytocin
21. pancreas
22. insulin
23. glucagon

24. pineal
25. melatonin
26. thymus
27. thymosin
28.. false, do not have
29. false, target
30. false, exocrine
31. false, are produced by
32. false, steroid
33. false, negative
34. false, increase
35. false, posterior
36. false, anterior
37. false, glucagon
38. false, decreases
38. false, sympathetic
40. false, away from
41. false, male
42. false, counteracts
43. a
44. b
45. b
46. a

47. b
48. c, d
49. b, f
50. e
51. b
52. g
53. f
54. h
55. f
56. a
57. f
58. a, d
59. c
60. a, c, d
61. e
62. c
63. b
64. e
65. b
66. c
67. d
68. a
69. d

70. An organ can respond to the presence of a hormone only if it has the proper receptor molecule in its cell membrane or internally. Only kidney cells have the specific receptor for ADH, while most cells have receptors for insulin.

Chapter 33: The Nervous System and the Senses

OVERVIEW

This chapter covers the structure and function of nerve cells, the nature of resting potentials in nerves, and how action potentials are generated and conducted. The authors discuss the human nervous system, including the central nervous system (brain and spinal cord), the peripheral nervous system, neurotransmitters, and neuromodulators. The ways animals perceive and respond to nervous stimulation are discussed, and the structures and functions of the major sense organs are covered. Learning, memory, and retrieval are briefly covered.

1) How Do Nervous and Endocrine Communication Differ?

Both hormone-producing cells and nerve cells make "messenger" chemicals that they release into extracellular spaces. However, there are four differences in how the nervous and endocrine systems use chemical messages: (1) Endocrine cells release hormones into the bloodstream, while nerve cells release their products (*neurotransmitters*) very close to the cells they influence; (2) bloodborne hormones bathe many cells indiscriminately, but a nerve cell releases its neurotransmitter onto one or a few specific cells; (3) nerve cells speed information from one place to another by electrical signals that travel within the cell itself, and release neurotransmitters only near the target cell, while hormones move slowly and usually are released far away from the target cells; and (4) the effects of messages sent by nerve cells (*neurons*) are of much shorter duration than are the effects of hormones. However, the endocrine and nervous systems are closely coordinated in their control of bodily functions.

2) What Are the Structures and Functions of Neurons?

Individual nerve cells are **neurons**, each of which has four functions besides normal metabolism: (1) receive information; (2) integrate information and produce an appropriate output signal; (3) conduct the signal to its terminal ending; and (4) transmit the signal to other nerve cells, glands, or muscles. A typical vertebrate neuron has four structural regions called dendrites, cell body, axon, and synaptic terminals. **Dendrites** (branched tendrils with large surface area that extend outward from the nerve body) receive signals from other neurons or from the environment. Dendrites of *sensory neurons* have specialized membranes to respond to specific stimuli (pressure, odor, light, heat). Dendrites of brain and spinal cord neurons respond to chemical neurotransmitters released by other neurons. Electrical signals travel down the dendrites to the neuron's integration center, the **cell body**. The cell body integrates all the signals received and, if this is sufficiently positive, the neuron will produce an **action potential** or electrical output signal. The cell body also performs typical metabolic activities. A long, thin fiber called an **axon** extends from the cell body. The action potential begins at the **spike initiation zone**, a site where the axon leaves the cell body. Some axons are three feet long. Axons carry action potentials to the *synaptic terminals* located at the far ends of each axon. Axons are normally bundled together into **nerves**, like bundles of wires in an electrical cable. Some axons are wrapped with insulation (**myelin**), which allows very rapid conduction of the electrical signal. Signals are conducted to other cells at

synaptic terminals, swellings at the branched ends of axons. Most synaptic terminals contain a **neurotransmitter** chemical that is released in response to the passage of an action potential down the axon. The output of the first cell's axon becomes the input of the next cell's dendrite. The site where synaptic terminals communicate with each other is called the **synapse**.

3) How Is Neural Activity Generated and Communicated?

Neurons create electrical signals across their membranes. Unstimulated, inactive neurons maintain a constant electrical difference, or *potential*, across their cell membranes. This **resting potential** is always negative inside the cell (–40 to –90 millivolts). If the neuron is stimulated, the inner negative potential can be altered. If it becomes sufficiently less negative, it reaches a **threshold** (usually 15 millivolts less negative than resting potential). At threshold, an action potential is triggered at a spike initiation zone and the neuron's potential rapidly rises to +50 millivolts inside, after which the cell's normal resting potential is restored. The positive charge of the action potential flows rapidly down the axon to the synaptic terminal, where the signal is communicated to another cell at the synapse. The signals transmitted at synapses are called **postsynaptic potentials** (PSPs). The **synaptic cleft** is a tiny gap separating the synaptic terminal of the first neuron (the **presynaptic neuron**) from the second or **postsynaptic neuron**. When an action potential reaches a synaptic terminal, the inside of the terminal becomes positively charged, triggering storage vesicles in the synaptic terminal to release a chemical neurotransmitter into the synaptic cleft. These molecules rapidly diffuse across the gap and bind briefly to membrane receptors of the postsynaptic neuron. Receptor proteins in the postsynaptic membrane bind to a specific type of neurotransmitter, causing ion-specific channels in the postsynaptic membrane to open to allow ions to flow across the plasma membrane along their concentration gradients. This causes a small, brief change in electrical charge called the PSP. Depending on which channels are opened and which ions flow, PSPs can be *excitatory* (EPSPs), making the neuron less negative inside and more likely to produce an action potential, or *inhibitory* (IPSPs), making it more negative and less likely to produce an action potential. The PSPs travel to the cell body where they determine whether an action potential will be produced. The dendrites and cell bodies of one neuron can receive EPSPs and IPSPs from the presynaptic terminals of thousands of presynaptic neurons. These are then "added up" or **integrated** in the cell body of the postsynaptic neuron, producing an action potential only if the EPSPs and IPSPs collectively raise the electrical potential inside the neuron above threshold.

The nervous system uses over 50 different neurotransmitters and **neuropeptides**. **Acetylcholine** is the only neurotransmitter released at the synapses between neurons and skeletal muscles, where it is always excitatory. Curare blocks acetylcholine receptors, preventing muscle contraction and causing paralysis. Degeneration of **dopamine**-producing neurons in the brain leads to Parkinson's disease (uncontrolled tremors). Schizophrenia is treated successfully with drugs that block dopamine receptors. **Serotonin** acts in the brain and spinal cord; it can inhibit pain sensory neurons in the spinal cord. Animals with blocked serotonin production are unable to sleep normally, and too little serotonin may cause depression. **Norepinephrine** (*noradrenaline*) is released by neurons of the sympathetic nervous system onto many body organs when fright occurs; both amphetamines and cocaine prolong the effect of norepinephrine. Examples of neuropeptides are **opiates** such as **endorphin**, morphine, opium, codeine, and heroin, all of which reduce pain.

4) How Is the Nervous System Designed?

Information processing requires four basic operations: (1) determine the type of stimulus; (2) signal the intensity of a stimulus; (3) integrate information from many sources; and (4) initiate and direct the response. The type of stimulus is distinguished by wiring patterns in the brain. The nervous system monitors which neurons are firing action potentials. The brain interprets optic nerve action potentials as light, and olfactory nerve action potentials as odors, etc. The **intensity** or strength of a stimulus is coded by the frequency of action potentials. The most intense the stimulus , the faster the neuron produces action potentials, and the larger is the number of neurons that respond. Thus loud noise stimulates a larger number of auditory nerves to fire more rapidly than does less intense sound. The nervous system processes information from many sources through **convergence** (many neurons funnel their signals to fewer decision-making neurons in the brain). Complex responses occur through **divergence** of signals, the flow of electrical signals from a relatively small number of decision-making cells in the brain onto many different neurons that control muscle or glandular activity.

Neuron-to-muscle pathways direct behavior and are composed of four elements: (1) **Sensory neurons** respond to a stimulus; (2) **Association neurons** receive signals from many sources and activate motor neurons; (3) **Motor neurons** receive instructions from association neurons and activate muscles or glands; and (4) **Effectors**, usually muscles or glands that perform the response directed by the nervous system. The simplest behavior is the **reflex**, a largely involuntary response produced by neurons in the spinal cord not requiring interaction with the brain, for example, knee jerk and pain-withdrawal reflexes.

Increasingly complex nervous systems are increasingly centralized. Cnidarians have radial symmetry without a head, and have a diffuse nervous system. They have a **nerve net** of neurons, with occasional clusters of neurons called **ganglia** but no real brain. More advanced animals with bilateral symmetry and **cephalization** (head region with a concentration of sense organs) have a centralized nervous system with a brain containing nearly all the cell bodies of the nervous system.

5) How Is the Human Nervous System Organized?

The human nervous system has two parts: the **central nervous system** (CNS with **brain** and **spinal cord**) extending down the dorsal part of the torso, and **peripheral nervous system** made of nerves that connect the CNS to the rest of the body. The peripheral nervous system (PNS) consists of **peripheral nerves**, which link the brain and spinal cord to the rest of the body. Peripheral nerves contain axons of sensory neurons that bring information to the CNS from the body parts, and axons of motor neurons that carry signals from the CNS to the organs and muscles. The motor portion of the PNS has two parts: (1) the **somatic nervous system** (control skeletal muscles and voluntary muscle movement; the motor neurons are located in the *gray matter* of the spinal cord and their axons go directly to the muscles they control), and (2) the **autonomic nervous system**, made up of motor neurons control involuntary responses, forming synapses on the heart, smooth muscles, and glands. The autonomic nervous system is controlled both by the *medulla* and the *hypothalamus* of the brain. It has two divisions, each making contact with the same organs but producing opposite results: (1) the **sympathetic division** (preparing the body for stressful or highly energetic activities such as "fight or flight") and (2) the **parasympathetic division** (maintains activities that can be done at leisure, such as digestion)

The CNS consists of the spinal cord and brain. The CNS receives and processes information, generates thoughts, and directs response. The CNS contains up to 100 billion association neurons. The brain and spinal cord are protected in three ways: the *skull* and *vertebral column*; a triple layer of connective tissue called **meninges**; and **cerebrospinal fluid** between the layers of the meninges.The spinal cord is a cable of axons protected by the backbone. Between vertebrae, nerves carrying axons of sensory neurons and motor neurons arise from the dorsal and ventral portions of the spinal cord, respectively, and merge to form the peripheral nerves of the spinal cord (part of the PNS). In the center of the spinal cord is the **gray matter** (cell bodies of neurons that control voluntary muscles and the autonomic nervous system, and neurons that communicate with the brain and other parts of the spinal cord). The gray matter is surrounded by **white matter** formed of myelin-coated axons of neurons that extend up and down the spinal cord.

The brain consists of many parts specialized for specific functions. Embryologically, the vertebrate brain begins as a simple tube that develops into three parts: **hindbrain** (autonomic behaviors such as breathing and heart rate), **midbrain** (vision), and **forebrain** (smell). The human hindbrain includes the **medulla** (controls autonomic functions such as breathing, heart rate, blood pressure, and swallowing), the **pons** (influences transitions between sleep and wakefulness and between stages of sleep, and the rate and patterns of breathing), and the **cerebellum** (coordinates movements of the body, and learning and memory storage for behaviors).

The midbrain in humans is small and contains an auditory relay center, a center that controls reflex movements of the eyes, and another relay center, the **reticular formation** (which plays a role in sleep and arousal, emotion, muscle tone, and certain movements and reflexes; it filters sensory inputs before they reach the conscious regions of the brain) passes through it.

The large human forebrain (**cerebrum**) includes the **cerebral cortex**, the **thalamus** (which carries sensory information to and from other forebrain regions), and **limbic system** (which produces our most basic and primitive emotions, drives, and behaviors such as fear, thirst, pleasure, sexual response, and memory formation). The limbic system includes the **hypothalamus** (containing neurons and neurosecretory cells that release hormones into the blood, control the release of hormones from the pituitary gland, and direct the activities of the autonomic nervous system), the **amygdala** (containing neurons producing the sensations of pleasure, punishment, fear, rage, and sexual arousal when stimulated), and the **hippocampus** (playing a role in emotions and in the formation of long-term memory needed for learning). The human cerebral cortex (the outer layer of the forebrain) is the largest part of the brain. It is divided into the **cerebral hemispheres** that communicate through axons making up the **corpus callosum**. To accommodate 100 billion neurons, the cortex forms folds called **convolutions** to increase its area. These neurons receive and process sensory information, store some of it as memory for future use, direct voluntary movements, and are responsible for thinking. The cerebral cortex is divided anatomically into four regions: the frontal, parietal, occipital, and temporal lobes. Damage to the cortex from trauma, stroke, or a tumor results in specific deficits.

6) How Does the Brain Produce the Mind?

The "mind-brain" problem has occupied philosophers and neurobiologists. Some interesting observations follow. Studies of accident, stroke, or surgery patients have revealed that the "left cerebrum" (speech, reading, writing, language comprehension, mathematical ability, and logical problem solving) and "right cerebrum" (musical skills, artistic ability, facial recognition, spatial visualization, and ability to recognize and express emotions) are specialized for different functions.

The mechanics of learning and memory are poorly understood. Memory may be brief or long lasting. Learning occurs in two phases: an initial **working memory** (temporary electrical or biochemical activity in the brain) followed by **long-term memory** (involving brain structure changes such as new long-lasting neural connections or the strengthening of existing connections). Working memory can be converted to long-term memory, perhaps involving the action of the hippocampus. Learning , memory , and retrieval may be controlled by separate regions of the brain. Learning and memory may be controlled by the hippocampus, while retieval seems to be controlled by the temporal lobes.

7) How Do Sensory Receptors Work?

Generally, a **receptor** is a structure that changes when acted upon by a stimulus. All receptors are **transducers** (structures that convert signals from one form to another). A **sensory receptor** may be an entire neuron specialized to produce an electrical response to particular stimuli. Stimulation of a sensory receptor causes a **receptor potential**, an electrical signal whose size is proportional to the strength of the stimulus. Intensity is conveyed to the nervous system by the frequency, not the size, of action potentials. **Thermoreceptors** are free nerve endings that respond to fluctuations in temperature. **Mechanoreceptors** produce a receptor potential in response to stretching of their plasma membranes. **Hair cells** are mechanoreceptors located in the inner ear; these receptors for sound, motion, and gravity produce an action potential when their hairs are bent. **Photoreceptors** (receptors for light), **chemoreceptors** (for chemicals which we perceive as tastes or smells), and **pain receptors** posses specialized membrane receptor proteins.

8) How Is Sound Sensed?

Sound is produced by any vibrating object. The ear structure helps capture, transmit, and transduce sound. The **outer ear** consists of the **external ear** and the **auditory canal**, which conducts sound waves to the **middle ear** consisting of the **tympanic membrane** (eardrum), three tiny bones called the hammer, the anvil, and the stirrup, and the **Eustachian tube** (connects to the pharynx and equalizes the air pressure between the middle ear and the atmosphere). In the middle ear, sound vibrates the tympanic membrane, which in turn vibrates the hammer, anvil, and the stirrup bones, which transmit vibrations to the **inner ear**. The fluid-filled hollow bones of the inner ear form a spiral-shaped **cochlea** and other structures that detect head movement and the pull of gravity. Sounds enters the inner ear when the stirrup vibrates the **oval window** membrane that covers a hole in the cochlea.

The central canal of the cochlea contains the **basilar membrane**, on top of which are receptor or hair cells. Protruding into the central canal is the gelatinous **tectorial membrane** in which the hairs are embedded. Sound perception occurs when the oval window passes vibrations to the fluid in the cochlea, which in turn vibrates the basilar membrane, causing it to move up and down. This bends the hairs embedded in the tectorial membrane, producing receptor potentials in the hair cells, which release transmitter onto neurons of the **auditory nerve**, whose action potentials travel to the brain. The inner ear also allows us to perceive *loudness* (magnitude of sound vibrations) and *pitch* (the frequency of sound vibrations).

9) How is Light Sensed?

All forms of vision use photoreceptors containing receptor molecules called colored **photopigments**, which absorb light and chemically change in the process to alter ion channels in the cell membrane, producing a receptor potential. The arthropods evolved **compound eyes** consisting of a mosaic of many individual light-sensitive subunits called **ommatidia**. Each ommatidium acts as a light detector, collectively producing a grainy image of the world and is particularly good at detecting movement and color perception.

The mammalian eye collects, focuses, and transduces light waves. Incoming light first encounters the transparent **cornea**, behind which is a chamber filled with watery nourishing fluid called **aqueous humor**. The adjustable **iris** (pigmented muscular tissue with a circular opening, the **pupil**) controls how much light enters. Light then strikes the **lens** (a flattened sphere of transparent protein fibers), which is suspended behind the pupil by ligaments and muscles that regulate its shape. Behind the lens is a chamber filled with clear jellylike **vitreous humor**. After passing through the vitreous humor, light reaches the **retina**, a multilayered sheet of photoreceptors and neurons, where light energy is converted into electrical nerve impulses transmitted to the brain. Behind the retina is the **choroid**, a darkly pigmented tissue. Surrounding the outer portion of the eyeball is the **sclera**, tough connective tissue forming the "white" of the eye. The adjustable lens allows focusing of both distant and nearby objects. The visual image is focused most sharply on a small area of the retina called the **fovea**. If your eyeball is too long, you are nearsighted (unable to focus on distant objects); if the eyeball is too short, you are farsighted.

Light striking the retina is captured by photoreceptors; the signal is processed by layers of overlying neurons. The retinal cells closest to the vitreous humor consists of **ganglion cells**, whose axons make up the **optic nerve**. Ganglion cell axons must pass through the retina to reach the brain at a spot called the **optic disc** or **blind spot**. Photoreceptors called **rods** and **cones** are present in the retina. Although cones are located throughout the retina, they are concentrated in the fovea. Human eyes have three types of cones, each containing a slightly different photopigment most strongly stimulated by a different wavelength of light (red, green, and blue).The brain distinguishes color according to the relative intensity of stimulation of different cones. Rods dominate the peripheral portions of the retina. Rods are more sensitive to light than cones and are largely responsible for our vision in dim light. Rods do not distinguish color.

Binocular vision allows depth perception. Most herbivores have one eye on each side of the head. Predators and omnivores (humans) have both eyes facing forward, with slightly different but extensively overlapping visual fields (**binocular vision**) that allows for depth perception. Herbivores have no depth perception but do have nearly 360-degree field of view, an advantage when looking out for predators.

10) How Are Chemicals Sensed?

Through chemical senses, animals may find food, avoid poisons, locate homes, and find mates. Terrestrial animals have two chemical senses: **olfaction** (smell) for airborne molecules and **taste** for molecules dissolved in water or saliva. In vertebrates, receptors for smell are nerve cells with hairy dendrites in a patch of mucus-covered epithelial tissue in the upper portion of each nasal cavity. Taste receptors are located in clusters called **taste buds** on the tongue. Each taste bud has up to 80 taste receptor cells surrounded by supporting cells in a small pit. The four major types of taste receptors are sweet, sour, salty, and bitter. What we call taste is mostly smell. Pain is actually a specialized chemical sense. Most pain is caused by tissue damage. When cells are damaged, their contents flow into the extracellular fluid. Potassium ions in the cell contents stimulate pain receptors. Damaged cells also release enzymes that produce **bradykinin**, another stimulus that activates pain receptors.

KEY TERMS AND CONCEPTS

Fill in the crossword puzzle with key terms, based on the following clues.

Across

1. _____ ear: composed of the bony, fluid-filled tubes of the cochlea and the vestibular apparatus.
5. part of the hindbrain of vertebrates that controls automatic activities such as breathing, swallowing, heart rate, and blood pressure.
6. a long process of a nerve cell, extending from the cell body to synaptic endings on other nerve cells or on muscles.
7. part of the hindbrain of vertebrates coordinating movements of the body.
8. part of the forebrain of vertebrates, involved in production of appropriate behavioral responses to environmental stimuli.

9. portion of the hindbrain just above the medulla containing neurons that influence sleep and the rate and pattern of breathing.
11. part of a nerve cell in which most of the common cellular organelles are located.
14. photoreceptor cell in the vertebrate retina, sensitive to dim light but not involved in color vision.
16. the part of the central nervous system of vertebrates enclosed within the skull.
17. _____canal: a canal within the outer ear that conducts sound from the external ear to the eardrum.
19. the pigmented muscular tissue of the vertebrate

eye that surrounds and controls the size of the pupil.

20. the central region of the vertebrate retina, upon which images are focused.

21. the clear outer covering of the eye in front of the pupil and iris.

Down

2. a single nerve cell.

3. a flexible or movable structure used to focus light upon a layer of photoreceptor cell in eyes.

4. photoreceptor cell in the vertebrate retina; the three types of cones are most sensitive to different colors of light and provide color vision.

6. a neurotransmitter found in the brain and in synapses of motor neurons onto skeletal

muscles.

9. the adjustable opening through which light enters the eye.

10. a bundle of axons of nerve cells, bound together in a sheath.

12. a chemical formed during tissue damage that binds to receptor molecules on pain nerve endings, giving rise to the sensation of pain.

13. branched tendrils that extend outward from the cell body, specialized to respond to signals from the external environment or from other neurons.

15. a transmitter in the brain whose actions are largely inhibitory; loss causes Parkinson's disease.

18. _____window: the membrane-covered entrance to the inner ear.

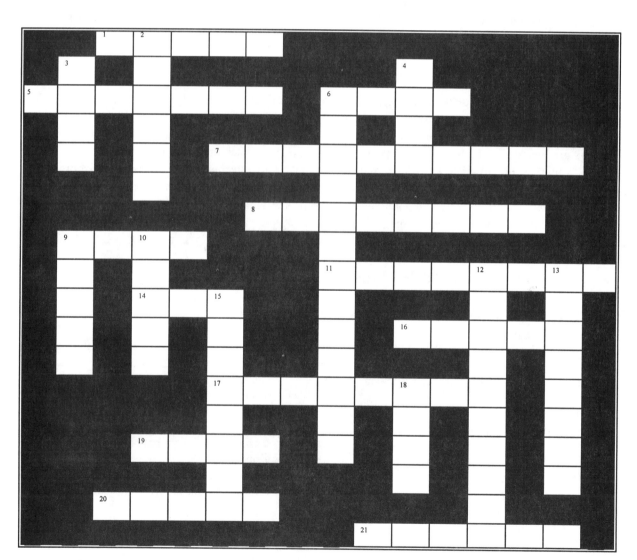

Key Terms and Definitions

acetylcholine: a neurotransmitter found in the brain and in synapses of motor neurons onto skeletal muscles.

action potential: a rapid change from a negative to a positive electrical potential in a nerve cell. This signal travels along an axon without change in size.

amygdala: part of the forebrain of vertebrates, involved in production of appropriate behavioral responses to environmental stimuli.

aqueous humor: the clear, watery fluid between the cornea and lens of the eye.

association neuron: in neural networks, a nerve cell that is postsynaptic to a sensory neuron and presynaptic to a motor neuron. In actual circuits, there maybe many association neurons between individual sensory and motor neurons.

auditory canal: a canal within the outer ear that conducts sound from the external ear to the eardrum.

auditory nerve: the nerve leading from the mammalian cochlea to the brain, carrying information about sound.

autonomic nervous system: part of the peripheral nervous system of vertebrates that synapses on glands, internal organs, and smooth muscle and produces largely involuntary responses.

axon: a long process of a nerve cell, extending from the cell body to synaptic endings on other nerve cells or on muscles.

basilar membrane: a membrane in the cochlea that bears hair cells that respond to the vibrations produced by sound.

binocular vision: the ability to see objects simultaneously through both eyes, providing greater depth perception and more accurate judgment of the size and distance of an object from the eyes.

blind spot: see optic disc.

bradykinin: a chemical formed during tissue damage that binds to receptor molecules on pain nerve endings, giving rise to the sensation of pain.

brain: the part of the central nervous system of vertebrates enclosed within the skull.

cell body: part of a nerve cell in which most of the common cellular organelles are located. Also often a site of integration of inputs to the nerve cell.

central nervous system: in vertebrates, the brain and spinal cord.

cephalization: the increasing concentration over evolutionary time of sensory structures and nerve ganglia at the anterior end of animals.

cerebellum: part of the hindbrain of vertebrates, concerned with coordinating movements of the body.

cerebral cortex: a thin layer of neurons on the surface of the vertebrate cerebrum, in which most neural processing and coordination of activity occurs.

cerebral hemisphere: one of two nearly symmetrical halves of the cerebrum, connected by a broad band of axons, the corpus callosum.

cerebrospinal fluid: a clear fluid produced within the ventricles of the brain that fills the ventricles and cushions the brain and spinal cord.

cerebrum: part of the forebrain of vertebrates concerned with sensory processing, direction of motor output, and coordination of most bodily activities. The cerebrum consists of two nearly symmetrical halves (the hemispheres) connected by a broad band of axons, the corpus callosum.

chemoreceptor: a sensory receptor that responds to chemicals from the environment; used in the chemical senses of taste and smell.

choroid: a darkly pigmented layer of tissue behind the retina that contains blood vessels and pigment that absorbs stray light.

cochlea: a coiled, bony, fluid-filled tube found in the mammalian inner ear, which contains receptors (hair cells) that respond to the vibration of sound.

compound eye: a type of eye found in arthropods, composed of numerous independent subunits, called ommatidia. Each ommatidium apparently contributes a single piece of a mosaic-like image perceived by the animal.

cone: a cone-shaped photoreceptor cell in the vertebrate retina, not as sensitive to light as the rods. The three types of cones are most sensitive to different colors of light and provide color vision. See also *rod*.

convergence: a condition in which a large number of nerve cells provide input to a smaller number of cells.

convolution: folding of the cerebral cortex of the vertebrate brain.

cornea: the clear outer covering of the eye in front of the pupil and iris.

corpus callosum: the band of axons that connect the two cerebral hemispheres of vertebrates.

dendrite: branched tendrils that extend outward from the cell body, specialized to respond to signals from the external environment or from other neurons.

divergence: a condition in which a small number of nerve cells provide input to a larger number of cells.

dopamine: a transmitter in the brain whose actions are largely inhibitory. Loss of dopamine-containing neurons causes Parkinson's disease.

dorsal root ganglion: a ganglion located on the dorsal (sensory) branch of each spinal nerve, containing the cell bodies of sensory neurons.

echolocation: use of ultrasonic sounds, which bounce back from nearby objects, to produce an "auditory image" of nearby surroundings; used by bats and porpoises.

effector: a part of the body (usually a muscle or gland) that carries out responses as directed by the nervous system.

electrolocation: the production of high-frequency electrical signals from an electric organ in front of the tail of weak electrical fish; used to detect and locate nearly objects.

endorphin: one of a group of peptide neuromodulators in the vertebrate brain that, by reducing the sensation of pain, mimics some of the actions of opiates.

Eustachian tube: a tube connecting the middle ear with the pharynx that allows pressure between the middle ear and the atmosphere to equilibrate.

external ear: the fleshy portion of the ear that extends outside the skull.

forebrain: during development, the anterior portion of the brain. In mammals, the forebrain differentiates into the thalamus, the limbic system, and the cerebrum. In humans, the cerebrum contains about half of all the neurons in the brain.

fovea: the central region of the vertebrate retina, upon which images are focused. The fovea contains closely packed cones.

ganglion: a cluster of neurons.

ganglion cell: a type of cell comprising the innermost layer of the vertebrate retina whose axons form the optic nerve.

gray matter: outer portions of the brain and inner region of the spinal cord composed largely of neuron cell bodies which give these areas a gray color.

hair cell: the type of receptor cell found in the inner ear. Hair cells bear hairlike projections, the bending of which between two membranes causes the receptor potential.

hindbrain: the posterior portion of the brain, containing the medulla, pons, and cerebellum.

hypothalamus: part of the forebrain of vertebrates, located just below the thalamus, involved in regulation of hormone production, autonomic nervous system responses, and many behaviors and emotions.

inner ear: the innermost part of the mammalian ear, composed of the bony, fluid-filled tubes of the cochlea and the vestibular apparatus.

integration: in nerve cells, the process of adding up electrical signals from sensory inputs or other nerve cells, to determine the appropriate outputs.

intensity: the strength of stimulation or response.

iris: the pigmented muscular tissue of the vertebrate eye that surrounds and controls the size of a central opening, the pupil, through which light enters.

lens: a clear object that bends light rays; in eyes, a flexible or movable structure used to focus light upon a layer of photoreceptor cells.

limbic system: a diverse group of brain structures, mostly in the lower forebrain, including the thalamus, hypothalamus, amygdala, hippocampus, and parts of the cerebrum, involved in basic emotion, drives, and behaviors, and in learning.

long-term memory: the second phase of learning; a more-or-less permanent memory formed by a structural change in the brain, brought on by repetition.

mechanoreceptor: a receptor that responds to mechanical deformation, such as is caused by pressure, touch, or vibration.

medulla: part of the hindbrain of vertebrates that controls automatic activities such as breathing, swallowing, heart rate, and blood pressure.

meninges: three layers of connective tissue that surround the brain and spinal cord.

midbrain: during development, the central portion of the brain, which contains an important relay center, the reticular formation.

middle ear: part of the mammalian ear composed of the tympanic membrane, the eustachian tube, and three bones that transmit vibrations from the auditory canal to the oval window.

motor neuron: a neuron that receives instructions from the association neurons and activates effector organs such as muscles or glands.

myelin: a wrapping of insulating membranes of specialized nonneural cells around the axon of a vertebrate nerve cell. Myelin increases the speed of conduction of action potentials.

nerve: a bundle of axons of nerve cells, bound together in a sheath.

nerve net: a simple form of nervous system consisting of a network of neurons that extend throughout the tissues of an organism, such as a cnidarian.

neuron: a single nerve cell.

neuropeptide: a small protein molecule with neurotransmitter-like actions.

neurotransmitter: a chemical released by a presynaptic cell at a synapse, which binds to receptors on the postsynaptic cell, causing changes in the electrical potential of the second cell.

norepinephrine: also called noradrenaline, a neurotransmitter which is released by neurons of the parasympathetic nervous system and prepares the body to respond to stressful situations.

olfaction: a chemical sense, the sense of smell; in terrestrial vertebrates the result of detection of airborne molecules.

ommatidium: individual light-sensitive subunit of a compound eye. Each ommatidium consists of a lens and several receptor cells.

opioid: a group of peptide neuromodulators found in the vertebrate brain that mimics some of the actions of opiates (such as opium) and also seems to influence many other processes, including emotion and appetite.

optic disk: the area of the retina at which the axons of the ganglion cell merge to form the optic nerve; the blind spot of the retina.

optic nerve: the nerve leading from the eye to the brain, carrying visual information.

outer ear: the outermost part of the mammalian ear, including the external ear and auditory canal leading to the tympanic membrane.

oval window: the membrane-covered entrance to the inner ear.

pain receptor: a receptor that has extensive areas of membranes studded with special receptor proteins that respond to light or to a chemical.

parasympathetic division: the division of the autonomic nervous system that produces largely involuntary responses related to maintenance of normal body functions, such as digestion.

peripheral nerve: a nerve that links the brain and spinal cord to the rest of the body.

peripheral nervous system: in vertebrates, that part of the nervous system connecting the central nervous system to the rest of the body.

photopigment: a chemical substance in photoreceptor cells that changes molecular conformation when struck by light.

photoreceptors: receptor cells that respond to light; in vertebrates, rods and cones.

pons: a portion of the hindbrain just above the medulla containing neurons that influence sleep and the rate and pattern of breathing.

postsynaptic neuron: the nerve cell at a synapse that changes its electrical potential in response to a chemical (the neurotransmitter) released by another (presynaptic) cell.

postsynaptic potential (PSP): an electrical signal produced in a postsynaptic cell by transmission across the synapse. It may be excitatory (EPSP), making the cell more likely to produce an action potential, or inhibitory (IPSP), tending to inhibit an action potential.

presynaptic neuron: a nerve cell that releases a chemical (the neurotransmitter) at a synapse, which causes changes in the electrical activity of another (postsynaptic) cell.

pupil: the adjustable opening in the center of the iris through which light enters the eye.

receptor: (1) a cell that responds to an environmental stimulus (chemicals, sound, light, pH, etc.) by changing its electrical potential; (2) a protein molecule in a cell membrane that binds to another molecule (hormone, neurotransmitter) triggering metabolic or electrical changes in a cell.

receptor potential: an electrical potential change in a receptor cell produced in response to reception of an environmental stimulus (chemicals, sound, light, heat, etc.). The size of the receptor potential is proportional to the intensity of the stimulus.

reflex: a simple, stereotyped movement of part of the body that occurs automatically in response to a stimulus.

resting potential: a negative electrical potential found in unstimulated nerve cells.

reticular formation: a diffuse network of neurons extending from the hindbrain, through the midbrain, and into the lower reaches of the forebrain, involved in filtering sensory input and regulating what information is relayed to higher brain centers for further attention.

retina: a multi-layered sheet of nerve tissue at the rear of camera-type eyes, composed of photoreceptor cells plus associated nerve cells that refine the photoreceptor information and transmit it to the optic nerve.

rod: a rod-shaped photoreceptor cell in the vertebrate retina, sensitive to dim light but not involved in color vision. See also *cone*.

sclera: a tough white connective tissue layer that covers the outside of the eyeball and forms the white of the eye.

sensory neuron: a nerve cell that responds to a stimulus, either from the internal or external environment.

sensory receptor: a cell specialized to respond to particular internal or external environmental stimuli by producing an electrical potential.

serotonin: a neurotransmitter in the central nervous system, which is involved in mood, sleep, and the inhibition of pain.

somatic nervous system: that portion of the peripheral nervous system that controls voluntary movement by activating skeletal muscles.

spike initiation zone: on a neuron, the site where the action potential begins; where the axon leaves the cell body.

spinal cord: part of the central nervous system of vertebrates, extending from the base of the brain to the hips, protected by the bones of the vertebral column; contains the cell bodies of motor neurons synapsing on skeletal muscles, the circuitry for some simple reflex behaviors, and axons communicating with the brain.

sympathetic division: the division of the autonomic nervous system that produces largely involuntary responses that prepare the body for stressful or highly energetic situations.

synapse: the site of communication between nerve cells. One cell (presynaptic) usually releases a chemical (the neurotransmitter) that changes the electrical potential of the second (postsynaptic) cell.

synaptic cleft: a tiny space that separates the pre- and postsynaptic cells.

synaptic terminal: swellings at the branched endings of axons, where the axon forms a synapse.

taste: a chemical sense for substances dissolved in water or saliva; in mammals, perceptions of sweet, sour, bitter, or salt produced by stimulation of receptors on the tongue.

taste bud: a cluster of taste receptor cells and supporting cells located in a small pit beneath the surface of the tongue and communicating with the mouth through a small pore. The human tongue has about 10,000 taste buds.

tectorial membrane: one of the membranes of the cochlea, in which the hairs of the hair cells are embedded. During sound reception, movement of the basilar membrane relative to the tectorial membrane bends the cilia.

thalamus: part of the forebrain, the thalamus relays sensory information to many parts of the brain.

thermoreceptor: a sensory receptor that responds to changes in temperature.

threshold: the electrical potential (less negative than the resting potential) at which an action potential is triggered.

transducer: a device that converts signals from one form to another. Sensory receptors are transducers that convert environmental stimuli, such as heat, light, or vibration, into electrical signals (such as action potentials) recognized by the nervous system.

tympanic membrane: the eardrum; a membrane stretched across the opening of the ear, which transmits vibration of sound waves to bones of the middle ear.

vitreous humor: a clear jellylike substance that fills the large chamber of the eye between the lens and retina.

white matter: portions of the brain and spinal cord consisting largely of myelin-covered axons that give these areas a white appearance.

working memory: the first phase of learning; short-term memory that is electrical or biochemical in nature.

THINKING THROUGH THE CONCEPTS

True or False: Determine if the statement given is true or false. If it is false, change the <u>underlined</u> word so that the statement reads true.

22. _____ <u>Dendrites</u> carry an impulse away from the nerve cell body and are the long extensions of a nerve cell.
23. _____ <u>Axons</u> initiate an impulse.
24. _____ Nerves pass on impulses with <u>undiminished</u> intensity.
25. _____ The resting potential inside a nerve cell is always <u>positive</u> within the cell.
26. _____ When the threshold level is reached, the <u>sodium</u> channel opens.
27. _____ The size of the action potential is <u>dependent on</u> the strength of the stimulus.
28. _____ Myelin covers some <u>dendrites</u>.
29. _____ Receptors for neurotransmitters are located on the <u>postsynaptic</u> neuron.
30. _____ A centralized nervous system is characteristic of <u>radially</u> symmetrical animals.
31. _____ The <u>cerebellum</u> controls coordination.
32. _____ The <u>thalamus</u> connects the two sides of the cerebrum.
33. _____ The <u>left</u> side of the brain is associated with creativity.
34. _____ Short-term memory is <u>electrical</u>.
35. _____ Long-term memory is <u>chemical</u>.
36. _____ Structures that change when acted upon by stimuli are <u>acceptors</u>.
37. _____ Structures that convert signals from one form to another are <u>reducers</u>.
38. _____ High frequency vibrations of air or water is <u>sound</u>.
39. _____ The cochlea is part of the human <u>middle</u> ear.
40. _____ The sense of vision involves <u>chemical</u> changes.
41. _____ The fovea contains virtually no <u>cones</u>.
42. _____ The retinas of nocturnal animals are made up of <u>cones</u>.
43. _____ Pain is a special kind of <u>chemical</u> sense.
44. _____ <u>Rods</u> are responsible for color vision.

Matching: Structural regions of neurons.

45. _____ carry action potentials to output terminals

46. _____ the cell's integration center

47. _____ receive information from the environment

48. _____ sites where signals are transmitted to other cells

49. _____ convert environmental information into electrical signals

50. _____ bundled together into "nerves"

51. _____ initiate action potentials

Choices:

    a. cell bodies

    b. synaptic terminals

    c. dendrites

    d. axons

Matching: Nerve cell function.

52. _____ always negative (-40 to -90 millivolts) within a nerve cell

53. _____ a sudden positive charge within a nerve cell

54. _____ occurs when a nerve cell becomes sufficiently less negative inside

Choices:

    a. threshold

    b. action potential

    c. resting potential

Matching: Autonomic nervous system.

55. _____ prepares the body for "flight or fight" situations

56. _____ conducts messages between the environment and the central nervous system

57. _____ regulates various internal organs and is regulated by the hypothalamus

58. _____ speeds up heart rate

59. _____ associated with "rest and rumination" activities

60. _____ opens the eye pupils

61. _____ increases urine production

Choices:

    a. sympathetic nervous system

    b. parasympathetic nervous system

    c. both of these

    d. neither of these

Matching: The human brain. There may be more than one correct answer.

62. _____ midbrain

63. _____ like an extension of the spinal cord

64. _____ channels sensory information to other forebrain parts

65. _____ hindbrain

66. _____ controls several autonomic functions

67. _____ hypothalamus

68. _____ forebrain

69. _____ receives input from all sense organs and "decides" which require attention

70. _____ controls learning, emotions, and the autonomic nervous system

71. _____ largest part of the brain

72. _____ coordinates body movements and body positions

73. _____ controls speech, reading, math ability, and musical skills

Choices:

a. cerebellum

b. cerebrum

c. limbic system

d. medulla

e. reticular activating formation center

f. thalamus

Matching: Hormone versus neuron control.

74. _____ indiscriminately affect millions of cells

75. _____ speed transmissions quickly using electrical signals

76. _____ release neurotransmitters very close to the cells they influence

77._____ their chemicals travel slowly to the sites of action

78. _____ release chemicals into the bloodstream that carries them great distances

79. _____ precisely affect small numbers of cells

80. _____ some release neurohormones

Choices:

a. nerve cells

b. hormone producing cells

c. both cell types

d. neither cell type

Multiple Choice: Pick the most correct choice for each question.

81. A rapid change from a negative to a positive electrical potential in a nerve cell is
    a. a reflex
    b. an amygdala
    c. an action potential
    d. a divergence
    e. a synapse

82. The gap between the axon of one neuron and the dendrite of another is called a
    a. synapse
    b. node of Ranvier
    c. cell body
    d. convergence
    e. dendritic junction

83. The resting potential of a neuron becomes negative when
    a. K^+ ions diffuse into the cytoplasm
    b. Na^+ ions diffuse into the cytoplasm
    c. Na^+ ions diffuse out of the cytoplasm
    d. K^+ ions diffuse out of the cytoplasm

84. The autonomic nervous system innervates all of the following **except**
    a. skeletal muscle
    b. heart
    c. stomach
    d. kidney

85. The sympathetic nervous system
    a. is part of the central nervous system
    b. prepares the body for stressful situations
    c. is under voluntary control
    d. is composed primarily of association neurons
    e. allows the body to relax

86. A mechanoreceptor is a receptor designed for
    a. touch
    b. light
    c. chemicals
    d. pain
    e. pleasure

87. The middle ear contains
    a. cochlea
    b. three bones (hammer, anvil and stirrup)
    c. semicircular canals
    d. receptor cells for hearing

88. The amount of light entering the human eye is regulated by the muscular
    a. iris
    b. pupil
    c. sclera
    d. cornea

89. Completely color-blind animals lack
    a. cones
    b. rods
    c. ommatidia
    d. irises
    e. lenses

CLUES TO APPLYING THE CONCEPTS

These practice questions are intended to sharpen your ability to apply critical thinking and analysis to biological concepts covered in this chapter.

90. There are many chemicals that can act on the nervous system and cause problems. Determine the effect each of the following has on a person subjected to their action. (1) Botulism toxin, from a certain type of bacteria, inhibits the release of acetylcholine. (2) Curare, used by South American Indians to coat the tip of their arrows, binds to the same receptors that normally bind acetylcholine. (3) Diisopropyl flourophosphate, a chemical that could be used as a nerve gas during war, blocks the enzyme that breaks down acetylcholine after it has crossed the synapse and becomes bound to the receiving neuron.

_____

_____

_____

_____

_____

_____

_____

_____

_____

91. Explain how each of the following medical problems could result in deafness. (1) injury to the auditory nerve; (2) arthritis of the middle ear bones; (3) a punctured eardrum; (4) too much earwax in the auditory canal; and (5) damage to the hair cells from going to rock concerts.

_____

_____

_____

_____

_____

_____

_____

_____

ANSWERS TO EXERCISES

1.	inner	22.	false, axons	45.	d	68.	b, c, f
2.	neuron	23.	false, dendrites	46.	a	69.	e
3.	lens	24.	true	47.	c	70.	c
4.	cone	25.	false, negative	48.	b	71.	b
5.	medulla	26.	true	49.	c	72.	a
6.	(across) axon	27.	false, independent of	50.	d	73.	b
6.	(down) acetylcholine	28.	false, axons	51.	a	74.	b
7.	cerebellum	29.	true	52.	c	75.	a
8.	amygdala	30.	false, bilaterally	53.	b	76.	a
9.	(across) pons	31.	true	54.	a	77.	b
9.	(down) pupil	32.	false, corpus callosum	55.	a	78.	b
10.	nerve	33.	false, right	56.	d	79.	a
11.	cell body	34.	true	57.	c	80.	a
12.	bradykinin	35.	false, morphological	58.	a	81.	c
13.	dendrite	36.	false, receptors	59.	b	82.	a
14.	rod	37.	false, transducers	60.	a	83.	d
15.	dopamine	38.	true	61.	b	84.	a
16.	brain	39.	false, inner	62.	e	85.	b
17.	auditory	40.	true	63.	d	86.	a
18.	oval	41.	false, rods	64.	f	87.	b
19.	iris	42.	false, rods	65.	a, d	88.	a
20.	fovea	43.	true	66.	d	89.	a
21.	cornea	44.	false, cones	67.	c		

90. Normally, acetylcholine stimulates receiving neurons to transmit action potentials to skeletal muscles, which then contract. (1) When botulism toxin inhibits the release of acetylcholine, transmitting cells will not be able to signal receiving cells, and nerve signals will be stopped at the synapse, resulting in paralysis of skeletal muscles. (2) When curare blocks the acetylcholine receptors, skeletal muscles cannot be stimulated to contract, resulting in paralysis. (3) Diisopropyl flourophosphate prevents acetylcholine from being removed from the receptors where it is bound, so that the receiving cells will continue to transmit signals to skeletal muscles, the result being prolonged contraction of the muscles.

91. (1) If the auditory nerve is injured, the brain will not be able to receive signals that sound is occurring. (2) If the middle ear bones are arthritic, they will be unable to move in response to sound vibrations, preventing them from transmitting vibrations to the cochlea. (3) A punctured eardrum will be unable to pick up vibrations from the air and transmit them to the ear bones. (4) Too much earwax will block sound waves from striking the eardrum and setting it in motion. (5) Damaged hair cells will not produce receptor potentials and, thus, will not release transmitter onto neurons of the auditory nerve.

Chapter 34: Action and Support: The Muscles and Skeleton

OVERVIEW

This chapter discusses the structures and tissues that allow the body to move, have form, and support itself. Muscle tissue is specialized to contract and relax. With contracting or relaxing, muscles move the body, move substances through organs, and allow the heart to beat. The human body takes its support from the bony skeleton. The muscles attached to the skeleton by connective tissue allow it to move; however, movement would not be possible if the skeleton were not articulated, or jointed. This chapter discusses how these structures function together.

1) How Do Muscles Work?

Muscle tissue is specialized according to its function. **Skeletal muscle**, or **striated muscle**, functions to move the skeleton and is usually under voluntary control. **Cardiac muscle** is found only in the heart. Cardiac muscle is stimulated by nerves and hormones, although it initiates its own contractions. **Smooth muscle** is found lining the walls of organs such as the uterus, stomach, intestines, and esophagus, as well as the walls of blood vessels. Smooth muscle contractions are involuntary.

Interactions between microfilaments of the proteins **actin** and **myosin** are involved in cellular movement. These interactions are the ancesters of today's muscle movement. Muscle cells are referred to as **muscle fibers** and contain many individual contractile subunits called **myofibrils**. Each myofibril is enclosed by a **sarcoplasmic reticulum**, storing high concentrations of calcium ions. The myofibrils are arranged in subunits called **sarcomeres**. Within each sarcomere, **thick filaments** of myosin lie between **thin filaments** of actin. The thin and thick filaments connect by small branches of myosin called **cross-bridges**. Each sarcomere is bounded by **Z lines** to which thin filaments attach. Extending into each muscle fiber are extensions of their plasma membrane called **transverse tubules (T tubules)**. When muscles contract, the thin filaments of actin are pulled past the thick filaments of myosin.

Muscle contraction is controlled by the nervous system. The point of attachment between a motor neuron and a muscle cell is called a **neuromuscular junction**. The strength and degree of contraction is regulated by the number of muscle fibers stimulated and the frequency at which it occurs. If rapid firing is prolonged, **tetany**, or a sustained maximal contraction, results. Motor neurons form synapses with a group of muscle fibers called a **motor unit**.

The heart is composed of cardiac muscle. It, too, contains sarcomeres of thin and thick filaments. Cardiac muscle fibers initiate their own contractions, particularly in the pacemaker, the S A node. Gap junctions allow the action potential to spread from one cardiac muscle cell to another, synchronizing the contraction.

Smooth muscle lines the walls of organs such as the uterus, stomach, intestines, and esophagus, as well as the walls of blood vessels. Smooth muscle does not have the regular arrangement of sarcomeres that skeletal and cardiac muscle exhibit. As in cardiac muscle, gap junctions allow the action potential to spread along the muscle, synchronizing the contraction in slow, sustained, or wavelike movements.

Contractions of smooth muscle is involuntary and may be stimulated by stretching, hormonal signals, nervous signals, or a combination of signals.

2) What Does the Skeleton Do?

The **skeleton** is a supporting framework for the body. **Hydrostatic skeletons** consist of a fluid-filled sac. This type of skeleton is found in worms, mollusks, and cnidarians. **Exoskeletons** support the body from the outside. Both thin, flexible and thick, rigid exoskeletons exist to support the bodies of arthropods. **Endoskeletons** support the body from the inside and are found in echinoderms and chordates. The vertebrate skeleton consists of the **axial skeleton** and the **appendicular skeleton**.

3) Which Tissues Compose the Vertebrate Skeleton?

Cartilage and bone are the tissues making up the vertebrate skeleton. The cells of both tissues are embedded in a protein matrix called **collagen**. **Cartilage** is found at the ends of bones at joints; it forms the ears and nose, the structures of the respiratory system, and makes up the "shock absorbing" pads, the **intervertebral discs**, between vertebrae. The cells of cartilage are **chondrocytes**. Since no blood vessels are found in cartilage, the chondrocytes metabolize very slowly and rely on diffusion to receive nutrients and remove wastes.

The collagen in bone tissue is hardened by calcium phosphate, forming a dense support structure. Bones have a hard outer covering of **compact bone** and a lightweight, porous interior of **spongy bone**. Bone marrow is found in the cavities of spongy bone. The cells of bone include **osteoblasts**, **osteocytes**, and **osteoclasts**. In the process of forming bone tissue (bone remodeling), concentric layers of bone embedded with osteocytes, **osteons**, are created. In the center of an osteon, a capillary passes, supplying the cells with nutrients. Bone remodeling slows as bone ages, resulting in more fragile bones.

4) How Does the Body Move?

The body moves through the use of **antagonistic muscles**. When one muscle is contracted, another is extended. Muscles in vertebrates move the bones of the skeleton at points where two bones meet, the **joints**. Muscles attach to bones by connective tissue forming **tendons**. Bones attach to other bones by connective tissue forming **ligaments**. In **hinge joints**, a bone on one side of the joint moves while the other bone maintains a fixed position. The antagonistic muscle pair involved in moving a hinged joint includes a **flexor** and an **extensor**. Each muscle is attached to an immovable bone. This attachment point is called the **origin**. The end attached to the moveable bone is the **insertion** point. **Ball-and-socket joints** consist of a rounded end of one bone that fits into a hollow depression of another. This type of joint allows movement in many directions and involves two pairs of muscles oriented at right angles to each other.

KEY TERMS AND CONCEPTS

Fill-In: From the following list of key terms fill in the blanks in the following statements.

actin	exoskeleton	myofibrils	skeletal muscle
antagonistic muscles	extensor	myosin	skeleton
appendicular skeleton	flexor	neuromuscular junction	smooth muscle
axial skeleton	hinge joints	origin	spongy bone
ball-and-socket joints	hydrostatic skeleton	osteoblasts	striated muscle
cardiac muscle	insertion	osteoclasts	tendons
cartilage	intervertebral discs	osteocytes	tetany
chondrocytes	joint	osteons	thick filaments
collagen	ligaments	osteoporosis	thin filaments
compact bone	motor unit	sarcomeres	transverse (T) tubules
cross-bridges	muscle fibers	sarcoplasmic reticulum	Z lines
endoskeleton			

(1) _____ is used to move the skeleton. It is also called (2) _____

because of its appearance under the microscope. (3) _____ is found only in the

heart, and (4)_____ lines the walls of hollow organs and blood vessels.

Movement within cells occurs because of protein microfilaments made of (5) _____.

These microfilaments interact with another protein, (6) _____, to change the shape

of the cell.

Individual muscle cells are called (7) _____, which contain individual contractile

subunits, the (8) _____. Within muscle cells, high concentrations of Ca $^{++}$ are stored

in fluid of the (9) _____. Deep indentations of muscle cell plasma membrane called

(10) _____ extend into the muscle fiber. A precise arrangement of actin and myosin

filaments make up (11) _____. Actin molecules make up (12) _____,

while myosin molecules make up (13) _____. Actin and myosin filaments interact using

connections called (14) _____. Between adjacent sarcomeres, thin filaments are

attached to (15) _____.

Motor neurons form synapses with muscle cells at (16) _____, which are always

excitatory. However, most motor neurons synapse with many muscle fibers. This group of fibers is a

(17) _____. If a motor neuron elicits many action potentials in rapid succession for a

prolonged period, the muscle produces a sustained maximal contraction called

(18) _____.

The supporting structure for the body is the (19) _____. A (20) _____ is a fluid-filled sac that is highly flexible. Arthropods, such as insects and crustaceans, are encased in an (21) _____, while humans and other chordates have an (22) _____.

The vertebrate skeleton is subdivided into the (23) _____, containing the skull, vertebral column, and rib cage, and the (24) _____, forming the appendages.

The cells of the tissues making up the skeleton are imbedded in a protein matrix of (25) _____. When the skeleton is first formed during embryological development, it is composed of (26) _____. The cells of this connective tissue are called (27) _____. Adult skeletons still rely on this connective tissue to form the larynx, trachea, and bronchi. It also protects vertebrae by serving as shock-absorbing pads, the (28) _____.

Bones have a hard outer covering called (29) _____, with lightweight, porous (30) _____ on the inside. The cells of bone include (31) _____, the bone forming cells, (32) _____, mature bone cells, and (33) _____, the bone dissolving cells. Hard bone is composed of tightly packed units, the (34) _____, which are concentric layers of bone embedded with osteocytes.

The skeleton moves because of attached muscle pairs. These are (35) _____ in that when one contracts, the other is extended. Muscles are attached to bones by (36) _____. Bones, however, are attached to other bones by (37) _____. When bones move at areas where two bones attach, a (38) _____ is formed. (39) _____, such as the knee and elbow, are movable in only two dimensions. The pair of muscles controlling this type of joint includes a(n) (40) _____ and a(n) (41) _____. The area where muscle attaches to an immovable bone is called the (42) _____. The area where muscle attaches to a movable bone is called the (43) _____.
 (44) _____ such as those forming the hip and shoulder allow movement in several directions.

Key Terms and Definitions

actin: one of the major proteins of muscle, whose interactions with myosin produce contractions; found in the thin filaments of the muscle fiber. See also *myosin*.

antagonistic muscles: a pair of muscles, one of which contracts and in so doing extends the other; an arrangement that makes movement of the skeleton at joints possible.

appendicular skeleton: that portion of the skeleton consisting of the bones of the extremities and their attachments to the axial skeleton; the pectoral and pelvic girdles, the arms, legs, hands, and feet.

axial skeleton: the skeleton forming the body axis, including the skull, vertebral column, and rib cage.

ball-and-socket joint: a joint in which the rounded end of one bone fits into a hollow depression in another, as in the hip; allows movement in several directions.

cardiac muscle: specialized muscle of the heart, able to initiate its own contraction independent of the nervous system.

cartilage: a form of connective tissue forming portions of the skeleton, consisting of chondrocytes and their extracellular secretion of collagen. Cartilage resembles flexible bone.

chondrocyte: the living cells of cartilage, which with its extracellular secretions of collagen, form cartilage.

collagen: a protein matrix in which cartilage and bone are embedded.

compact bone: the hard and strong outer bone; composed of osteons (or Haversian systems).

cross-bridge: in muscles, an extension of myosin that binds to and pulls upon actin to produce contraction of the muscles.

endoskeleton: a rigid *internal* skeleton with flexible joints to allow for movement.

exoskeleton: a rigid *external* skeleton that supports the body, protects the internal organs and has flexible joints to allow for movement.

extensor: a muscle that straightens a joint.

flexor: a muscle that flexes (decreases the angle of) a joint.

hinge joint: a joint at which one bone is moved by muscle and the other bone remains fixed, such as in the knee, elbow, or fingers; allows movement in only two dimensions.

hydrostatic skeleton: a skeleton composed of fluid contained within a flexible, usually tubular, covering.

insertion: the site of attachment of a muscle to the relatively moveable bone on one side of a joint.

intervertebral discs: pads of cartilage found between the vertebrae that act as shock absorbers.

joint: a flexible region between two rigid units of an exoskeleton or endoskeleton, to allow for movement between the units.

ligament: a tough connective tissue band connecting two bones.

motor unit: a single motor neuron and all the muscle fibers on which it synapses.

muscle fiber: an individual muscle cell.

myofibril: a cylindrical subunit of each muscle cell, consisting of a series of sarcomeres. Myofibrils are surrounded by sarcoplasmic reticulum.

myosin: one of the major proteins of muscle, whose interaction with actin produces contraction; found in the thick filaments of the muscle fiber. See also *actin*.

neuromuscular junction: the synapse formed between a motor neuron and muscle fiber.

origin: the site of attachment of a muscle to the relatively stationary bone on one side of a joint.

osteoblast: a cell type that produces bone.

osteoclast: a cell type that dissolves bone.

osteocyte: a mature bone cell.

osteon: a unit of hard bone consisting of concentric layers of bone matrix, with embedded osteocytes, surrounding a small central canal that contains a capillary.

osteoporosis: a condition in which bones become porous, weak, and easily fractured; most common in elderly women.

sarcomere: the unit of contraction of a muscle fiber; a subunit of the myofibril, consisting of actin and myosin filaments and bounded by Z-lines.

sarcoplasmic reticulum: specialized endoplasmic reticulum found in muscle cells and forming interconnected hollow tubes. The sarcoplasmic reticulum stores calcium ions and releases them into the interior of the muscle cell to initiate contraction.

skeletal (striated) muscle: the type of muscle that is attached to and moves the skeleton and is under the direct, usually voluntary, control of the nervous system.

skeleton: a supporting structure for the body, upon which muscles act to change the body configuration.

smooth muscle: type of muscle found around hollow organs, such as the digestive tract, bladder, and blood vessels, normally not under voluntary control.

spongy bone: porous, lightweight bone tissue in the interior of bones; the location of bone marrow.

striated muscle: see *skeletal muscle.*

tendon: a tough connective tissue band connecting a muscle to a bone.

tetany: smooth, sustained maximal contraction of a muscle in response to rapid firing by its motor neuron.

thick filaments: bundles of myosin protein within the sarcomere that interact with thin filaments to produce muscle contraction.

thin filaments: protein strands within the sarcomere that interact with thick filaments to produce muscle contraction. Composed primarily of actin, with accessory proteins.

transverse (T) tubules: deep infoldings of the muscle cell membrane that conduct the action potential inside the cell.

Z lines: fibrous protein structures to which the thin filaments of skeletal muscle are attached, forming the boundaries of sarcomeres.

THINKING THROUGH THE CONCEPTS

True or False: Determine if the statement given is true or false. If it is false, change the underlined word so that the statement reads true.

45. _____ Moving food through the digestive system involves muscle action.

46. _____ Muscles are active only during the extension phase.

47. _____ Striated muscle contractions are under conscious control.

48. _____ Cardiac muscle contractions are under conscious control.

49. _____ Smooth muscle contractions are under conscious control.

50. _____ Actin and myosin work against one another to contract a muscle.

51. _____ One muscle cell may be 35 centimeters long.

52. _____ Actin and myosin filaments are arranged in subunits called motor units.

53. _____ An action potential causes the sarcoplasmic reticulum to release Ca^{++}, allowing muscle fibers to contract.

54. _____ Skeletal muscle is called striated muscle because of the appearance of the thick and thin filaments in sarcomeres.

55. _____ ATP provides the energy for myosin cross bridges to move along the thin filaments, contracting the muscle.

56. _____ An action potential causes the contraction of only one muscle cell.

57. _____ <u>Neuromuscular junctions</u> allow action potentials to travel from one cardiac muscle cell to another, synchronizing their contractions.

58. _____ Bones immobilized in a cast <u>remain strong and high in calcium</u>.

59. _____ Bones produce <u>red blood cells</u>.

60. _____ Cartilage is <u>living tissue</u>.

61. _____ <u>Spongy bone</u> provides sites for muscle attachment.

62. _____ Calcium levels in blood need to remain constant. If the blood level drops, calcium <u>is taken from bone</u> and retained in the blood.

63. _____ <u>Ligaments</u> connect muscle to bone.

64. _____ The <u>insertion point</u> of a muscle attaches to the mobile bone on the far side of a joint.

Matching: Correctly match the following characteristics with the appropriate cell type.

65. _____ bone dissolving cells

66. _____ cartilage cell

67. _____ mature bone cell

68. _____ bone forming cell

69. _____ found in osteons

70. _____ relies on diffusion for nutrients

71. _____ capillaries bring nutrients

72. _____ secrete flexible, elastic matrix

73. _____ dissolve cartilage

74. _____ replace cartilage with bone

75. _____ create channels during bone remodeling

76. _____ fill channels during bone remodeling

77. _____ become osteocytes

Choices:

a. chondrocyte

b. osteoblast

c. osteocyte

d. osteoclast

Identify the following: Are these bones part of the **axial** skeleton or the **appendicular** skeleton?

78. _____ skull

79. _____ humerus

80. _____ rib

81. _____ vertebra

82. _____ patella

83. _____ tarsals

84. _____ mandible

85. _____ sternum

86. _____ femur

87. _____ tibia

88. _____ coccyx

89. _____ ulna

Short answer:

90. List the five functions of the vertebrate skeleton.

_____

_____

_____

_____

_____

91. Use the following terms to explain how muscle contraction occurs. Include the differences between skeletal, cardiac, and smooth muscle in how contractions are induced

thin filament   motor neuron   myosin   action potential   T tubules   sarcoplasmic reticulum
calcium ions   accessory proteins   skeletal muscle   cardiac muscle   smooth muscle

_____

_____

_____

_____

_____

_____

_____

_____

_____

92. Which bones are involved in transmitting sound?

_____

CLUES TO APPLYING THE CONCEPTS

These practice questions are intended to sharpen your ability to apply critical thinking and analysis to biological concepts covered in this chapter.

93. Damage to a knee often involves loss of cartilage. A new procedure using chondrocyte transplants is being used to repair joints damaged in accidents (but not due to arthritis). This procedure involves culturing the chondrocytes of the injured individual and placing them at the injured site. Why is such a procedure necessary? That is; why is it unlikely that damaged cartilage will repair itself, while bone readily heals?

_____

_____

_____

_____

94. Using what you have read about bone formation and bone remodeling, what would you predict to be the effect weightlessness has on astronauts in space with respect to their bone strength?

_____

_____

_____

_____

_____

_____

ANSWERS TO EXERCISES

1. skeletal muscle
2. striated muscle
3. cardiac muscle
4. smooth muscle
5. actin
6. myosin
7. muscle fibers
8. myofibrils
9. sarcoplasmic reticulum
10. transverse (T) tubules
11. sarcomeres
12. thin filaments
13. thick filaments
14. cross bridges
15. Z lines
16. neuromuscular junctions
17. motor unit
18. tetany
19. skeleton
20. hydrostatic skeleton
21. exoskeleton
22. endoskeleton
23. axial skeleton
24. appendicular skeleton
25. collagen
26. cartilage
27. chondrocytes
28. intervertebral discs
29. compact bone
30. spongy bone
31. osteoblasts

32. osteocytes
33. osteoclasts
34. osteons
35. antagonistic muscle
36. tendons
37. ligaments
38. joint
39. Hinge joints
40. flexor
41. extensor
42. origin
43. insertion
44. ball-and-socket joints
45. true
46. false, contraction
47. true
48. false, nerve and hormone control
49. false, unconscious control
50. false, together
51. true
52. false, sarcomeres
53. true
54. true
55. true
56. false, a motor unit
57. false, gap junctions
58. false, weaken and lose calcium
59. true

60. true
61. false, compact bone
62. true
63. false, tendons
64. true
65. d
66. a
67. c
68. b
69. c
70. a
71. c
72, a
73. d
74. b
75. d
76. b
77. b
78. axial
79. appendicular
80. axial
81. axial
82. appendicular
83. appendicular
84. axial
85. axial
86. appendicular
87. appendicular
88. axial
89. appendicular

90. provides rigid framework that supports the body and protects the organs; allows animals to move; produces red blood cells, white blood cells. and platelets; stores calcium and phosphorous, sensory transduction (transmits sound vibrations)

91. A skeletal muscle cell is stimulated by a motor neuron creating an action potential. The action potential travels to the muscle cell's interior by passing down the T tubules. When the action potential reaches the sarcoplasmic reticulum, calcium ions are released into the cytoplasm of the muscle cell. The calcium ions bind to the accessory proteins of the thin filament. This changes the shape of the accessory proteins so that the myosin binding sites are exposed contracting the muscle fiber. Contraction of cardiac muscle occurs when an action potential travels through the T tubules releasing calcium ions from the sarcoplasmic reticulum. Calcium ions enter the cell from the extracellular fluid as well as from the sarcoplasmic reticulum. Gap junctions connect cardiac muscle cells, enabling the contractions to be synchronized. Smooth muscle cells do not have sacroplasmic reticula. Calcium ions, initiating contraction of smooth muscle cells, flow into the cell from the extracellular fluid. Gap

junctions between muscle cells synchronize contractions here, also.

92. The bones of the middle ear, the hammer, anvil, and stirrup, are involved in conducting sounds.

93. Damaged cartilage is unlikely to repair itself because this tissue lacks capillaries. It relies on diffusion to provide the necessary nutrients and to remove wastes. This process is very slow and the metabolism of cartilage is, therefore, very slow also. Bone, on the other hand, is supplied with a network of capillaries and active cells to heal damaged tissue.

94. Normal stress on bone increases its thickness and its strength. Simply immobilizing a bone in a cast decreases its strength. NASA is studying the effects of weightlessness on astronaut's bone strength. Astronauts experiencing extended periods without gravity putting stress on their bones have experienced bone loss.

Chapter 35: Animal Reproduction

OVERVIEW

This chapter presents the mechanisms by which animals reproduce. These mechanisms are the result of the effects of natural selection on generations of organisms. The chapter, however, focuses on human reproduction. Included is a summary of the reproductive organs and the hormones regulating sexual development and functioning. Information concerning embryological development and the changes that occur in the mother's body through pregnancy is also presented. Using the information learned about the human reproductive system, methods to limit or induce fertility are explored.

1) How Do Animals Reproduce?

Reproduction involving gametes produced through meiosis is considered **sexual reproduction**. The offspring produced will have a genome that combines those of the parents. However, some animals produce offspring through mitotic divisions in an area of the body. Through **asexual reproduction**, the offspring are genetically identical to the parent. Asexual reproduction may involve **budding**. These animals, such as hydra and sponges, produce a **bud** that is a miniature version of the "parent". The bud will eventually break off and continue independently. Reproduction by **regeneration** is rare, but does occur. Annelid and flatworm species may spontaneously break into two or more parts. Each part then regrows the missing body parts. Asexual reproduction may involve **fission**. Some corals divide in half, lengthwise, producing two smaller individuals. This type of reproduction usually involves some regeneration as well. Some species can produce eggs that will develop into adults without being fertilized. These adults are haploid and are produced by **parthenogenesis**.

Species with both male and female individuals are **dioecious**. Females produce **eggs** and males produce **sperm**. In **monoecious** species, an individual produces both egg and sperm. These individuals are **hermaphrodites**.

If the union of egg and sperm occurs outside the body of the parents, **external fertilization** has occurred. This is known as **spawning**. Animals that spawn must synchronize their release of egg and sperm. Female mussels and sea stars achieve this by releasing **pheromones** into the water when they release their eggs. The males detect this chemical and release sperm into the area. Other species rely on specialized mating rituals to bring males and females together. Frogs assume a specialized mating posture called **amplexus**. During amplexus, the male mounts the female and prods her side, causing her to release eggs that he then fertilizes by releasing sperm into the water.

If the union of the egg and sperm occur inside the female's body, **internal fertilization**, has occurred. Internal fertilization uses **copulation** behavior and increases the chances that the sperm will reach the egg. Some species deposit a packet of sperm, a **spermatophore**, which the female picks up and inserts into her reproductive cavity. Regardless of the method of getting sperm into the female body, fertilization will only occur if the female has released a mature egg. Mature eggs are released during ovulation.

2) How Does the Human Reproductive System Work?

The organs that produce the sex cells are called **gonads**. The gonads in the male are the **testes** which are contained outside the body within a sac called the **scrotum**. Within each **testis**, sperm are produced by the **seminiferous tubules**. **Interstitial cells** between the tubules produce the hormone testosterone. Inside the wall of the seminiferous tubules are the **spermatogonia**, the diploid cells from which sperm arise. Also within the wall of the tubules are **sertoli cells** which regulate **spermatogenesis**, the development of sperm. During spermatogenesis, spermatogonia grow and differentiate into **primary spermatocytes**. The primary spermatocytes divide by meiosis I, producing two **secondary spermatocytes**. Secondary spermatocytes divide by meiosis II, each producing two **spermatids**. It is the spermatids that differentiate into sperm nourished by the sertoli cells. The organization of sperm cells is unlike that of any other cell. Within the head of the sperm is the DNA and an **acrosome**. The acrosome contains enzymes that digest protective layers around the egg. Behind the head of the sperm is the midpiece containing numerous mitochondria to provide energy for the whiplike movement of the sperm's flagellum. Males do not begin producing sperm until the hypothalamus releases **gonadotropin-releasing hormone (GnRH)** at puberty. GnRH stimulates the anterior pituitary to produce **luteinizing hormone (LH)** and **follicle-stimulating hormone (FSH)**. LH stimulates interstitial cells to produce **testosterone**. Testosterone is needed to maintain an erection of the **penis** for successful intercourse.

Sperm are conducted outside the male body through accessory structures of the reproductive system. The seminiferous tubules merge to form a continuous tube, the **epididymis** which leads to the **vas deferens**. The vas deferens merges with the **urethra** forming a path shared with the urinary system to the tip of the penis. Sperm leaving the body (ejaculation) is mixed in a fluid called **semen**. Semen is formed from three glands, the **seminal vesicles**, the **prostate gland**, and the **bulbourethral glands**.

The gonads in the female are the **ovaries**. Within the ovaries, eggs are formed through **oogenesis**. Oogenesis begins in the ovaries of a developing fetus when the **oogonia** are formed. By the end of the third month, the oogonia have matured into **primary oocytes**. The primary oocytes begin meiosis I, but the process is halted at prophase. It is not until after puberty that development will continue, and then only a few primary oocytes each month will continue oogenesis. Surrounding each primary oocyte are accessory cells; together, they form a **follicle**. After puberty, pituitary hormones stimulate a few follicles to continue development. Within each follicle, the first meiotic division is completed, forming one **secondary oocyte** and one **polar body**. The polar body contains chromosomes to be discarded. The accessory cells of the follicle secrete **estrogen**. A mature follicle containing a secondary oocyte and the polar body is released from the ovary into the **oviduct** (or **fallopian tube**). If fertilization occurs, the secondary oocyte will undergo meiosis II. Meanwhile, accessory cells left behind in the ovary enlarge forming the **corpus luteum**. The corpus luteum secretes both estrogen and the hormone **progesterone**. The secondary oocyte (now refered to as the egg) is conducted into a fallopian tube by a current created by the ciliated **fimbriae**. Fertilization, if it occurs, usually takes place in the fallopian tube, forming a **zygote**. The zygote is carried through the fallopian tube by ciliated cells into the **uterus**. The inner wall of the uterus is well supplied with blood vessels and forms the **endometrium**, the mother's portion of the **placenta**. The outer wall of the uterus is the muscular **myometrium**, which will contract during delivery. The development of the endometrium is stimulated by estrogen and progesterone secreted by the corpus luteum. If the egg is not fertilized, the corpus luteum will disintegrate causing estrogen and progesterone levels to decrease. When the hormone levels drop, the endometrium disintegrates and is expelled from the uterus during **menstruation** as part of the **menstrual cycle**. At the outer end of the uterus is a ring of connective tissue forming the **cervix**. On the other side of the cervix is the birth canal, or **vagina**.

The menstrual cycle is regulated by interactions of hormones from the hypothalamus, pituitary gland,

and the ovary. Cells in the hypothalamus spontaneously release GnRH all the time. GnRH stimulates the secretion of FSH and LH by the anterior pituitary. FSH and LH circulating in the blood stream stimulate the development of several follicles in the ovaries as well as estrogen production by the follicles. One or two follicles complete development over a two-week period. As estrogen production increases, follicle development continues, endometrium growth in the uterus is stimulated, and FSH and LH are released in a surge by the hypothalamus and pituitary gland. The sudden increase in LH stimulates the primary oocyte to resume meiosis I, forming the secondary oocyte and first polar body; it initiates ovulation; and it transforms the follicle cells remaining in the ovary into the corpus luteum. The corpus luteum secretes estrogen and progesterone that turns off the release of FSH and LH, continues endometrial growth in the uterus, and turns off GnRH release. If fertilization occurs, during pregnancy the embryo secretes the hormone **chorionic gonadotropin (CG)**. CG prevents the breakdown of the corpus luteum. Therefore, estrogen and progesterone secretion continues, maintaining the endometrial lining that nourishes the developing embryo. If fertilization does not occur, the corpus luteum breaks down, the amount of estrogen and progesterone in the blood drops, the endometrial lining breaks down and is shed, and the production of GnRH resumes.

Traditionally, in order for a sperm to fertilize an egg, copulation (intercourse, in humans) must occur. With the influences of psychological and physical stimulation, blood flow into tissue spaces of the penis increases. When the tissues swell, the vessels draining blood from the penis become blocked. As pressure increases, the penis becomes erect, so that it can be inserted into the vagina. Further stimulation causes the muscles surrounding the epididymis, vas deferens, and the urethra to contract, resulting in ejaculation. During an ejaculation, 300-400 million sperm are released. Similarly, psychological and physical stimulation increases blood flow into the vagina and external reproductive tissues. The external reproductive tissues include folds of skin called the **labia** and a rounded projection, the **clitoris**. Since the clitoris develops from the same embryological tissue as the tip of the penis, it becomes erect when blood flow into the area increases. Sperm, released into the vagina following an ejaculation, swim up the female reproductive tract from the vagina, through the cervix, into the uterus, and up into the fallopian tubes. If an egg has been released in the past 24 hours or so, one sperm may succeed in fertilizing it.

The follicle cells that remain around the egg form the **corona radiata**, a barrier between the egg and sperm. Between the corona radiata and the egg is a second barrier, a jellylike **zona pellucida**. The hundreds of sperm reaching the egg release enzymes from their acrosomes, weakening the corona radiata and the zona pellucida, allowing one sperm to wiggle into the egg. The plasma membranes of the sperm and egg fuse and the head of the sperm is drawn into the egg's cytoplasm. As this occurs, the egg releases chemicals into the zona pellucida that reinforce it so that no other sperm can enter. The egg now begins meiosis II, finally producing a haploid gamete. The two haploid nuclei fuse forming a diploid cell, the zygote. Couples unable to successfully fertilize an egg are increasingly relying on artificial insemination or in vitro fertilization to help them out.

The zygote divides by mitosis. One week after fertilization a **blastocyst**, or hollow ball of cells, has formed. The **inner cell mass** will develop into the embryo while the cells on the outside are sticky and will adhere to the endometrium of the uterus. When this occurs, **implantation** has occurred. The embryo first obtains its nutrients from nearby cells of the endometrium. Within a few weeks of implantation, the placenta develops from cells of the embryo and the endometrium. Meanwhile, changes in the mother's breasts occur to prepare for nursing the baby. The increased amount of estrogen and progesterone in the blood stimulates the **mammary glands** to grow and develop the capacity to produce milk. Lactation, the production of milk, is promoted by the hormone prolactin. The substance first available to a nursing infant is **colostrum**. Colostrum is high in protein and antibodies from the mother. Mature milk, high in fats and lactose and lower in protein, gradually replaces the colostrum.

Birth of the fetus is initiated by stretching of the uterus and by hormones from both the fetus and the mother causing **labor**, a positive feedback reaction. The contractions of the uterine smooth muscle are triggered by the stretching caused by the growing fetus. Toward the end of development, the fetus produces steroid hormones that increase estrogen and prostaglandin production by the uterus and placenta. The increased hormone production, in combination with the stretching, increases the contractions of the uterus. As the baby's head pushes against the cervix, it dilates. This stretching signals the hypothalamus to release oxytocin. With oxytocin and prostaglandins circulating in the blood, the uterus contracts even more intensely, pushing the baby from the birth canal (the vagina). Soon after the baby's birth, contractions begin again to expel the placenta, the afterbirth. The umbilical cord produces prostaglandins that cause muscles around fetal umbilical vessels to contract, shutting off blood supply between the mother and the baby. The baby's circulatory system takes over.

3) How Can Fertility Be Limited?

Fertility can be limited by various methods of **contraception**. The most reliable method of contraception is, of course, abstinence. **Sterilization** provides a rather permanent means of contraception. Sterilization in men is achieved through a **vasectomy**. This rather minor operation, performed under local anesthetic, cuts the vas deferens. Sperm are still produced but are not expelled during ejaculation. A vasectomy has no known physical side effects. Sterilization in women is achieved through a **tubal ligation**. This operation, performed under general anesthetic, cuts the fallopian tubes. Ovulation still occurs but the sperm do not reach the eggs, nor can eggs reach the uterus. More temporary means of contraception prevent ovulation, prevent the sperm from reaching the egg, or prevent a fertilized egg from implanting in the uterus. **Birth control pills** prevent ovulation by suppressing LH release. The continuous estrogen and prostaglandin that the pill provides suppresses LH release. Barrier methods prevent sperm from reaching the egg. The **diaphragm** and the **cervical cap** fit securely over the cervix, preventing sperm entry. When used in conjunction with **spermicides**, these devices are very effective with no known side effects. As an alternative barrier method, a male can wear a **condom** over his penis. Less effective methods of contraception include **withdrawal** of the penis from the vagina before ejaculation and **douching** in an attempt to wash sperm from the vagina before they reach the uterus. Both of these methods are extremely unreliable. Another method that has a high failure rate is the **rhythm method**. This method involves abstinence from intercourse just before, during, and after ovulation. Users of this method often have difficulty determining when ovulation occurs each month or are undisciplined in their habits. An alternative method involves using devices, such as the **intrauterine device (IUD)**, that prevent a fertilized egg from implanting in the uterus. The "morning after pill," by providing a large dose of estrogen, essentially has the same effect When contraception fails, and an unplanned and unwanted pregnancy results, the pregnancy can be terminated by **abortion**. Abortion procedures typically involve dilation of the cervix followed by suction to remove the fetus and placenta. An alternative to the surgical procedure can be used within the first month of pregnancy. The RU-486 drug, in pill form, is used in some European countries and China. RU-486 blocks the action of progesterone, terminating the pregnancy.

KEY TERMS AND CONCEPTS

Fill-In: From the following list of key terms, fill in the blanks in the following statements.

abortion	epididymis	menstruation	secondary oocyte
acrosome	estrogren	monoecious	semen
amplexus	external fertilization	myometrium	seminal vesicles
asexual reproduction	fallopian tube	oogenesis	seminiferous tubules
blastocyst	fimbriae	oogonia	sertoli cells
budding	fission	ovaries	sexual reproduction
bulbourethral gland	follicle	oviduct	spawning
cervix	gonads	ovulation	spermatogenesis
clitoris	hermaphrodites	parthenogenesis	spermatagonia
colostrum	implantation	penis	spermatophore
contraception	inner cell mass	pheromone	testes
copulation	internal fertilization	placenta	urethra
corona radiata	interstitial cells	polar body	uterus
corpus luteum	labia	primary oocytes	vagina
dioecious	labor	prostrate gland	vas deferens
eggs	mammary glands	regeneration	zona pellucida
endometrium	menstrual cycle	scrotum	zygote

During (1) _____ haploid gametes are produced by meiosis. During
(2) _____, repeated mitosis of cells of some part of the body produces an exact copy
of the parent.

The process of (3) _____ produces a bud off the body of the adult that breaks off as
a new individual. The ability to regrow lost body parts is called (4) _____. Some
species are capable of reproducing by this method. Corals divide lengthwise, producing two smaller
copies of the parent coral. This type of asexual reproduction is (5) _____. When
haploid egg cells develop into adults, as with the development of male honey bees,
(6) _____ has occurred.

When species have both male and female individuals, the species is (7) _____.
Females produce (8) _____ and males produce (9) _____.
When individuals of a species produce both egg and sperm, the species is (10) _____
and the individuals are (11) _____.

If the union of the egg and sperm occurs outside the body, (12) _____ has taken place. (13) _____ involves egg and sperm being released into water. The sperm then swim to the eggs. The female may communicate to males that eggs are being released by releasing a (14) _____ into the water. Courtship rituals help ensure that both a male and a female are in the area at the right time. Frogs include a mating posture called (15) _____ in their mating behavior.

If sperm are taken into the female's body, (16) _____ occurs. Typically, this involves (17) _____, in which the male directly deposits the sperm into the female's body. The males of other species may package their sperm as a (18) _____, which the female picks up and places in her reproductive cavity. However sperm are to reach the egg, fertilization can only occur if a mature egg has been produced and (19) _____ has occured.

The paired organs that produce sex cells in mammals are called (20) _____. The male structures that produce sperm are the (21) _____ which are enclosed in the (22) _____ outside the body. Coiled within the testes are the (23) _____. The male hormone testosterone is produced by (24) _____. The diploid cells that give rise to sperm are the (25) _____, while the cells that will nourish the sperm are the (26) _____. The process by which sperm are formed is called (27) _____. The human sperm contains enzymes in the specialized lysosome, the (28) _____ and DNA in the head region. Sperm is released from the male body through the (29) _____ during ejaculation.

The seminiferous tubules merge to form the (30) _____, which leads to the (31) _____. This tube finally merges with the (32) _____ of the urinary system. The fluid ejaculated from the penis is called (33) _____ and contains sperm mixed with secretions from the (34) _____, the (35) _____, and the (36) _____.

The female structures that produce eggs are the (37) _____. The process by which eggs are formed is called (38) _____. The diploid cells that give rise to the eggs are the (39) _____. At three months of development, a female fetus has none of these precursor cells left. They have all matured into (40) _____, each of which will begin meiosis I. Surrounding each primary oocyte are accessory cells making up the (41) _____. After puberty, a few primary oocytes will initiate the completion of meiosis I. One oocyte will complete meiosis I producing one (42) _____ and one

(43) _____. The accessory cells of the follicle secrete the hormone
(44) _____. After being released from the ovary, the secondary oocyte may
complete meiosis II in the (45) _____. Accessory cells left behind in the ovary
form the (46) _____, which produces both estrogen and progesterone.

The oviduct, also known as (47) the _____, has an open end, almost surrounding
the ovary, with ciliated "fingers" called (48) _____. After the egg is fertilized,
the (49) _____ is swept along the oviduct to the (50) _____.
This organ has a two- layered wall. The inner layer is dense with blood vessels and forms the
(51) _____, the mother's contribution to the (52) _____. If
fertilization of the egg does not occur, this layer is shed during (53) _____, as part
of the (54) _____. The outer layer of the uterus is muscular.   This
(55) _____ contracts strongly during birth.   The outer end of the uterus is almost
closed off by a ring of connective tissue, the (56) _____. On the other side of the
ring of connective tissue is the (57) _____, which serves as the birth canal.

Sexual excitement in a female increases blood flow to external reproductive tissues. These include
external folds of skin, the (58) _____, and a rounded projection, the
(59) _____, which becomes erect.

The egg released from the ovary is surrounded by follicle cells that develop into the
(60) _____. Between this cell layer and the egg is a clear jellylike layer called
the (61) _____. Both layers protect the egg. About a week after a sperm has
broken through the protective layers and fertilized the egg, a hollow ball of cells, called a
(62) _____, has developed. The (63) _____ of the "ball"
will become the embryo. The outer layer of the "ball" will adhere to and burrow into the endometrium
during (64) _____.

When pregnancy occurs, estrogen and progesterone stimulate the (65) _____ to
develop the capacity to secrete milk. The first substance that is secreted for the newborn is
(66) _____, which is high in protein and antibodies.

The growing fetus causes the uterus to stretch. The stretching, in combination with hormones the fetus
and mother secrete, triggers contractions of the uterus and (67) _____ has begun.

The prevention of pregnancy involves various methods of (68) _____. When these
methods fail, a pregnancy may be terminated by (69) _____.

Key Terms and Definitions

abortion: the procedure for terminating pregnancy in which the cervix is dilated and the embryo and placenta removed by suction.

acquired immune deficiency syndrome (AIDS): an infectious disease caused by the human immunodeficiency virus (HIV); attacks and destroys T cells, weakening the immune system.

acrosome: a vesicle located at the tip of an animal sperm, which contains enzymes needed to dissolve protective layers around the egg.

amplexus: a form of external fertilization found in amphibians, in which the male holds the female during spawning and releases his sperm directly onto her eggs.

asexual reproduction: reproduction not involving the union of genetic material from two different organisms. Usually, asexual reproduction produces genetically identical copies of the parent organism.

birth control pill: a temporary contraceptive method that prevents ovulation by providing a continuing supply of estrogen and progesterone, which in turn suppresses LH release; these must be taken daily, usually for 21 days each menstrual cycle.

blastocyst: an early stage of human embryonic development, consisting of a hollow ball of cells, enclosing a mass of cells attached to its inner surface, which becomes the embryo.

bud: in animals, a small copy of an adult that develops on the body of the parent. It eventually breaks off and becomes independent.

budding: asexual reproduction by growth of a miniature copy, or bud, of the adult animal on the body of the parent. The bud breaks off to begin independent existence.

bulbourethral gland: in male mammals, a gland that secretes a basic, mucus-containing fluid that forms part of the semen.

cervical cap: a birth control device consisting of a rubber cap that fits over the cervix, preventing sperm form entering the uterus.

cervix: a ring of connective tissue at the outer end of the uterus, leading into the vagina.

chlamydia (kla-mid´-e-uh): a sexually transmitted disease, caused by a bacterium, that causes inflammation of the urethra in males and of the urethra and cervix in females.

chorionic gonadotropin (CG): a hormone secreted by the chorion (one of the fetal membranes), which maintains the integrity of the corpus luteum during early pregnancy.

clitoris: an external structure of the female reproductive system; composed of erectile tissue; a sensitive point of stimulation during sexual response.

colostrum: a yellowish fluid high in protein and containing antibodies that is produced by the female breasts before milk secretion begins.

condom: a contraceptive sheath that may be worn over the penis during intercourse, preventing sperm from being deposited in the vagina.

contraception: the prevention of pregnancy.

copulation: reproductive behavior in which the male penis is inserted into the female body, where it releases sperm.

corona radiata: the layer of cells surrounding an egg after ovulation.

corpus luteum: in the mammalian ovary, a structure derived from the follicle after ovulation, which secretes the hormones estrogen and progesterone.

crab lice: an arthropod parasite that can infest humans; can be transmitted by sexual contact.

diaphragm: a rubber cap that fits snugly over the cervix, preventing the sperm from entering the uterus and thereby preventing pregnancy.

dioecious: pertaining to organisms in which male and female gametes are produced by separate individuals.

douching: washing the vagina after intercourse in an attempt to remove sperm before they enter the uterus; an ineffective contraceptive method.

egg: the haploid female gamete, usually large and nonmotile, containing food reserves for the developing embryo.

endometrium: the nutritive inner lining of the uterus.

epididymis: tubes that connect with and receive sperm from the seminiferous tubules of the testis.

estrogen: in vertebrates, a female sex hormone produced by follicle cells of the ovary, that stimulates follicle development, oogenesis, development of secondary sex characteristics, and growth of the uterine lining.

external fertilization: union of sperm and egg outside the body of either parental organism.

fallopian tube: see *oviduct*.

fertilization: the fusion of male and female haploid gametes, forming a zygote.

fimbria (pl. fimbriae): in female mammals, the ciliated, fingerlike projections of the oviduct that sweep the ovulated egg from the ovary into the oviduct.

fission: asexual reproduction by dividing the body into two smaller, complete organisms.

follicle: in the ovary of female mammals, the oocyte and its surrounding accessory cells.

follicle-stimulating hormone (FSH): a hormone produced by the anterior pituitary gland that stimulates spermatogenesis in males and development of the follicle in females.

genital herpes: a sexually transmitted disease, caused by a virus, than can cause painful blisters on the genitals and surrounding skin.

gonad: organ where reproductive cells are formed; in males, the testes, and in females, the ovaries.

gonadotropin-releasing hormone (GnRH): a hormone produced by the neurosecretory cells of the hypothalamus, which stimulates cells in the anterior pituitary to release FSH and LH. GnRH is involved in the menstrual cycle and spermatogenesis.

gonorrhea (gon-a-re´-uh): a sexually transmitted bacterial infection of the reproductive organs; if untreated, can result in sterility.

hermaphrodite: an organism that produces both male and female gametes.

implantation: the process whereby the early embryo embeds itself within the lining of the uterus.

inner cell mass: the thickened region of the blastocyst that develops into the embryo.

internal fertilization: union of sperm and egg inside the body of the female.

interstitial cells: in the vertebrate testis, testosterone-producing cells located between the seminiferous tubules.

intrauterine device (IUD): a small copper or plastic loop, squiggle, or shield that is inserted in the uterus; a contraceptive method that works by irritating the uterine lining so that it cannot receive the embryo.

labium (pl., labia): one of a pair of folds of skin of the external structures of the mammalian female reproductive system.

labor: a series of contractions of the uterus that result in birth.

lactation: the secretion of milk from the mammary glands.

luteinizing hormone (LH): a hormone produced by the anterior pituitary gland that stimulates testosterone production in males and development of the follicle, ovulation, and production of the corpus luteum in females.

mammary gland (mam´-uh-re): a milk-producing gland used by female mammals to nourish their young.

menstrual cycle: in human females, a complex 28-day cycle during which hormonal interactions among the hypothalamus, pituitary gland, and ovary coordinate ovulation and the preparation of the uterus to receive and nourish the fertilized egg. If pregnancy does not occur, the uterine lining is shed during menstruation.

menstruation: in human females, the monthly discharge of uterine tissue and blood from the uterus.

monoecious: pertaining to organisms in which male and female gametes are produced in the same individual.

myometrium: the muscular outer layer of the uterus.

oogenesis: the process by which egg cells are formed.

oogonium (pl. oogonia): a diploid cell in female animals that gives rise to a primary oocyte.

ovary: the gonad of female animals.

oviduct: in mammals, the tube leading from the ovary to the uterus.

ovulation: the release of a secondary oocyte, ready to be fertilized, from the ovary.

parthenogenesis: a specialization of sexual reproduction, in which a haploid egg undergoes development without fertilization.

penis: an external structure of the male reproductive and urinary systems; serves to deposit sperm into the female reproductive system and delivers urine to the exterior.

pheromone (fer´-uh-mon): a chemical produced by an organism that alters the behavior or physiological state of another member of the same species.

placenta: in mammals, a structure formed of both embryonic and maternal tissues, which serves to exchange nutrients and wastes between embryo and mother.

polar body: in oogenesis, a small cell containing a nucleus but virtually no cytoplasm, produced by the first meiotic division of the primary oocyte.

primary oocyte: a diploid cell, derived from the oogonium by growth and differentiation, which undergoes meiosis to produce the egg.

primary spermatocyte: a diploid cell, derived from the spermatogonium by growth and differentiation, which undergoes meiosis to produce four sperm.

progesterone: a hormone produced by the corpus luteum that promotes development of the uterine lining.

prostate gland: a gland that produces part of the fluid component of semen. The prostate fluid is basic and contains a chemical that activates sperm movement.

regeneration: (1) regrowth of a body part after loss or damage; (2) asexual reproduction by regrowth of an entire body from a fragment.

rhythm method: a contraceptive method involving abstinence from intercourse during ovulation.

scrotum: the pouch of skin containing the testes of male mammals.

secondary oocyte: a large haploid cell derived from the first meiotic division of the diploid primary oocyte.

secondary spermatocyte: a large haploid cell derived by meiosis I from the diploid primary spermatocyte.

semen: the sperm-containing fluid produced by the male reproductive tract.

seminal vesicle: in male mammals, a gland that produces a basic, fructose-containing fluid that forms part of the semen.

seminiferous tubules: a series of tubes in the vertebrate testis in which sperm are produced.

Sertoli cell: a large cell in the seminiferous tubule that regulates spermatogenesis and nourishes the developing sperm.

sexually transmitted disease (STD): a disease that is passed from person to person by sexual contact.

sexual reproduction: a form of reproduction in which genetic material from two parental organisms is combined in the offspring. Usually, two haploid gametes fuse to form a diploid zygote.

spawning: a method of external fertilization in which male and female parents shed gametes into the water, and sperm must swim through the water to reach the eggs.

sperm: the haploid male gamete, small, motile, and containing little cytoplasm.

spermatid: a haploid cell derived from the secondary spermatocyte by meiosis II. The mature sperm is derived from the spermatid by differentiation.

spermatogenesis: the process by which sperm cells are formed.

spermatogonium (pl. spermatogonia): a diploid cell lining the walls of the seminiferous tubules that gives rise to a primary spermatocyte.

spermatophore: in some animals, in a variation on internal fertilization, the males package their sperm in a container that can be inserted in the female reproductive tract.

spermicide: a sperm-killing chemical; used for contraceptive purposes.

sterilization: the method of contraception in which the pathways through which the sperm (vas deferens) or egg (oviducts) must travel are interrupted. Generally permanent, it is the most effective form of contraception.

syphilis (si´-ful-is): a sexually transmitted bacterial infection of the reproductive organs; if untreated, can damage the nervous and circulatory systems.

testis (pl. testes): the gonad of male mammals.

testosterone: in vertebrates, a hormone produced by the interstitial cells of the testis; stimulates spermatogenesis and the development of male secondary sex characteristics.

trichomoniasis (trik-o-mo-ni´-uh-sis): a sexually transmitted disease, caused by the protist *Trichomonas*, that causes inflammation of the mucous membranes that line the urinary tract and genitals.

tubal ligation: surgical procedure in which a woman's oviducts are cut so that the egg cannot reach the uterus, which makes her infertile.

urethra: the tube leading from the urinary bladder to the outside of the body; in males, the urethra also receives sperm from the vas deferens and conducts both sperm and urine (at different times) to the tip of the penis.

uterus: in female mammals, the part of the reproductive tract that houses the embryo during pregnancy.

vagina: the passageway leading from the outside of the body to the cervix of the uterus.

vas deferens: the tube connecting the epididymis of the testis with the urethra.

vasectomy: a surgical procedure in which a man's vas deferens are cut, preventing sperm from reaching the penis during ejaculation, which makes him infertile.

withdrawal: removal of the penis from the vagina just before ejaculation in an attempt to avoid pregnancy; an ineffective contraceptive method.

zona pellucida: a clear, noncellular layer between the corona radiata and the egg.

zygote: the fertilized egg.

THINKING THROUGH THE CONCEPTS

True or False: Determine if the statement given is true or false. If it is false, change the <u>underlined</u> word so that the statement reads true.

70. _____ Meiosis occurs in <u>asexual</u> reproduction.

71. _____ A flatworm reproducing by regeneration would produce <u>identical</u> "offspring".

72. _____ Male honey bees develop from <u>unfertilized</u> eggs.

73. _____ Some species of fish <u>do not produce males; all members are female.</u>

74. _____ Due to the evolution of <u>sexual reproduction,</u> genetic variation occurs, enabling the action of natural selction to take place.

75. _____ Species with individuals of both sexes are <u>monoecious.</u>

76. _____ In order for <u>external fertilization</u> to be successful, egg and sperm must be released at the same time and in the same area.

77. _____ Copulation occurs during <u>external fertilization.</u>

78. _____ The testes are located in the scrotum, outside the body, keeping the them 4 C <u>warmer</u> than the body's core temperature.

79. _____ Primary spermatocytes are <u>diploid.</u>

80. _____ After <u>meiosis I,</u> spermatids are formed.

81. _____ <u>Oogenesis</u> begins at puberty.

82. _____ <u>Follicle-stimulating hormone</u> functions in males with testosterone to initiate spermatogenesis.

83. _____ A method of contraception that could be developed would block FSH production, resulting in <u>an infertile male who would not be impotent.</u>

84. _____ In males, the reproductive tract merges with the <u>urinary tract.</u>

85. _____ The egg is, in actuality, a <u>primary</u> oocyte.

86. _____ Fertilization occurs in the <u>vagina.</u>

87. _____ Every month the uterus prepares for <u>implantation</u>.

88. _____ The presence of a <u>corpus luteum</u> maintains the endometrium.

89. _____ The birthing process is a process regulated by <u>negative feedback</u>.

90. _____ The embryo secretes CG which <u>maintains the pregnancy</u>.

91. _____ <u>External fertilization</u> increases the chances for transmission of disease.

92. _____ The hormone prolactin <u>inhibits</u> milk production by the mammary glands.

93. _____ Contractions resume after the birth of a baby to expel the <u>placenta</u>.

94. _____ Birth control pills prevent <u>sperm from uniting with an egg</u>.

95. _____ Douching after intercourse is an <u>effective</u> method of contraception.

96. _____ Condom use not only prevents pregnancy, but it also <u>reduces</u> disease transmission.

97. _____ Successful use of the <u>diaphragm</u> requires diligent observations of changes in body temperature and mucus production from the cervix.

98. _____ <u>Condoms</u> are available for women.

Matching: Birth control methods.

99. _____ IUD

100. _____ condom

101. _____ cervical cap

102. _____ birth control pill

103. _____ tubal ligation

104. _____ spermicides

105. _____ douching

106. _____ vasectomy

107. _____ withdrawl

108. _____ rhythm method

Choices:

a. sterilization

b. barrier method

c. prevent ovulation

d. prevent implantation

e. ineffective

Identify the following: Are these sexually transmitted diseases caused by a **bacterium, virus, protistan,** or **arthropod**?

109. _____ gonorrhea

110. _____ AIDS

111. _____ trichomoniasis

112. _____ genital herpes

113. _____ syphilis

114. _____ crab lice

115. _____ chlamydia

Short Answer:

116. Trace the path of spermatogenesis from spermatagonium to mature sperm. Include where in the pathway meiosis I and meiosis II occur.

Spermatagonium ----> _____ ----> _____ ---->

_____ ----> _____ ----> _____ ----> _____

117. What is the purpose of the polar bodies?

_____

_____

_____

_____

118. Complete the following table by identifying the function of the given hormone and whether or not it is active in males, females, or both.

Hormone	Function	Active in males, females, or both
Gonadotropin-releasing Hormone (GnRH)		
Lutenizing Hormone (LH)		
Follicle-stimulating Hormone (FSH)		
Testosterone		
Estrogen		
Progesterone		
Chorionic gonadotropin (CG)		

CLUES TO APPLYING THE CONCEPTS

These practice questions are intended to sharpen your ability to apply critical thinking and analysis to the biological concepts covered in this chapter.

119. Explain how the birth process is controlled by a positive feedback mechanism.

_____
_____
_____
_____
_____

120. If a couple is unsuccessful in becoming pregnant, a physician may suggest a sperm count be conducted. Why would the number of sperm produced affect the success of attempts to become pregnant?

_____
_____
_____
_____
_____

ANSWERS TO EXERCISES

1. sexual reproduction
2. asexual reproduction
3. budding
4. regeneration
5. fission
6. parthenogensis
7. dioecious
8. eggs
9. sperm
10. monoecious
11. hermaphrodites
12. external fertilization
13. Spawning
14. pheromone
15. amplexus
16. internal fertilization
17. copulation
18. spermatophore
19. ovulation
20. gonads
21. testes
22. scrotum
23. seminiferous tubules
24. interstitial cells
25. spermatagonia
26. sertoli cells
27. spermatogenesis
28. acrosome
29. penis
30. epididymis
31. vas deferens
32. urethra
33. semen
34. seminal vesicles
35. prostate gland
36. bulbourethral gland
37. ovaries
38. oogenesis
39. oogonia
40. primary oocytes

41. follicle
42. secondary oocyte
43. polar body
44. estrogen
45. oviduct
46. corpus luteum
47. fallopian tube
48. fimbriae
49. zygote
50. uterus
51. endometrium
52. placenta
53. menstruation
54. menstrual cycle
55. myometrium
56. cervix
57. vagina
58. labia
59. clitoris
60. corona radiata
61. zona pellucida
62. blastocyst
63. inner cell mass
64. implantation
65. mammary glands
66. colostrum
67. labor
68. contraception
69. abortion
70. false, sexual
71. true
72. true
73. true
74. true
75. false, dioecious
76. true
77. false, internal fertilization
78. false, cooler

79. true
80. false, meiosis II
81. false, spermatogenesis
82. true
83. true
84. true
85. false, secondary oocyte
86. false, oviduct (fallopian tube)
87. true
88. true
89. false, positive feedback
90. true
91. false, internal fertilization
92. false, stimulates
93. true
94. false, ovulation
95. false, very ineffective
96. true
97. true
98. true
99. d
100. b
101. b
102. c
103. a
104. e
105. e
106. a
107. e
108. e
109. bacterium
110. virus
111. protistan
112. virus
113. bacterium
114. arthropod
115. bacterium

116. Spermatagonium ----> primary spermatocyte ----> meiosis I ----> secondary spermatocyte ----> meiosis II ----> spermatid ----> sperm

117. Polar bodies serve to hold discarded chromosomes. Meiosis produces haploid cells. Since the egg requires as much of the cytoplasm of the dividing cells as possible and "extra" chromosomes must be discarded to produce the haploid state, polar bodies are the result.

118.

Hormone	Function	Active in males, females, or both
Gonadotropin-releasing hormone (GnRH)	stimulates anterior pituitary to produce LH and FSH	both
Lutenizing hormone (LH)	stimulates interstitial cells to produce testosterone; initiates follicle development; surge triggers meiosis I to continue, ovulation, development of corpus luteum	both
Follicle-stimulating hormone (FSH)	stimulates sertoli cells and spermatagonia, causes spermatogenesis; initiates follicle development	both
Testosterone	stimulates sertoli cells and spermatagonia, causes spermatogenesis, triggers development of secondary sex characteristics, sexual drive, required for intercourse	males
Estrogen	stimulates endometrium production; decrease triggers menstruation; increase causes increase in LH and FSH	females
Progesterone	maintains endometrium production; decrease triggers menstruation and increases production of GnRH	females
Chorionic gonadotropin (CG)	prevents breakdown of corpus luteum	females

119. A positive feedback reaction is defined as a situation in which a change initiates events that tend to amplify the original change. Birth of the fetus is initiated by stretching of the uterus and by hormones from both the fetus and the mother, causing labor. The contractions of the uterine smooth muscle are triggered by the stretching caused by the growing fetus. Toward the end of development, the fetus produces steroid hormones that increase estrogen and prostoglandin production by the uterus and placenta. The increased hormone production, in combination with the stretching, increases the contractions of the uterus. As the baby's head pushes against the cervix, it dilates. This stretching signals the hypothalamus to release oxytocin. With oxytocin and prostoglandins circulating in the blood, the uterus contracts even more intensely, pushing the baby from the birth canal (the vagina). The changes that occur in the uterus cause further changes to occur until the process reaches a climax, the birth.

120. The number of sperm produced is important since a large number of sperm are needed to break through the corona radiata and the zona pellucida around the egg. If an insufficient number of sperm is produced to break down the barriers, no sperm can enter the egg to fertilize it.

Chapter 36: Animal Development

OVERVIEW

This chapter presents the processes that occur after a zygote has been produced. As the development of an embryo progresses, the cells differentiate so that the specialized body functions can take place. This chapter presents information relating to the indirect development of animals that undergo metamorphosis; however, it provides more detail relating to the direct development humans and other mammals follow.

1) How Do Cells Differentiate During Development?

Development is the process an organism goes through from a fertilized egg to maturity to death. During this process, cells that begin as identical cells from the zygote become specialized in form and function. The specialization that occurs is called **differentiation**. Differentiation is directed by the genes that are used or activated, transcribed, and translated. The genes that are used are controlled by regulatory molecules that bind to a gene, blocking or promoting transcription. The regulatory molecules are concentrated in the cytoplasm of the egg. When the fertilized egg divides, the daughter cells that are produced receive different regulatory molecules. The specialization pathway that each daughter cell will take is determined by which regulatory molecules it received. Cells may also differentiate according to chemicals received from other body cells.

2) How Do Indirect and Direct Development Differ?

An organism in its early stages of development is an **embryo**. A developing embryo is nourished by a protein and lipid rich **yolk**. The amount of yolk present in an egg depends on the way in which an animal develops. **Indirect development** occurs in animals that hatch from an egg in a form different from that of the adult. The maturation process involves drastic changes in the form of the animal. Amphibians and invertebrates, such as echinoderms and insects, undergo indirect development. The eggs that are laid by these animals are numerous, but they contain only a small amount of yolk. Since the amount of yolk is small, the embryo within an egg rapidly develops into a sexually immature **larva**. The larva will eventually undergo **metamorphosis** and become a sexually mature adult. **Direct development** occurs in animals that are born resembling miniature adults. The maturation process does not involve drastic changes in body form. These young may be provided large amounts of nourishing yolk within their eggs or they may be nourished within the mother's body before birth. Since both development strategies demand a large input of energy from the mother, a small number of offspring is produced.

The successful transition from aquatic life to terrestrial life occurred in animals after the **amniote egg** evolved in reptiles. Reptiles, birds, and mammals produce amniote eggs consisting of four **extraembryonic membranes**. The **chorion** is the outermost membrane and conducts gas exchange through the shell. The **amnion** is a membrane that encloses the embryo in a watery environment. The **allantois** surrounds wastes, and the **yolk sac** stores food for the developing embryo. Although the mammalian egg does not produce much yolk, the yolk sac membrane still exists.

3) How Does Animal Development Proceed?

Animal development begins with a series of mitotic divisions called **cleavage**. Although the number of cells increases during cleavage, the size of the overall structure does not increase. As daughter cells are produced, gene-regulating substances are distributed among them. As cleavage continues, a ball of cells about the same size as the zygote, called a **morula**, forms. The morula progresses to the development of a hollow ball of cells called a **blastula**. The division of frog zygotes involves cleavage through a pigmented area, the **gray crescent**, such that each cell contains half of the crescent. Based on experiments separating developing frog embryos, the gray crescent contains gene-regulating substances that are required for the normal development of a tadpole. Further divisions of the blastula result in the formation of the **blastopore**. During the formation of the blastopore, cell movement occurs, called **gastrulation**. Three embryonic tissue layers form during gastrulation. The **endoderm** forms from the cells lining the blastopore. This eventually becomes the digestive tract. The **ectoderm** forms from the cells lining the outside of the **gastrula**. These cells give rise to the epidermis and the nervous system. Cells that move to the area between the endoderm and ectoderm are called the **mesoderm**. Mesoderm cells develop into muscles, the skeleton, including the **notochord**, and the circulatory system.

   Through a process called **induction**, chemicals produced by cells influence the development of other cells. **Organogenesis**, the process of organ development, is controlled by induction. Such signals may include "survival" signals or "death" signals. Some cells will die unless they receive a survival signal from surrounding cells, while other cells will live unless they receive a death signal from surrounding cells. In this way, motor neurons synapse with muscle cells properly and separate fingers and toes form in humans. Development does not stop at birth, however. The development of mature reproductive organs, for example, occurs at a genetically controlled age. The genes regulating this development may be triggered by environmental or social cues. As cells age they function less efficiently. They may be programmed to divide a certain number of times and then die. Or a cell's longevity may depend on its ability to repair damage done to its DNA. As cells die, the organism ages. Cancer cells, however, have not only lost the mechanism controlling division, but they also have an indefinite life span.

4) How Do Humans Develop?

Human development reflects our evolutionary ancestry. The fertilized egg undergoes cleavage as it travels through the fallopian tubes to the uterus. It is a **blastocyst**, rather than a blastula, that implants into the uterine wall. The blastocyst has a thick **inner cell mass** on one side of its hollow ball structure. It is the inner cell mass that will develop into the embryo and three of the extraembryonic membranes. The thin outer wall becomes the fourth extraembryonic membrane, the embryo's portion of the placenta, the chorion. The **embryonic disc** separates two fluid-filled sacs. One sac is enclosed by the amnion and contains the amniotic fluid. The amnion grows around the fetus so that it is maintained within the watery environment all animal embryos need. The second sac is the yolk sac but it contains no yolk. As gastrulation begins, the **primitive streak** forms in the embryonic disc. This corresponds to the blastopore mentioned above. The mesoderm of the embryo is formed from cells migrating through the primitive streak to the interior of the embryo. The cell layer above the primitive streak forms the ectoderm. The cell layer below the primitive streak forms the endoderm. As the embryo grows, the endoderm forms a tube that later becomes the digestive tract. The notochord, formed from mesoderm, causes the ectoderm to form a groove. The groove closes over, creating the predecessor to the brain and spinal cord, the **neural tube**. Continued development generates a beating heart and rapid growth of the brain. By the second month, all of the major organs have developed, including the gonads. Sex hormones are secreted by the testes or ovaries, which influence the development of embryonic organs and certain regions of the brain. The embryo is called a **fetus** at the end of the second month.

The development of the **placenta** begins as the embryo burrows into the endometrium. The outer cells of the embryo form the chorion. The chorion sends fingerlike projections called **chorionic villi** into the endometrium, intricately linking the two.   The placenta secretes estrogen to stimulate the growth of the uterus and mammary glands, and progesterone to stimulate the mammary glands and prevent premature uterine contractions. The placenta also regulates the exchange of substances between the mother and the fetus. The membranes of the fetal capillaries and the chorionic villi are very selective with respect to the substances that are exchanged; however, some disease-causing organisms and damaging chemicals can pass through. The next seven months of fetal development involves, for the most part, growth of the already-formed structures. The brain and spinal cord grow and the fetus begins to respond to stimuli. The respiratory, digestive, and urinary tracts enlarge and begin functioning. During the last two months, the fetus positions itself head downward, resting against the cervix in preparation for birth. At birth, the baby suddenly must breath in oxygen and exhale carbon dioxide, it must attempt to regulate its body temperature, and it must rely on suckling to receive nourishment. Suddenly, it is out of the safe environment of the uterus and out in the world.

KEY TERMS AND CONCEPTS

Fill-In: From the following list of key terms, fill in the blanks in the following statements.

allantois	differentiation	gastrulation	notochords
amnion	direct development	gray crescent	organogenesis
amniote egg	ectoderm	indirect development	placenta
blastocyst	embryo	induction	primitive streak
blastopore	embryonic disc	inner cell mass	yolk
blastula	endoderm	larva	yolk sac
chorion	extraembryonic membrane	mesoderm	
chorionic villus	fetal alcohol syndrome (FAS)	metamorphosis	
cleavage	fetus	morula	
development	gastrula	neural tube	

(1) _____ is the process of an organism developing from a fertilized egg, through adulthood, to death. Throughout this process cells specialize by (2) _____.

Animal development begins with an egg containing (3) _____, rich in protein and lipids needed by the (4) _____.

Animals progressing through (5) _____ undergo drastic changes from a sexually immature (6) _____ to a sexually mature adult. A caterpillar will undergo (7) _____ to become a butterfly. Animals progressing through (8) _____ are born as miniature versions of the adult.

The evolution of the (9) _____ enabled animals to live away from standing water. This structure contains four (10) _____. The (11) _____ allows gas exchange through the shell. The (12) _____ surrounds the embryo, maintaining it in a watery environment. The (13) _____isolates wastes from the embryo. The (14) _____stores food for the developing embryo.

A series of mitotic divisions initiated after the zygote forms is called (15) _____. After several division cycles, a solid ball of cells forms, called the (16) _____. With continued divisions, a cavity develops, and this hollow ball of cells is called a (17) _____. Research determining the presence of different gene-regulating substances in the cytoplasm involved experiments using the (18) _____ of frog eggs.

An indentation forms in the blastula, called the (19) _____. Three embryonic tissues form as cells migrate through the enlarging indentation. Cells lining the indentation develop the (20) _____, eventually becoming the digestive tract. Cells to the outside form the (21) _____, eventually becoming the epidermis and the nervous system. Cells that migrate between the two layers form the (22) _____, eventually becoming the muscles and skeleton including the (23) _____. The movement of the cells is called (24) _____ and the structure that is formed is the (25) _____.

Chemical messages received from other cells often determines the fate of cell development. This developmental process is (26) _____. The formation of organs from rearrangement of cells is (27) _____.

During human development, instead of a blastula forming, a (28) _____ forms. The thick (29) _____ will become the (30) _____, while the thin outer wall will become the chorion. The cells destined to become the embryo grow and split so that two fluid-filled sacs are separated by a double layer of cells, the (31) _____. The double layer of cells splits apart slightly, forming the (32) _____ (or blastopore in other animals). During the third week of development, the (33) _____ is generated, the precursor of the brain and spinal cord. At the end of the second month, the embryo is referred to as a (34) _____.

The chorion from the embryo penetrates the endometrium of the uterus using (35) _____. The chorion, interacting with the endometrium, generates the (36) _____,which is selective as to which substances pass into the bloodstream of the fetus. Not all harmful substances are blocked however. For instance, alcohol readily passes from the mother's bloodstream into that of the fetus, often strongly affecting the fetus. Alcoholic mothers give birth to children displaying (37)_____.

(38) _____ genes code for proteins that turn on or turn off other genes, determining the differentiation of embryonic cells.

Key Terms and Definitions

allantois: one of the embryonic membranes of reptiles, birds, and mammals. In reptiles and birds, the allantois serves as a waste-storage organ. In mammals, the allantois forms most of the umbilical cord.

amnion: one of the embryonic membranes of reptiles, birds, and mammals, enclosing a fluid-filled cavity that envelops the embryo.

amniote egg (am-ne-ot′): the egg of reptiles and birds; contains an amnion that encloses the embryo in a watery environment, allowing the egg to be laid on dry land.

blastocyst: in mammalian embryonic development, a hollow ball of cells formed at the end of cleavage.

blastopore: the site at which a blastula indents to form a gastrula.

blastula: in animals, the embryonic stage attained at the end of cleavage, in which the embryo usually consists of a hollow ball with a wall one or several cell layers thick.

chorion: the outermost embryonic membrane in reptiles, birds, and mammals. In birds and reptiles, the chorion functions mostly in gas exchange. In mammals, the chorion forms most of the embryonic part of the placenta.

chorionic villus (kor-e-on-ik; pl., chorionic villi): in mammalian embryos, a fingerlike projection of the chorion that penetrates the uterine lining and forms the embryonic portion of the placenta.

cleavage: the early cell divisions of embryos, in which little or no growth occurs between divisions. Cleavage reduces the cell size and distributes gene-regulating substances to the newly formed cell.

development: the process by which an organism proceeds from fertilized egg through adulthood to eventual death.

differentiation: the process whereby relatively unspecialized cells, especially of embryos, become specialized into particular tissue types.

direct development: a developmental pathway in which the offspring is born as a miniature version of the adult and does not radically change its body form as it grows and matures.

ectoderm: the outermost embryonic tissue layer, that gives rise to structures such as hair, the epidermis of the skin, and the nervous system.

embryo: stages of animal development that begin with fertilization of the egg and end with hatching or birth. In mammals, embryo refers to the early stages when the developing animal does not yet resemble the adult of the species.

embryonic disk: in human embryonic development, the flat two-layered group of cells that separates the amniotic cavity from the yolk sac.

endoderm: the innermost embryonic tissue layer, which gives rise to structures such as the lining of the digestive and respiratory tracts.

extraembryonic membranes: in reptile, bird, and mammal embryonic development, the chorion, amnion, allantois, and yolk sac that function in gas exchange, provision of the watery environment needed for development, waste storage, and storage of the yolk, respectively.

fetal alcohol syndrome (FAS): a cluster of symptoms including retardation and physical abnormalities that occur in infants born to mothers who consumed large amounts of alcoholic beverages during pregnancy.

fetus: the later stages of mammalian embryonic development (after the second month for humans), when the developing animal has come to resemble the adult of the species.

gastrula: in animal development, a three-layered embryo with ectoderm, mesoderm, and endoderm cell layers. The endoderm layer usually encloses the primitive gut.

gastrulation: the process whereby a blastula develops into a gastrula, including the formation of endoderm, ectoderm, and mesoderm.

gray crescent: in frog embryonic development, an area of intermediate pigmentation in the fertilized egg containing gene-regulating substances required for the normal development of the tadpole.

homeobox: a sequence of DNA coding for special, 60 amino acid proteins, that activate or inactivate genes that control development; homeoboxes thus specify embryonic cell differentiation.

indirect development: a developmental pathway in which an offspring goes through radical changes in body form as it matures.

induction: the process by which a group of cells causes other cells to differentiate into a specific tissue type.

inner cell mass: in human embryonic development, the cluster of cells on one side of the blastocyst, that will develop into the embryo.

larva: a small, sexually immature form of an animal with indirect development, often much different in body form from the adult.

mesoderm: the middle embryonic tissue layer, lying between the endoderm and ectoderm, and usually the last to develop. Mesoderm gives rise to structures such as muscle and skeleton.

metamorphosis: in animals with indirect development, a radical change in body form from larva to sexually mature adult.

morula: in animals, an embryonic stage during cleavage, when the embryo consists of a solid ball of cells.

neural tube: a structure derived from ectoderm during early embryonic development that later becomes the brain and spinal cord.

notochord (not´-o-kord): a stiff but somewhat flexible, supportive rod found in all members of the phylum Chordata at some stage of development.

organogenesis: the process by which the layers of the gastrula (endoderm, ectoderm, mesoderm) rearrange themselves into organs.

placenta: in mammals, a structure formed by a complex interweaving of the uterine lining and the embryonic membranes, especially the chorion; functions in gas, nutrient, and waste exchange between embryonic and maternal circulatory systems and secretes hormones.

primitive streak: in reptiles, birds, and mammals, the region of the ectoderm of the two-layered embryonic disk through which cells migrate to form mesoderm.

yolk: protein- or lipid-rich substances contained in eggs as food for the developing embryo.

yolk sac: one of the embryonic membranes of reptile, bird, and mammalian embryos. In birds and reptiles, the yolk sac is a membrane surrounding the yolk in the egg. In mammals, the yolk sac is empty but forms part of the umbilical cord and the digestive tract.

THINKING THROUGH THE CONCEPTS

True or False: Determine if the statement given is true or false. If it is false, change the underlined word so that the statement reads true.

39. _____ Cells specialize because underlined{unnecessary genes are lost}.
40. _____ Chemical messages from other cells change how a cell differentiates.
41. _____ Animals whose young will develop indirectly produce eggs with a large amount of yolk.
42. _____ Larvae are sexually mature.
43. _____ Human eggs contain a large amount of yolk.
44. _____ During cleavage, embryonic cells do not grow.
45. _____ The presence of a large yolk may actually hinder cleavage.
46. _____ The grey crescent of a frog egg is necessary for proper development of an embryo.

47. _____   Gastrulation involves the migration of cells, forming three layers.

48. _____   Cells transplanted to a new region of the embryo differentiate according to the area into which they were transferred.

49. _____   Distinct fingers develop because webbing cells receive a "death signal" from other cells.

50. _____   Cell differentiation and development stops after birth.

51. _____   Cell death is preprogrammed.

52. _____   Long-lived cells can divide damaged DNA.

53. _____   The neural tube is the precursor structure to the digestive tract.

54. _____   The X chromosome determines if the gonads will develop into ovaries or testes.

55. _____   Testosterone and estrogen affect certain areas of the brain.

56. _____   The amnion and the endometrium create the placenta.

57. _____   The kidneys in a fetus are functional before birth.

58. _____   Most drugs can pass through the placenta to the fetus.

59. _____   Women who smoke are more likely to have a miscarriage than women who do not smoke.

Matching: Adult tissue types and the embryonic layers from which they developed.

60. _____   epidermis of skin   Choices:

61. _____   muscle   a. endoderm

62. _____   skeleton   b. mesoderm

63. _____   digestive tract   c. ectoderm

64. _____   nervous system

65. _____   circulatory system

Identify the Following: **Can** or **cannot** the following substances cross the placental barrier from the mother's circulatory system to that of the fetus?

66. _____ alcohol   72. _____ syphilis

67. _____ most cells   73. _____ oxygen

68. _____ nicotine   74. _____ large proteins

69. _____ carbon monoxide   75. _____ HIV

70. _____ Thalidomide   76. _____ German measles

71. _____ aspirin   77. _____ nutrients

Short Answer:

78. The cells undergoing cleavage do not increase in size. Why not?

_____

_____

_____

79. Place the following terms in the proper developmental order.

   blastula   zygote   fetus   embryo   gastrula   morula

_____

80. Complete the following table comparing the extraembryonic membranes produced in reptilian and human eggs.

Extraembryonic membrane	Structure and location in reptilian egg	Function in reptilian egg	Structure and location in human egg	Function in human egg
Chorion				
Amnion				
Allantois				
Yolk sac				

CLUES TO APPLYING THE CONCEPTS

These practice questions are intended to sharpen your ability to apply critical thinking and analysis to the biological concepts covered in this chapter.

81. Identify the two ways cells initiate differentiation.

_____

_____

82. Compare indirect development with direct development. Identify two types of animals that develop indirectly and two that develop directly.

_____

_____

_____

_____

_____

_____

_____

_____

_____

_____

ANSWERS TO EXERCISES

1. Development
2. differentiation
3. yolk
4. embryo
5. indirect development
6. larva
7. metamorphosis
8. direct development
9. amniote egg
10. extraembryonic membranes
11. chorion
12. amnion
13. allantois
14. yolk sac
15. cleavage
16. morula
17. blastula
18. gray crescent
19. blastopore
20. endoderm
21. ectoderm
22. mesoderm
23. notochord
24. gastrulation
25. gastrula
26. induction
27. organogenesis

28. blastocyst
29. inner cell mass
30. embryo
31. embryonic disc
32. primitive streak
33. neural tube
34. fetus
35. chorionic villi
36. placenta
37. Fetal Alcohol Syndrome (FAS)
38. Homeobox
39. false, gene-regulating substances
40. true
41. false, small amount of yolk
42. false, sexually immature
43. false, no yolk
44. true
45. true
46. true
47. true
48. true
49. true
50. false, continues
51. true

52. false, repair
53. false, brain and spinal cord
54. false, Y chromosome
55. true
56. false, chorion
57. true
58. true
59. true
60. c
61. b
62. b
63. a
64. c
65. b
66. can
67. cannot
68. can
69. can
70. can
71. can
72. can
73. can
74. cannot
75. can
76. can
77. can

78. The zygote is a relatively large cell, containing a great deal of cytoplasm. In order for body cells to be an appropriate size for efficient functioning, the first divisions during development produce smaller and smaller cells with less cytoplasm.

79. zygote to morula to blastula to gastrula to embryo to fetus

80.

Extraembryonic membrane	Structure and location in reptilian egg	Function in reptilian egg	Structure and location in human egg	Function in human egg
Chorion	lines inside of shell	regulates exchange of O_2, CO_2, H_2O	forms placenta with endometrium	regulates exchange of gases, nutrients, and wastes
Amnion	surrounds embryo	encloses embryo in watery environment	surrounds embryo	encloses embryo in watery environment
Allantois	surrounds wastes, connects to embyro's circulation	stores wastes	provides umbilical cord vessels	carries blood between embryo and placenta
Yolk sac	surrounds yolk	yolk nourishes embryo; becomes digestive tract	empty	forms digestive tract

81. Cells differentiate as directed by genes that are turned on by gene-regulating substances in the cytoplasm or by chemicals released by other cells.

82. The amount of yolk present in an egg is related to the way in which an animal develops. Indirect development occurs in animals that hatch from an egg in a form different from that of the adult. The maturation process involves drastic changes in the form of the animal. Amphibians and invertebrates, such as echinoderms and insects, undergo indirect development. The eggs that are laid by these animals are numerous but they contain only a small amount of yolk. Since the amount of yolk is small, the embryo within the egg rapidly develops into a sexually immature larva. The larva will eventually undergo metamorphosis and become a sexually mature adult. Direct development occurs in animals that are born resembling miniature adults such as reptiles, birds, and mammals. The maturation process does not involve drastic changes in body form. These young may be provided with large amounts of nourishing yolk within their eggs or they may be nourished within the mother's body before birth. Since both developmental strategies demand a large input of energy from the mother, a small number of offspring is produced.

Chapter 37: Animal Behavior

OVERVIEW

This chapter considers various aspects of behavior. The authors cover instinctive and learned behaviors, as well as the mechanisms of communication, competitive behaviors within species, cooperative behavior in various animal societies, and the study of human behavior.

1) How Do Innate and Learned Behaviors Differ?

Behavior is any observable response to external or internal stimuli. **Innate (instinctive) behaviors** can be performed in reasonably complete form even the first time an animal of the right age and motivational state encounters a particular stimulus. Many simple orientation behaviors are innate. Simple **kinesis**, a behavior in which an organism changes the speed of random movements, enables pillbugs to reach the moist areas they need to survive. A **taxis** is a directed movement toward (positive taxis) or away from (a negative taxis) a stimulus. For example, *Euglena* shows a positive taxis toward dim light but a negative taxis toward bright light. **Fixed action patterns** are complex stereotyped innate behaviors. The **releaser** (appropriate stimulus) for a fixed action pattern is typically a very specific characteristic of the animal toward which the behavior is directed. For example, when a female red-winged blackbird raises her tail to a particular angle, this releases copulatory behavior in the male.

Learned behaviors are modified by experience. The capacity to make changes in behavior based on experience is called **learning**. A common form of learning is **habituation** (a decline in response to a repeated stimulus). The ability to habituate prevents an animal from wasting energy and attention on irrelevant stimuli. For example, the protist *Stentor* retracts when touched but gradually stops retracting if touching is continued. Humans habituate to many stimuli. Conditioning is a more complex form of learning. In **classical conditioning**, an animal learns to perform a response, normally caused by one stimulus, to a new stimulus. For example, Pavlov rang a bell while feeding dogs, and later the dogs salivated to the sound of the bell alone. During **operant conditioning**, an animal learns to perform a behavior (such as pecking at a button) in order to receive a reward or to avoid punishment. For example, Skinner used special boxes in which animals learned to press a lever for food. By **trial-and-error learning**, animals acquire new and appropriate responses to stimuli through experience. For example, a hungry toad learns to avoid bees after its tongue is stung by a bee it happens to capture. Much learning of this type occurs during play in animals with complex nervous systems. **Insight learning** occurs when animals seem to solve problems suddenly, without the benefit of prior experience. For example, a hungry chimpanzee can stack boxes to reach a banana hanging from the ceiling. In practice, few behaviors are unambiguously instinctive or unequivocally learned but rather are a mixture. Seemingly innate behavior can be modified by experience. Habituation can fine-tune an organism's innate response to environmental stimuli.

Learning may be governed by innate constraints. **Imprinting** is a special form of learning in which learning is rigidly programmed to occur only at a certain critical period of development (the **sensitive period**). For example, mallard ducks learn to follow the animal or object that they most frequently encounter during the period from about 13–16 hours after hatching. All behavior arises out of interaction between genes and the environment.

2) How Do Animals Communicate?

Communication is the production of a signal (sound, movement, chemical emitted) by one organism that causes another organism to change its behavior in a way beneficial to one or both. Most communication occurs between members of the same species to resolve conflicts for food, space, or mates with minimal damage. For animals with well-developed eyes, visual communication is most effective over short distances. Visual signals may be active (a specific movement or posture) or passive (the size, shape, or color of an animal). Active and passive signals can be combined, as is seen in many courtship rituals. Visual signals can be advantageous (they are instantaneous, can be rapidly revised, quiet, and unlikely to alert distant predators) or disadvantageous (ineffective in darkness or dense vegetation, and limited to close-range communication).

Communication by sound is effective over longer distances. Sound can be transmitted instantaneously, through darkness, dense forests, and water, and over long distances, and can be varied to convey rapidly changing messages quickly.

Chemical messages persist longer but are hard to vary. **Pheromones** are chemical substances produced by an individual that influence the behavior of others of its species. Chemicals may carry messages over long distances for long periods of time, take very little energy to produce, and may not be detected by other species. Fewer message are communicated with chemicals than with sight or sound, and pheromone signals lack the diversity and gradation of auditory or visual signals. Pheromones act in two ways. **Releaser pheromones** cause an immediate, observable behavior in the animal that detects them. **Primer pheromones** stimulate a physiological change in the animal that detects them, normally in its reproductive state. For example, queen bees produce a primer pheromone called **queen substance**, which is eaten by hivemates and prevents other females from becoming sexually mature.

Communication by touch helps establish social bonds among group members. Touch can also influence human well-being.

3) How Do Animals Interact?

Sociality is a widespread feature of animal life. Competition for resources underlies many forms of social interaction. Aggressive behavior helps secure resources such as food, space, or mates. **Aggression** is antagonistic but usually harmless behavior, typically involving symbolic displays or rituals between members of the same species. During aggressive displays, animals may exhibit weapons (claws or fangs) and often behave in ways that make them appear larger (stand upright, fluff their feathers or fur, and extend their ears or fins), and emit intimidating sounds. Actual fighting tends to be a last resort.

Dominance hierarchies reduce aggressive interactions. In a dominance hierarchy, each animal establishes a rank that determines its access to resources. Although aggressive encounters occur while the hierarchy is being established, disputes are minimized after each animal learns its place.

Animals may defend territories that contain resources. **Territoriality** is the defense of an area where important resources (such as places to mate and raise young, feed, or store food) are located. Territorial behavior is most commonly seen in adult males, and territories are normally defended against members of the same species, who compete most directly for the resources being protected. Once a territory is established through aggressive interaction, relative peace prevails as boundaries are recognized and respected ("good fences make good neighbors"). For males of many species, successful territorial defense has a direct impact on reproductive success. Territoriality limits the population size of some species, helping keep it within the limits set by the available resources. Territories are advertised through sight, sound, and smell.

Sexual reproduction commonly involves social interactions between mates called courtship behavior. Before mating can occur, animals must identify one another as belonging to the same species, as members of the opposite sex, and as being sexually receptive. Vocal and visual signals encode sex, species, and individual quality. The intertwined functions of sex recognition and species recognition, advertisement of individual quality, and synchronization of reproductive behavior commonly require a complex series of signals, both active and passive, by both sexes. Chemical signals (pheromones) bring both sexes together. Aggression can be suppressed by submissive signals during reproductive interactions, such as a head-down posture, mimicking juvenile behavior, or presenting "gifts."

Social behavior within animal societies requires cooperative interactions. Social grouping may have disadvantages and advantages. Some disadvantages include: (1) increased competition for limited resources; (2) increased risk of infection; (3) increased risk that offspring will be killed; and (4) increased risk of being detected by predators. Some benefits include: (1) increased protection from predators; (2) increased hunting efficiency; (3) advantages from division of labor; (4) conservation of energy; and (5) increased likelihood of finding mates. Some types of animals cooperate on the basis of changing needs, such as coyotes being solitary when food is abundant, but hunting in packs when food is scarce.

Some cooperative societies are based on behavior that seems to sacrifice the individual for the good of the group, such as worker ants who die in defense of their nest. These behaviors characterize **altruism**, behavior that decreases the reproductive success of one individual to the benefit of another. The *selfish gene theory* proposes that genes are preserved by self-sacrificing behavior: An individual would promote the survival of its own type of genes through behaviors that maximize the survival of its close relatives. This concept is called **kin selection**.

The most difficult of all animal societies to explain are those of the bees, ants, and termites, where most individuals never breed but labor slavishly to feed and protect the offspring of a different individual. Honeybees form complex insect societies, with one reproductive queen, male drones, and sterile female workers who bring food to other bees, construct and clean the hive, and forage for pollen and nectar, communicating the location of these resources to other workers by means of the **waggle dance**. Pheromones play a major role in regulating the lives of social insects.

With the exception of human society, vertebrate societies are not as complex as insect societies. Bullhead catfish illustrate a simple vertebrate society based almost entirely on pheromones. Naked mole rats form a complex vertebrate society not unlike that of an ant or termite colony.

4) Can Biology Explain Human Society?

Some scientists hypothesize that because humans are animals whose behaviors have an evolutionary history, the techniques of **ethology** (the study of behavior) can be used to understand human behavior. But human ethologists cannot experiment with people as animal ethologists do with animals.

The behavior of newborn infants has a large innate component. The rhythmic movement of a baby's head searching for mother's breast is a fixed action pattern. Suckling, smiling, walking movements when the body is supported, and grasping with the hands and feet also are innate actions. Innate tendencies can be revealed by exaggerated human releasers. Simple behaviors shared by diverse groups, such as the facial expressions for pleasure and rage and the "eye flash" greeting, may be innate. People may respond to pheromones, perhaps through detection by a small pitlike structure inside the nose called the *vomeronasal organ* (VNO).

Comparison of identical and fraternal twins reveal genetic components of behavior. When twins are reared together, environmental influences on behavior are very similar for each member of the pair, so behavioral differences between twins must have a large genetic component. If a particular behavior is heavily influenced by genetic factors, we would expect to find similar expression of that behavior in identical twins but not in fraternal twins. Such studies have shown a significant genetic component for traits such as activity level, alcoholism, sociability, intelligence, dominance, and even political attitudes. On the basis of tests to measure many aspects of personality, identical twins are about twice as similar in personality as are fraternal twins. The field of human behavior genetics is controversial, because it challenges the long-held belief that environment is the most important determinant of human behavior.

KEY TERMS AND CONCEPTS

Fill-In: From the following list of key terms, fill in the blanks in the following statements.

altruism	habituation	pheromone
behavior	imprinting	releaser
classical conditioning	innate (instinctive)	sensitive period
communication	insight learning	taxis
dominance hierarchy	kinesis	territoriality
ethology	learning	trial-and-error learning
fixed action pattern	operant conditioning	

(1)_____is any observable response to external or internal stimuli.

(2)_____ behaviors can be performed in reasonably complete form even the first time an animal of the right age and motivational state encounters a particular stimulus.

Simple (3)_____is a behavior in which an organism changes the speed of random movements, enabling pillbugs, for instance, to reach the moist areas they need to survive.

A (4)_____is a directed movement toward or away from a stimulus.

A (5)_____ is a complex stereotyped innate behavior. The (6)_____, or appropriate stimulus, for a fixed action pattern is typically a very specific characteristic of the animal toward which the behavior is directed.

The capacity to make changes in behavior based on experience is called (7)_____.

(8)_____is a decline in response to a repeated stimulus.

In (9)_____, an animal learns to perform a response, normally caused by one stimulus, to a new stimulus.

During (10)_____, an animal learns to perform a behavior (such as pecking at a button) in order to receive a reward or to avoid punishment.

By (11)_____, animals acquire new and appropriate responses to stimuli through experience.

(12)_____occurs when animals seem to solve problems suddenly, without the benefit of prior experience.

(13)_____is a special form of learning in which learning is rigidly programmed to occur only at a certain critical period of development called the (14)_____.

(15)_____is the production of a signal (sound, movement, or emitted chemical) by one organism that causes another organism to change its behavior in a way beneficial to one or both.

A (16)_____is a chemical substance produced by an individual that influences the behavior of others of its species.

In a (17)_____, each animal establishes a rank that determines its access to resources while reducing aggressive interactions.

(18)_____is the defense of an area where important resources, such as places to mate and raise young, feed, or store food, are located.

(19)_____is behavior that decreases the reproductive success of one individual to the benefit of another.

(20)_____is the study of behavior.

Key Terms and Definitions

aggression: antagonistic behavior, usually between members of the same species, often resulting from competition for resources.

altruism: a type of behavior that may decrease the reproductive success of the individual performing it but benefits that of other individuals.

behavior: as any observable response to external or internal stimuli.

classical conditioning: a training procedure in which an animal learns to perform a response (such as salivation) to a new stimulus that did not elicit that response originally (such as a sound). This is accomplished by pairing a stimulus that elicits the response automatically (in this case, food) with the new stimulus.

communication: the act of producing a signal that causes another animal, usually of the same species, to change its behavior in a way beneficial to one or both participants.

dominance hierarchy: a social arrangement in which animals, usually through aggressive interactions, establish a rank for some or all of the members that determines access to resources.

ethology: the study of animal behavior under natural or near-natural conditions.

fixed action pattern: stereotyped, rather complex behavior that is genetically programmed (innate); often triggered by a stimulus called a releaser.

habituation: simple learning characterized by a decline in response to a harmless, repeated stimulus.

imprinting: the process by which an animal forms an association with another animal or object in the environment during a sensitive period.

innate: inborn; instinctive; determined by the genetic makeup of the individual.

insight learning: a complex form of learning that requires manipulation of mental concepts to arrive at adaptive behavior.

instinctive: innate; inborn; determined by the genetic makeup of the individual.

kin selection: the concept that natural selection selects for altruistic behaviors that benefit close relatives of the individual performing the behavior.

kinesis: an innate behavior in which an organism changes the speed of its random movement in response to an environmental stimulus.

learning: an adaptive change in behavior as a result of experience.

operant conditioning: a laboratory training procedure in which an animal learns to perform a response (such as pressing a lever) through reward or punishment.

pheromone: a chemical produced by an organism that alters the behavior or physiological state of another of the same species.

primer pheromone: a chemical produced by an organism that alters the physiological state of another of the same species.

queen substance: a chemical produced by a queen bee that can act as both a primer and a releaser pheromone.

releaser: a stimulus that triggers a fixed action pattern.

releaser pheromone: a chemical produced by one organism that alters the behavior of another of the same species.

sensitive period: the particular stage in an animal's life during which it imprints.

taxis (pl. taxes): innate behavior that is a directed movement of an organism toward or away from a stimulus such as heat, light, or gravity.

territoriality: the defense of an area in which important resources are located.

trial-and-error learning: a process by which adaptive responses are learned through rewards or punishments provided by the environment.

waggle dance: symbolic communication used by honeybee foragers to communicate the location of a food source to their hivemates.

THINKING THROUGH THE CONCEPTS

True or False: Determine if the statement is true or false. If it is false, change the <u>underlined</u> word so that the statement reads true.

21. _____   <u>Some</u> behavior has some genetic basis.

22. _____   Flexible behaviors are <u>learned</u>.

23. _____   Direction of movement is influenced by the direction of stimulus in <u>kinesis</u>.

24. _____   A <u>taxis</u> is characterized by random movement.

25. _____   The brain <u>is not</u> involved in reflex actions.

26. _____   The trigger for a fixed action pattern is known as a <u>stimulus</u>.

27. _____   The <u>more complex</u> the animal, the more it relies on instinct.

28. _____   In insect societies, almost all behavior is <u>innate</u>.

29. _____   <u>Habituation</u> is characterized by a crucial time during which it can become part of an animal's behavior.

30. _____   <u>Interspecific</u> communication is more common and important.

31. _____   If an animal assumed a particular posture to communicate with another animal, this behavior is considered <u>active</u>.

32. _____   Both sound and visual signals <u>cannot</u> be of graded intensity.

33. _____   Competition is greatest between members of <u>different</u> species.

34. _____   Most aggressive encounters between members of the same species are <u>real</u>.

35. _____   Territories are more commonly laid out by <u>females</u>.

36. _____   Territories are usually defended against invasion by members of <u>the same</u> species.

37. _____   Established territories promote <u>conflict</u> among members of the same species.

38. _____   The majority of encounters between <u>social</u> animals are competitive and aggressive.

39. _____   Vertebrates have <u>less</u> flexible behavior patterns than invertebrates.

40. _____   An animal that exposes its neck is displaying <u>aggressive</u> behavior.

41. _____   Suckling by babies is <u>instinctive</u> behavior.

42. _____   Menstrual cycles of roommates are usually <u>unsynchronized</u>.

43. _____   Women isolated from men tend to have <u>longer</u> menstrual cycles.

Matching: Innate (instinctive) behavior.

44._____ movement of a body part in response to a stimulus

45._____ entirely programmed by the genes

46._____ moths fly toward light

47._____ performed correctly without learning or prior experience

48._____ nest-cleaning in birds

49._____ an organism changes its speed of random movement in response to a stimulus

50._____ eye-blinking

51._____ directed movement of the entire body towards or away from a stimulus

52._____ nut-burying in squirrels

53._____ pillbugs seeking a moist environment

54._____ a fish swims with its dorsal surface towards light

55._____ complex series of movements in response to the appropriate releaser stimulus

Choices:
a. kinesis
b. taxis
c. reflex
d. fixed action pattern
e. all of these
f. none of these

Matching: Categories of learning.

56._____ decline in response to a harmless, repeated stimulus

57._____ modification of behavior by experience

58._____ a strong association learned during a sensitive period in life

59._____ training pigeons for sea search and rescue missions

60._____ primarily instinctive

61._____ humans ignoring night sounds while asleep

62._____ an animal learns to associate an old behavior with a new stimulus

63._____ a toad will avoid bees if it is stung on the tongue

64._____ the "following response" of young birds

65._____ manipulating concepts in the mind to arrive at adaptive behavior

66._____ an animal learns to perform a behavior to receive a reward

67._____ dogs salivating at the sound of a can opener

68._____ acquiring new and appropriate responses to stimuli by experience

69._____ a hungry chimp will stack boxes to reach bananas suspended from the ceiling

Choices:
a. classical conditioning
b. habituation
c. insight
d. imprinting
e. operant conditioning
f. trial and error learning
g. all of these
h. none of these

Matching: Modes of communication. There may be more than one correct answer.

70._____ dog urine

71._____ easily alerts predators

72._____ uses pheromones

73._____ may be active or passive

74._____ persists after the animal has departed

75._____ can establish social bonds among group members

76._____ limited to close-range communications

77._____ ignored by other species

78._____ grooming in primates

Choices:

a. visual

b. sound

c. chemical

d. touch

Matching: Mechanisms of competition.

79._____ includes behavior that makes the animal appear larger

80._____ defense of an area where important resources are located

81._____ instinctive behavior

82._____ establishing a rank that determines social status

83._____ harmless symbolic displays or rituals for resolving conflicts without fighting

84._____ adult males defend an area against members of the same species

85._____ scent-marking boundaries with pheromones

86._____ exhibiting fangs, claws, or teeth

87._____ pecking orders in chickens

88._____ "good fences make good neighbors"

89._____ the sheep with the biggest horns gets most access to necessary resources

Choices:

a. aggression

b. dominance hierarchies

c. territoriality

d. all of these

e. none of these

Multiple Choice: Pick the most correct choice for each question.

90. The basis of all social behavior is
    a. communication
    b. reproduction
    c. obtaining food
    d. competition
    e. aggression

91. Chemicals produced by an individual that influence the behavior of members of the same species are called
    a. hormones
    b. enzymes
    c. stimuli
    d. pheromones

92. The dominance hierarchy within a group of
animals functions to
   a.   eliminate competition
   b.   limit population numbers
   c.   minimize aggression
   d.   increase competition
   e.   ensure reproduction

93. An example of innate behavior is
   a.   habituation
   b.   conditioning
   c.   taxis
   d.   imprinting

94. Pillbugs that find damp leaves and rotten
logs by chance and stay there because they
need moisture for survival exhibit
   a.   kinesic behavior
   b.   taxic behavior
   c.   reflexive behavior
   d.   fixed action behavior

95. All the following are innate behaviors **except**
   a.   kinesis
   b.   imprinting
   c.   taxis
   d.   reflex

96. A stereotyped, complex series of movements
performed correctly the first time an
appropriate stimulus is presented is termed a
   a.   fixed action
   b.   reflex
   c.   kinesis
   d.   conditional action

97. Evidence that a fixed action is not learned
was obtained by
   a.   testing hybrids for intermediate
behavior patterns
   b.   showing that animals without prior
experience will produce the same
response as those with experience
   c.   using exaggerated releasers called
supernormal stimuli
   d.   all the above choices are correct

98. An adaptive change in behavior as a result of
experience is called
   a.   instinct
   b.   learning
   c.   innate behavior
   d.   fixed action behavior

Short Answer:

99. List three advantages and three disadvantages of visual communication.

Advantages: Disadvantages:

_____   _____

_____   _____

_____   _____

_____

100. List two advantages of sound communication over visual communication.

_____   _____

CLUES TO APPLYING THE CONCEPTS

This practice question is intended to sharpen your ability to apply critical thinking and analysis to biological concepts covered in this chapter.

101. The powerful drugs given to cancer patients often have nausea and loss of appetite among their side effects. One unintentional, and seemingly bizarre, additional side effect is that former patients suffer bouts of nausea many weeks after treatments cease if they eat the same food they ate just before the long-ago drug treatment. Eating a piece of chocolate cake, always a delightful treat, can thus become a dreaded experience. Explain this "food aversion" based on what you have learned in this chapter. Give an example of how this sort of learning can be helpful to animals living in the wild.

_____

_____

_____

_____

_____

_____

_____

_____

_____

_____

ANSWERS TO EXERCISES

1. behavior	26. false, releaser	51. b	75. d
2. innate (instinctive)	27. false, simpler	52. d	76. a, d
3. kinesis	28. true	53. a	77. c
4. taxis	29. false, imprinting	54. b	78. d
5. fixed action pattern	30. true	55. d	79. a
6. releaser	31. true	56. b	80. c
7. learning	32. false, can	57. g	81. d
8. habituation	33. false, the same	58. d	82. b
9. classical conditioning	34. false, symbolic	59. e	83. a
10. operant conditioning	35. false, males	60. h	84. c
11. trial-and-error learning	36. true	61. b	85. c
12. insight learning	37. false, harmony	62. a	86. a
13. imprinting	38. false, solitary	63. f	87. b
14. sensitive period	39. false, more	64. d	88. c
15. communication	40. false, submissive	65. c	89. b
16. pheromone	41. true	66. e	90. a
17. dominance hierarchy	42. false, synchronized	67. a	91. d
18. territoriality	43. true	68. f	92. c
19. altruism	44. c	69. c	93. c
20. ethology	45. e	70. c	94. a
21. false, all	46. b	71. b	95. b
22. true	47. e	72. c	96. a
23. false, taxis	48. d	73. a	97. d
24. false, kinesis	49. a	74. c	98. b
25. true	50. c		

99. Advantages: instantaneous, rapidly revised, quiet, and unlikely to attract predators. Disadvantages: ineffective in darkness, ineffective in dense vegetation, limited to close-range communication, may not be noticed by distracted animals.

100. Sound communication is effective in darkness and through dense forests, in water, and over long distances.

101. This is an example of classical conditioning or learning by association. The cancer patient has inadvertently learned to associate the chocolate cake with the nausea actually caused by the chemotherapy. Animals use associations to avoid foods that cause digestive distress. For instance, birds quickly learn to avoid distasteful butterflies that eat alkaloids harmful to birds.

Chapter 38: Population Growth and Regulation

OVERVIEW

This chapter examines the factors that control the size and rate of growth of populations. It covers how the environment plays a role in controlling populations and how individual interactions among members of the same species, as well as among members of different species, influence population size. How a population grows may depend on how its members are distributed within a given area, or it may depend on the number of offspring that survive to reach maturity. These factors are applied to the human population as well.

1) How Do Populations Grow?

The members of a species that live in an area that allows interbreeding define a **population**. The size of the population changes depending on the number of births and deaths, the number leaving (**emigration**), and the number coming in (**immigration**). If life in the ecosystem is ideal, the population will increase according to its **biotic potential**, that is, its maximum rate. However, the environment cannot sustain the population at this rate since resources, such as food and space, are limited and organisms interact with one another. Therefore, the population's size is limited according to **environmental resistance**.

The rate at which a population size changes, the **rate of growth,** is determined by: $b - d = r$ (number of births - number of deaths = rate of growth). However, if the *number of individuals* that are new to a population within a certain time period is in question, then the rate of growth (r) is multiplied by the number of members in the population at the beginning of the time period (N): rN = population growth within a given time period. If a population is growing at an ever-accelerating rate, then the population is experiencing **exponential growth**. Exponential growth is graphed as a **J-curve**.

2) How Is Population Growth Regulated?

Populations that grow exponentially may suddenly undergo substantial deaths due to disease, to seasonal changes, or to resource availability. Some populations go through exponential growth and massive die-offs on a cyclical basis; these are referred to as **boom-and-bust cycles**. Other populations that undergo exponential growth may do so because there is no natural predator in the area. This often occurs when **exotic** species have been introduced to an ecosystem. The size of most populations, however, is controlled by the ecosystem. Any given area can support only a certain population size indefinitely. This size is the **carrying capacity** of the ecosystem. Population numbers that have reached carrying capacity can be graphed as an **S-curve**.

Population numbers are affected by the course of nature in ways that may or may not be due to the size of the population. Populations that get too crowded or dense may be adversely affected by **density-dependent factors** such as predation, parasitism, disease, or intense competition. On the other hand, **density independent factors**, such as weather, fire events, or human activities, impact a population regardless of its size.

When an animal kills and eats another organism, **predation** has occurred, and the animal doing the killing and eating is the **predator**. Predation is an important mechanism in natural population control. However, predation not only controls the size of the prey populations, but it also serves to control the size of the predator populations. As predators reduce the number of prey available, they are, in effect, reducing their own food resource. This results in a reduction in the predator population. When predator numbers are reduced, the prey population will increase again. Thus, predator populations and prey populations undergo **population cycles**. When an animal feeds on another organism without killing it, the animal is a **parasite**, and the organism on which it is feeding is its **host**.

When population numbers increase, **competition** for the resources on which the organisms depend becomes more intense. If the competition occurs among members of different species, **interspecific competition** is occurring. However, if the competition is among members of the same species, the more intense **intraspecific competition** is occurring. In order to survive, organisms have evolved mechanisms to gather the scarce resources. **Scramble competition** results in the survival of the individuals best able to accrue the most resources. **Contest competition** uses social or chemical interactions to gather resources. Territory protection and defense is a type of contest competition.

3) How Are Populations Distributed?

If members of a population cluster together in herds, packs, or flocks, or around an area of a plentiful resource, these organisms are **clumped** in their distribution pattern. Other organisms defend territories that are relatively evenly spaced and thus are **uniform** in their distribution within an area. Rarely, organisms distribute themselves **randomly** throughout the ecosystem. This may occur when resources are equally available and plentiful throughout the area.

4) What Survivorship Patterns Do Populations Exhibit?

Graphing the number of individuals in an age group against their age illustrates a pattern of survivorship, a **survivorship curve**. Different species have different patterns of survivorship. Species that have low offspring mortality and many members who tend to reach maturity before death show a "late loss" curve. Species that have an equal chance of survival as they do death at any age show a "constant loss" when graphed. Species that have high offspring deaths with few members reaching maturity have an "early loss" curve.

5) How Is The Human Population Changing?

The human population grew slowly for over one million years. During that time, fire was discovered, tools and weapons were fashioned, shelters were built, and clothing was made to protect individuals. Each of these "inventions" lead to a **cultural revolution** as the populations adapted to these innovations. The domestication of crops and animals lead to an **agricultural revolution**, providing a more dependable food supply. Once advances in medicine and health care occurred the human death rate was reduced dramatically. This **industrial-medical revolution** lead to an increase in population. In developed countries, the industrial-medical revolution also lead to reduced birth rates, stabilizing their population growth. In developing countries, however, reduced birth rates have not occurred, primarily due to social traditions and a lack of access to education and contraceptives. The differences in population growth between developing and developed countries can be illustrated using **age structure** diagrams. These diagrams graph the distribution of males and females in each age group. If there is a large portion of the population below reproductive years, even if people entering their reproductive years had only the number of children needed to replace themselves [**replacement level fertility (RLF)**], the population would continue to grow *due to the number* of people entering their reproductive years. The United States is experiencing rapid growth because the "baby boom" generation has reached their reproductive years. However, continued immigration to the United States also contributes to the population growth.

KEY TERMS AND CONCEPTS

Fill-In: From the following list of key terms, fill in the blanks in the following statements.

age structure	ecology	parasite
agricultural revolution	ecosystem	parasitism
biotic potential	emigration	population
boom-and-bust cycle	environmental resistance	predation
carrying capacity	exponential	predator
clumped distribution	host	randomly distributed
community	immigration	replacement-level fertility
competition	industrial-medical revolution	S-curve
contest competition	interspecific competition	scramble competition
cultural revolution	intraspecific competition	survivorship curve
density dependent	J-curve	uniformly distributed
density independent		

(1) _____ is the study of the interrelationships among living organisms and their nonliving environment. The term (2) _____ , however, is used to describe all the organisms and their non-living environment within a specified area. All the potentially interbreeding members of a species within an ecosystem is referred to as a (3) _____.

The size of a population changes as new members are born, move into the area (called (4) _____), leave the area (called (5) _____), or die. The actual rate at which a population changes in size depends on the (6) _____, which is the maximum rate a population could increase with unlimited resources, and the (7) _____, which includes the limited availability of the resources such as food, water, space etc. However, to measure the change in population size, the number of deaths per person is subtracted from the number of births per person that occur in a given time period. This is the (8) _____ or (r).

When a population's size increases at a continuously accelerating rate the population is experiencing (9) _____ growth; an increasing number of individuals is added to the population with each generation. When the population numbers are graphed for this type of growth, a J-shaped curve or a (10) _____ is the result.

When resources for a species are temporarily abundant, its population may grow exponentially until it is limited by an environmental factor (temperature, for example). This pattern of rapid growth followed by a massive die-off is called a (11) _____. Species that have been introduced to an area from a foreign location are termed (12) _____ species. Populations of these introduced species often exhibit exponential growth since their natural predators are missing from their new ecosystem.

In populations were birth rates equal death rates, the population stabilizes. These populations exhibit S-shaped curves or (13) _____ when population growth rates are graphed. Populations often stabilize when the population numbers reach the maximum size that the ecosystem can support indefinitely; the (14) _____ of the ecosystem.

Human activities and environmental factors such as weather events or forest fires may limit population size regardless of its density. These are examples of (15) _____ factors. In contrast, parasites, disease, competition for limited resources, and predation are examples of (16) _____ factors that affect population size more intensely as population density increases.

Animals capturing, killing, and eating other organisms is called (17) _____. This act naturally helps control population size. The animal that captures, kills, and eats the organism (the prey) is referred to as the (18) _____. As prey population numbers increase, the predator population numbers will increase. Then, as more predators capture and eat more prey, the prey numbers decline. This will cause the predator population numbers to decrease as well. This effect, the (19) _____ , is always out of phase.

If an organism feeds within or on a larger organism but does not kill it outright, this is called (20) _____. The organism feeding is the (21) _____, while the organism that is being fed upon is the (22) _____.

When resources in an area are limited, the organisms using the resources are in (23) _____ for the resources. If the organisms vying for the resources belong to different species then (24) _____ is occurring. If the organisms vying for the resources belong to the same species then (25) _____ is occurring and its effect is more intense. One way that some species have evolved to deal with the intensity of this type of competition involves a "free-for-all," the winner of (26) _____ gets the resource. Other species defend territories when population numbers exceed available resources. These organisms engage in (27) _____ where only the most fit are able to defend their territory.

The populations of species in any given area may be distributed in specific patterns. If the members of a population live in groups such as in herds or flocks or along resource lines, the population has a (28) _____. If the organisms live a rather consistent distance from each other, then these organisms are (29) _____. Rarely, if the members of the population do not form social groups or do not use territorial spacing, their dispersal may not have any observable pattern and they are (30) _____.

Over time, populations tend to show patterns in the numbers that die or survive. The pattern that any given species exhibits is called its (31) _____.

Humans have developed ways to increase the carrying capacity of the ecosystem of which they are a part. By using fire and tools and developing protective shelter and clothing, a (32) _____ occurred, allowing previously uninhabitable areas to be habitable. As farming and animal husbandry evolved, a dependable food supply came about with an (33) _____. The human population grew slowly until advances were made that reduced the number of deaths. This (34) _____ began during the mid-eighteenth century and continues today.

A diagram of a population representing the number of individuals in specific age categories is the (35) _____ of the population.

When individuals in a population of reproductive age bear only the number of children needed to replace themselves, then (36) _____ is occurring.

Key Terms and Definitions

age structure: the distribution of males and females in a population according to age groups.

biotic potential: the maximum rate at which a population could increase, assuming ideal conditions that support a maximum birth rate and minimum death rate.

boom-and-bust cycle: a population cycle characterized by rapid exponential growth followed by a sudden massive die-off seen in seasonal species and some populations of small rodents, such as lemmings.

carrying capacity: the maximum population size that an ecosystem can support indefinitely, determined primarily by the availability of space, nutrients, water, and light.

clumped distribution: The distribution characteristic of populations in which individuals are clustered into groups; may be social or based on the need for a localized resource.

competition: interaction among individuals when both attempt to use a resource (for example, food or space) that is limited relative to the demand for it.

contest competition: a mechanism for resolving intraspecific competition using social or chemical interactions.

density-dependent: any factor, such as predation, that limits population size more effectively as the population density increases.

density-independent: any factor, such as freezing weather, that limits a population's size and growth regardless of its density.

ecology: the study of the interrelationships of organisms with each other and with their nonliving environment.

ecosystem: all the organisms and their nonliving environment within a defined area.

emigration: migration of individuals out of an area.

environmental resistance: any factor that tends to counteract biotic potential, limiting population size.

exotic: referring to a foreign species introduced into an ecosystem

exponential growth: a continuously accelerating increase in population size.

growth rate: a measure of the change in population size per individual per unit of time.

host: the prey organism on or in which a parasite lives; is harmed by the relationship.

immigration: migration of individuals into an area.

interspecific competition: competition among individuals of different species.

intraspecific competition: competition among individuals of the same species.

J-curve: the J-shaped growth curve of an exponentially growing population, in whichincreasing numbers of individuals join the population during each succeeding time period.

parasite: an organism that lives in or on a larger prey organism, called a host, weakening it.

parasitism: the process of feeding on a larger organism without killing it immediately or directly.

population: all the members of a particular species within an ecosystem.

population cycle: out-of-phase cyclical patterns of predator and prey populations.

predation: the act of killing and eating another living organism.

predator: an organism that kills and eats other organisms.

prey: organisms that are killed and eaten by another organism.

random distribution: the distribution characteristic of a population in which the probability of finding an individual is equal in all parts of an area.

replacement-level fertility (RLF): the average birth rate at which a reproducing population exactly replaces itself during its lifetime.

scramble competition: a free-for-all scramble for resources among individuals of the same species.

S-curve: the S-shaped growth curve that describes a population of long-lived organisms introduced into a new area. It consists of an initial period of exponential growth, followed by decreasing growth rate, and finally, relative stability.

survivorship curve: a curve resulting when the number of individuals of each age in apopulation is graphed against their age, usually expressed as a percentage of their maximum life span.

uniform distribution: a distribution characteristic of a population with a relatively regular spacing of individuals, commonly a result of territorial behavior.

THINKING THROUGH THE CONCEPTS

True or False: Determine if the statement given is true or false. If it is false change the underlined word so that the statement reads true.

37. _____ A population is made up of all the members of a species in a certain area that has the potential to interbreed.
38. _____ The availability of food and space serve to limit the biotic potential of a population.
39. _____ In nature, exponential growth occurs for prolonged periods of time.
40. _____ Carrying capacity of an ecosystem is determined in part by renewable resources.
41. _____ The effect of density-dependent factors is unaffected by population size.
42. _____ As a population increases and becomes more dense, the result is less die-off from disease and parasites.
43. _____ Predators not only exert an influence on the size of their prey populations, but prey populations also affect predator population size.
44. _____ Humans are likely to cause extinctions because impacts such as pollution and habitat alteration are not density dependent.
45. _____ Parasites living in their host quickly weaken and kill their host.
46. _____ Intraspecific competition is less intense then interspecific competition.
47. _____ Populations may exhibit a clumping pattern due to localized resources.
48. _____ A species' population will show a characteristic pattern of survivorship; either "early loss," "constant loss," or "late loss."
49. _____ Humans have found ways to overcome environmental resistance.
50. _____ In age structure diagrams, if the number of children (ages 0–14) exceeds the number of reproducing individuals (ages 15–45), the population is decreasing.
51. _____ Delayed childbearing slows population growth.
52. _____ The U.S. population is currently growing exponentially.

Matching:

Choices:

53. _____ illustrates exponential growth   a. age structure diagram

54. _____ illustrates a stable population   b. population growth

55. _____ $r = b - d$ c. S - curve

56. _____ rN d. growth rate

57. _____ illustrates patterns of death   e. survivalship curve

58. _____ illustrates distribution of males and females   f. J - curve

Short answer:

59. Population change = (_____ - _____) + (_____ - _____)

60. Identify 3 of the 5 factors influencing the biotic potential of a species.

   _____

   _____

   _____

   _____

61. Briefly discuss three factors that can lead to boom-and-bust cycles. Include examples of the types of organisms that are susceptible to these cycles.

   _____

   _____

   _____

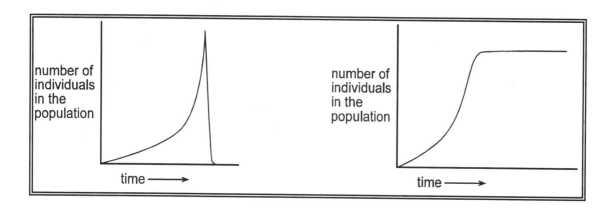

Identify the Following: Using the diagram above determine if the following are characteristics of a J-shaped population curve, an S-shaped population curve, or both. (Write **J**, **S**, or **b** on the line provided)

62. _____ Initial population growth is small

63. _____ Population growth accelerates with time

64. _____ Population growth accelerates with time, then levels off

65. _____ Population grows indefinitely, exceeding carrying capacity

66. _____ Population is limited at carrying capacity

67. _____ Population size is probably limited by environmental factors

68. _____ Density-dependent factors influence population size

69. _____ Population may suffer a sudden crash

70. Explain why exotic or introduced species tend to display exponential growth. Discuss the subsequent effects on the ecosystem.

_____

_____

_____

_____

_____

_____

_____

_____

_____

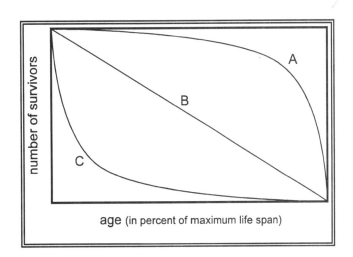

Identify the Following: Using the survivalship curves **A**, **B**, and **C**, graphed above identify the following.

71. _____ Which curve shows continued loss?

72. _____ Which curve shows early loss?

73. _____ Which curve shows late loss?

74. _____ Which curve would diagram the survivorship of a Maple tree population?

75. _____ Which curve would diagram the survivorship of an elephant population?

76. _____ Which curve would diagram the survivorship of a population of American Robins?

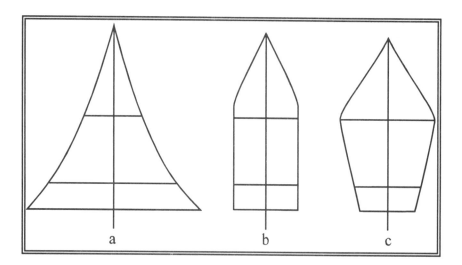

Using the age structure diagrams **a, b,** and **c,** above answer the following.

77. _____   Which diagram shows a growing population ?

78. _____   Which diagram shows a shrinking population?

79. _____   Which diagram shows a stable population?

80. _____   Which would diagram the population of a developing country?

81. _____   Which would diagram the population of a country with replacement level fertility ?

82. _____   Which would diagram the population of a country with fewer children than reproducing adults?

ANSWERS TO EXERCISES

1. Ecology
2. ecosystem
3. population
4. immigration
5. emigration
6. biotic potential
7. environmental resistance
8. growth rate
9. exponential
10. J-curve
11. boom-and-bust cycle
12. exotic
13. S-curve
14. carrying capacity
15. density independent
16. density dependent
17. predation
18. predator
19. population cycle
20. parasitism

21. parasite
22. host
23. competition
24. interspecific competition
25. intraspecific competition
26. scramble competition
27. contest competition
28. clumped distribution
29. uniformly distributed
30. randomly distributed
31. survivorship curve
32. cultural revolution
33. agricultural revolution
34. industrial-medical revolution
35. age structure
36. replacement-level fertility
37. true
38. true
39. false; short

40. true
41. false; density independent
42. false; more
43. true
44. true
45. false; slowly
46. false; more intense
47. true
48. true
49. true
50. false; increasing
51. true
52. true
53. f
54. c
55. d
56. b
57. e
58. a

59. Population change = (births - deaths) + (immigrants - emigrants).
60. Age at which the organism first reproduces, frequency with which reproduction occurs, average number of offspring produced each time, length of reproductive life span of an organism, death rate of individuals under ideal conditions.
61. Short-lived, rapidly reproducing organisms such as bacteria, algae, and insects have population cycles that are often dependent on amount of rainfall (flood or drought), temperature (intense heat or killing frost), or nutrient availability.

62. b
63. b
64. S
65. J
66. S
67. b
68. S
69. J

70. Introduced foreign, exotic, species often invade new habitats where conditions are favorable, food or nutrients are plentiful, and competition is scarce. Predators, parasites, or disease have little or no effect on the population of the introduced species. This leads to exponential growth of the introduced species which may seriously damage the ecosystem by displacing, out competing, or preying on native species.

71. B
72. C
73. A
74. C
75. A
76. B
77. a
78. c
79. b
80. a
81. b
82. c

Chapter 39: Community Interactions

OVERVIEW

This chapter looks at the how organisms interact with members of their own species, as well as with members of different species. These interactions include mechanisms that have evolved to deal with competition for limited resources, predator/prey relationships, and symbiotic relationships. The species that are present in a community depends on the ecosystem that is present. This chapter also deals with how an ecosytem changes over time (succession).

1) Why Are Community Interactions Important?

The interactions among populations within a **community** serve to maintain a balance between available resources (i.e., food, water, shelter) and the number of individuals using them. As the interactions among the populations serve to limit population size, they also lead to changes in characteristics and behaviors, increasing the fitness of the total population (evolution). When changes in one species results in adaptive changes in an interacting species, **coevolution** has occurred.

2) What Are the Effects of Competition Among Species?

Competition among species, or **interspecific competition**, has such a strong effect on the species involved that each evolves ways to reduce any overlap in needs. In other words, each species specializes within the community, developing its own well-defined, **ecological niche**, and thus **partitioning resources**. If one of the competing species is removed from the community, the other species may expand its niche since the competition pressure has been reduced. Gause showed this in his experiments with two species of *Paramecium*, as did ecologist J. Connell in his study of the barnacles *Chthamalus* and *Balanus*.

3) What Are the Results of Interactions Between Predators and Their Prey?

Predation interactions have intense effects on the species involved. Predators have evolved ways to best capture their prey, while the prey have evolved mechanisms to elude their predators. This coevolution has resulted in some very complex physical characteristics and behaviors. Bats and their moth prey have developed complex "cat and mouse" behaviors, while other species **camouflage** themselves to avoid predators or detection by prey. In contrast to camouflaged species, others stand out with bright coloration. These species advertize their presence, safe in the knowledge that they will be left alone. Their bright colors (or **warning coloration**) warn potential predators that they are poisonous or otherwise distasteful and are to be avoided. Species with common characteristics may share warning patterns as well; for example, stinging insects tend to be bright yellow with black stripes, and poisonous frogs from the tropics display very colorful skin pigments. Some harmless species have evolved to **mimic** their poisonous relatives, thus taking advantage of the effect of the warning pattern on potential predators. Some stingless wasps, for example, are bright yellow with black stripes. Devious predators exists as well: **Aggressive mimicry** has evolved among species that resemble harmless species, yet are truly waiting to take a bite out of an unsuspecting prey. Predators, however, may be caught off-guard. Some prey make use of color patterns that mimic a larger organism. These species use their **startle coloration** to scare their predator and make a safe getaway. Some prey species have the ultimate defense: "chemical warfare." Coevolution, however, has also lead to a few predator species that are not harmed by the chemical produced and may even use it as its own defense mechanism.

4) What Is Symbiosis?

Within a community, interacting with other species is unavoidable; however, some species have such close interactions that they have developed **symbiotic** relationships. When one species of the relationship benefits and the other is unaffected, the relationship is **commensalistic**. If one species benefits and the other is harmed, the relationship is **parasitic**. If both species benefit, the relationship is **mutualistic**.

5) How Do Keystone Species Influence Community Structure?

The influence of species on community structure is not necessarily equal. When one species has a role that is out of proportion to its population size, that species is a **keystone** in the community. Often, a keystone species cannot be identified until it has actually been removed from the community. At this point it may be too late to reduce the impact its absence will have on the community.

6) How Does a Community Change over Time?

The interactions among members of a community lead to structural changes within that community; changes that are identified as stages in **succession** of the community. **Primary succession** begins with **pioneer species** such as lichen and mosses establishing a hold on bare rock. As soil slowly forms, additional species move into the young community in a recognizable pattern. **Secondary succession** occurs after an established community has been disturbed perhaps by fire, wind storm, or farming. If left undisturbed, succession will continue to a stable endpoint, the **climax**, determined in a large part by the geography and climate of the area. If a community is regularly disturbed, it will be maintained at a succession point below the climax, a **subclimax**. Climax communities covering broad geographical regions are **biomes**. Biomes are distinguished by specific climatic conditions and characterized by specific plant communities.

KEY TERMS AND CONCEPTS

Fill-In: From the following list of key terms, fill in the blanks in the following statements.

aggressive mimicry	exotic species	primary succession
biomes	interspecific competition	resource partitioning
camouflaged	intertidal zone	secondary succession
climax community	keystone species	startle coloration
coevolution	mimicry	subclimax
community	mutualism	succession
competitive exclusion principle	parasites	symbiosis commensalism
ecological niche	pioneers	warning coloration

All the interacting populations within an ecosystem make up an ecological (1) _____.

In order for (2) _____ to occur, two interacting species serve as agents of natural selection on one another over evolutionary time.

When two or more species try to use the same limited resource, each species is harmed. This type of species interaction is called (3) _____.

In order to reduce competition among species, each species has evolved its own (4) _____; each species has its own physical environmental factors necessary for its survival, as well as a specific "occupation" within its habitat. With this in mind, the (5) _____ states that no two species can inhabit the same ecological niche; eventually one species would eliminate the other through competition. However, examples of species with very similar niche requirements exist. Through evolutionary adaptations, these organisms have reduced the overlap of their niches. This is referred to as (6) _____.

The barnacles *Chthamalus* and *Balanus* live along rocky ocean shores and are exposed to flood and drought conditions as the tide comes in and recedes. This area of the coastline is referred to as the (7) _____.

Many animals have evolved colors, patterns, or shapes to avoid a predator or to avoid being noticed by their prey. These animals are (8) _____. Other animals display bright (9) _____ announcing that they are poisonous or otherwise nonpalatable. Some harmless species take full advantage of the brightly colored species by resembling them in appearance. This (10) _____ saves the harmless, tasty species from predation.

Predators may also deceive. Through (11) _____ a predator resembles a harmless species, allowing its prey to unknowingly come into range for an attack.

Certain moths and caterpillars have evolved "eye-spots" and other color patterns that resemble the eyes of a larger animal. When the prey uses its (12) _____, the predator is frightened, allowing the prey to escape.

(13) _____ is defined as a close interaction between members of different species for an extended period of time. In a relationship such as (14) _____, one species benefits while the other is unaffected. However, (15) _____ live on or in their hosts, usually harming the host in some way but not killing it. When both species in the relationship benefit from the interaction, the relationship is called (16) _____.

Some communities contain one species that plays a major role in determining the community's structure. If this (17) _____ is removed, the structure of the community would be dramatically altered.

When a community changes structurally over time, (18) _____ has occurred. The changes begin with the first species to invade an area; these species are called the (19) _____. As time and climate allow, the community will reach a relatively stable endpoint, the (20) _____. When there has been no previous community in existence, the ecosystem begins with the process of (21) _____. If, however, an ecosystem has been disturbed by fire, storm, or farming, a new ecosystem develops through the process of (22) _____. If an ecosystem is periodically and regularly disturbed, the potential climax community may not be reached, instead, a (23) _____ community is maintained. Large geographical regions consisting of climax communities and characterized by specific plant species make up the (24) _____ of the world.

The balance of a community may be severely disrupted when an (25) _____ species is introduced to the community.

Key Terms and Definitions

aggressive mimicry: the evolution of a predatory organism to resemble a harmless animal or part of the environment, thus allowing it to gain access to its prey.

biome: a general type of ecosystem occupying extensive geographical area characterized by similar plant communities, for example, deserts.

camouflage: coloration and/or shape that renders an organism inconspicuous in its environment.

climax community: a diverse and relatively stable community that forms the endpoint of succession.

coevolution: the process by which two interacting species act as agents of natural selection on one another over evolutionary time.

commensalism: a symbiotic relationship in which one species benefits while the other is neither harmed nor benefitted.

community: all the interacting populations within an ecosystem.

competitive exclusion principle: the concept that no two species can simultaneously and continuously occupy the same ecological niche.

ecological niche: the role of a particular species within an ecosystem, including all aspects of its interaction with the living and nonliving environments.

exotic species: species introduced into an ecosystem where it did not evolve; such species may flourish and out compete native species.

interspecific competition: competition among members of different species.

intertidal zone: an area of the ocean shore that is alternately covered and exposed by the tides.

keystone species: species whose influence on community structure is greater than its population size would suggest.

mimicry: the situation in which a species has evolved to resemble something else, typically another type of organism.

mutualism: a symbiotic relationship in which both participating species benefit.

parasite: an organism that feeds in or on its prey (called a host).

parasitism: a symbiotic relationship in which one organism (normally smaller and more numerous than its host) benefits by feeding on the other, which is usually harmed but not killed.

pioneer: an organism that is among the first to colonize an unoccupied habitat in the first stages of succession.

primary succession: succession that occurs in an environment, such as bare rock, in which no trace of a previous community was present.

resource partitioning: the coexistence of two species with similar niche requirements, each occupying a smaller niche than it would by itself, as means of minimizing their competitive interactions.

secondary succession: succession that occurs after an existing community is disturbed, for example, after a forest fire; much more rapid than primary succession.

startle coloration: a form of mimicry in which a color pattern (in many cases resembling large eyes) can be displayed suddenly by a prey organism when approached by a predator.

subclimax: a community where succession is stopped before the climax community is reached and is maintained by regular disturbances; for example, tallgrass prairie is maintained by periodic fires.

succession: a structural change in the community and its nonliving environment over time. Community changes alter the ecosystem in ways that favor competitors, and species replace one another in a somewhat predictable manner until a stable; self-sustaining climax community is reached.

symbiosis: a close interaction between organisms of different species over an extended period; includes parasitism, mutualism, and commensalism.

warning coloration: bright coloration that warns predators that the potential prey is distasteful or even poisonous.

THINKING THROUGH THE CONCEPTS

True or False: Determine if the statement given is true or false. If it is false, change the underlined word so that the statement reads true.

26. _____ The interacting populations of a community influence one another's ability to survive and reproduce, leading to a system that results in coevolution.

27. _____ A species' niche in the environment can be described as its occupation or role in the community.

28. _____ Species may reduce competition by partitioning the available resources resulting in each species occupying a larger niche than if there were no competition.

29. _____ Many poisonous or harmful species display bright coloration to standout to a predator; in this way, the predator is sure to eat the bright organism.

30. _____ Common coloration patterns of equally poisonous or harmful species aid in a predator learning to avoid these prey items.

31. _____ Prey species have evolved predator avoidance mechanisms, while predator species have responded by evolving deceptive mechanisms to catch their prey.

32. _____ Plants are defenseless against species that feed on them.

33. _____ Parasites, generally, do not kill their hosts.

34. _____ If elephants disappear from the African savanna, the grasslands will eventually succeed to forest. This is because the elephant is a prey species.

35. _____ Following a forest fire, the new community that will develop will do so through primary succession.

36. _____ The organisms that invade bare rock to begin a new community are the biomes of the community.

37. _____ If a farmer allows a field to lie fallow, or abandons the field altogether, secondary succession will quickly establish a new community structure.

38. _____ A pond that is left undisturbed will eventually fill in with silt, forming a marsh; a meadow may eventually be formed as the marsh dries.

39. _____ Climax communities undergo constant change, resulting in diverse populations inhabiting numerous ecological niches.

40. _____ Fields maintained for agriculture represent communities held at a specific subclimax by humans.

Matching: Nature's "Chemical warfare."

Choices:

41. _____ flowering lupines
42. _____ squid, octopus
43. _____ spiders, snakes
44. _____ bombardier beetle
45. _____ milkweed plants
46. _____ grasses

a. produce toxic, distasteful chemicals
b. produce silicon
c. produce a boiling hot, toxic spray
d. produce ink clouds
e. produce a paralyzing venom
f. produce alkaloids

Identify the Following: Are these interactions examples of **competition**, **parasitism**, **commensalism**, or **mutualism**?

47. _____ Lichens are a growth form that occurs when fungi and algae live together. The fungus absorbs nutrients for the alga while the alga photosynthesizes providing carbohydrates for the fungus.

48. _____ Bromiliads, "air plants," grow in the notches of tropical trees. The trees are not harmed nor do they benefit from the bromiliads; however, the bromiliads absorb water and nutrients collected in rainwater in the notch.

49. _____ Hyenas and vultures both feed from animal remains after lions have finished feeding.

50. _____ A tick that has imbedded into the hide of a deer feeds off the deer, possibly weakening or infecting the deer.

51. _____ Egrets can often be seen following cattle through a field. As the cattle disturb insects in the grass the egrets eat the insects; the cattle are unaffected.

52. _____ Roundworms are often found in the intestines of feral cats. The roundworm feeds off the nutrients the cat has ingested, leaving the cat malnurished.

53. _____ An oak seedling and a maple seedling are growing in an opening in the forest. They both require sunlight, water, and nutrients from their environment.

54. _____ Flowers are pollinated by insects. The insects, in turn, receive pollen and nectar for nourishment from the flowers.

Identify the Following: Are the following examples of **primary** or **secondary** succession?

55. _____ Large areas of Australia were burned during the southern hemisphere's summer season of 1997. However, new groundcover soon germinated in the open areas and rich nutrients of the fire ashes.

56. _____ When Mt. St. Helens volcano in Washington State, USA, erupted in 1980, the existing communities were destroyed. Within two years subalpine flowers could be seen blooming and mountain meadows had formed; the communities were reforming.

57. _____ Tornados during the northern hemisphere's spring season of 1985 leveled acres of virgin forests in northwest Pennsylvania Commonwealth, USA. This gave researchers first-hand information on regeneration patterns in forests never touched by human development.

58. _____ As global temperatures rise and glaciers recede, the stratum left behind will be inhabited by lichen, followed by mosses. As plant matter decays and weathering continues, soil will slowly develop which will sustain tundra grasses and wildflowers.

Short Answer:

59. A predator's prey may avoid being eaten by using body parts that have evolved to camouflage it with the environment in which the it evolved. Identify three examples of how a prey species may camouflage itself. A predator may also have evolved camouflage patterns, concealing it from its prey. Identify three examples.

Camouflaged prey: Camouflaged predator:

_____   _____

_____   _____

_____   _____

60. Warning colorations are commonly displayed by poisonous organisms and are often mimicked by equally harmful species and by harmless species as well. Identify three toxic organisms that display warning coloration. Match these organisms with their equally distasteful or harmless mimics.

Poisonsous organism: Mimic:

_____   _____

_____   _____

_____   _____

61. What is the key difference between primary succession and secondary succession?

_____
_____
_____

CLUES TO APPLYING THE CONCEPTS

These practice questions are intended to sharpen your ability to apply critical thinking and analysis to the biological concepts covered in this chapter.

62. Kangaroo of Australia fill a niche very similar to that of deer in North America. If a deer population from North America were introduced to Australia, the kangaroo and deer would be in direct competition for many resources. Using what you have learned about competition, the competitive exclusion principle, and introduced species, explain what would happen between the two populations.

_____

_____

_____

_____

_____

_____

_____

_____

_____

63. The organisms in a community are tied together either directly or indirectly. This is illustrated best by studying keystone species. Briefly explain how the removal of a keystone species from an ecosytem affects the balance of the communities found there.

_____

_____

_____

_____

_____

_____

_____

64. The snowberry fly has evolved specific behavorial patterns to ward off its predator, a jumping spider. Explain how the fly's behavior is its defense against predation.

_____

_____

_____

_____

_____

_____

ANSWERS TO EXERCISES

1. community
2. coevolution
3. interspecific competition
4. ecological niche
5. competitive exclusion principle
6. resource partitioning
7. intertidal zone
8. camouflaged
9. warning coloration
10. mimicry
11. aggressive mimicry
12. startle coloration
13. symbiosis
14. commensalism
15. parasites
16. mutualism
17. keystone species
18. succession
19. pioneers

20. climax community
21. primary succession
22. secondary succession
23. subclimax
24. biomes
25. exotic
26. true
27. true
28. false, smaller
29. false, sure to avoid
30. true
31. true
32. false, may produce defense chemicals
33. true
34. false, keystone
35. false, secondary
36. false, pioneers
37. true
38. true

39. false, are stable
40. true
41. f
42. d
43. e
44. c
45. a
46. b
47. mutualism
48. commensalism
49. competition
50. parasitism
51. commensalism
52. parasitism
53. competition
54. mutualism
55. secondary
56. secondary
57. secondary
58. primary

59. Camouflaged prey - dappled fawns; grasshoppers; prey resembling leaves, twigs, thorns, bird droppings; plants resembling rocks. Camouflaged predators - spotted cheetah; striped tiger, frogfish resembling algae covered rocks.
60. Yellow jacket, hornets, bees; coral snake, mountain king snake; monarch butterfly, viceroy butterfly.
61. Primary succession - bare rock, secondary succession - community existed previously
62. The competitive exclusion principle states that if two species with the same niche are placed together and forced to compete for limited resources, one will out compete the other. The deer, as the introduced species would have few or no natural predators. This would give the deer a competitive edge. The deer would most likely out compete and eventually replace the kangaroo in Australia.
63. If the keystone species is removed, the species that it preyed upon will increase in number since the keystone species kept the prey population under control. The prey species may out compete the other species in the community, possibly eliminating them. With the species distribution altered, the balance of the community has changed.
64. When approached by a predatory jumping spider, the snowberry fly mimics the behavior and appearance of a jumping spider protecting its territory. Seeing this specific behavior pattern, the predator retreats, leaving the fly alone.

Chapter 40: How Do Ecosystems Work?

OVERVIEW

This chapter traces the pathways of energy and nutrients through the ecosystem. Energy follows a one-way path through the ecosystem, passing from the sun through the organisms in a community and being lost as heat as it is transferred. Nutrients, on the other hand, cycle through the ecosystem using the natural recycling properties of each community. When the cycles are influenced by man, as is currently the case with the carbon cycle, the natural balance is disturbed. The result has led to acid deposition and global warming.

1) What Are the Pathways of Energy and Nutrients through Communities?

Energy flows through communities from the sun to the organisms inhabiting them. As the energy flows, some is lost to the environment as heat. The energy supplied must be continuously replenished from the sun. Nutrients, on the other hand, are recycled. Thus, they remain in the ecosystem.

2) How Does Energy Flow through Communities?

Plants, algae and a few other protists, and cyanobacteria absorb sunlight energy using light-absorbing pigments within them. Using the process of photosynthesis, sunlight energy is converted to chemical energy and is stored as sugar and structures making up the photosynthetic organism. These organisms are **autotrophs**. Since they produce their "food" themselves, they are also called **producers**. Organisms that feed on other organisms are **heterotrophs**. Since these organisms consume other organisms, they are called **consumers**. The stored energy in photosynthetic organisms is available to the other members of a community. The amount of energy that has been stored is the **net primary productivity**. Net primary productivity is often measured as the **biomass** of the producers that is added to the ecosystem.

As energy flows through the community, it passes from one **trophic level** to the next. Trophic levels begin with producers, then progress to consumers. The number of trophic levels in a system depends on the level of consumers involved. Organisms that feed directly on the producers are **herbivores** and are also **primary consumers**; they form the second trophic level. Organisms that feed on the primary consumers are **carnivores** (flesh-eaters). The carnivores are **secondary consumers**; they form the third trophic level. A fourth trophic level is formed when carnivores eat other carnivores; these are **tertiary consumers**. When trophic levels are traced to diagram relationships between producers, primary consumers, secondary consumers, and tertiary consumers, a **food chain** has been developed. In nature, however, relationships are not so simple. Instead of a food chain, a **food web** illustrates how food chains merge in a complex relationship. Rounding out the food web is the group of organisms that break down and decompose plant and animal matter. **Detritus feeders**, including earthworms, centipedes and millipedes, and vultures, feed on dead organic matter such as fallen leaves, cast-off exoskeletons, carcasses, and bodily wastes. **Decomposers**, such as bacteria and fungi, further break down dead organic matter, releasing any remaining nutrients.

As energy flows through the trophic levels, approximately 10 percent is transferred from one level to the next . **Energy pyramids** are used to diagram this transfer of energy.

A caterpillar that feeds on leaves sprayed with a pesticide may not die from ingesting the chemical. Instead, the chemical may be stored in its body. A bird that feeds on caterpillar may eat several caterpillars with the chemical stored in their bodies. This means that the bird is consuming a great deal more of the pesticide. It is also stored in the bird's body. Any organism that feeds on the bird will ingest a concentrated amount of the chemical each time it feeds on such birds. This organism will store even greater amounts of the pesticide in its body. This is known as **biological magnification**. It occurs because the chemicals used are not **biodegradable**, that is they are not easily broken down by decomposers into harmless substances.

3) How Do Nutrients Move within Ecosystems?

Nutrients, as defined here, are elements and small molecules that are used to make the chemical building blocks of life. Molecules needed by organisms in large amounts are the **macronutrients**; those needed only in very small amounts are the trace nutrients or **micronutrients**. Nutrients do not flow through communities but cycle through. These are the **nutrient cycles**, also referred to as the **biogeochemical cycles**. The nutrients tend to be stored in nonliving, or **abiotic**, **reservoirs** such as CO_2 and nitrogen in the atmosphere and phosphorous in rocks. Other major reservoirs for CO_2 include the dissolved form in the oceans and the remains of ancient plants and animals transformed by heat and pressure into **fossil fuels**.

Although the atmosphere is composed of 79 % nitrogen gas, plants and animals can not use nitrogen in its gaseous form. Atmospheric nitrogen is converted to more usable forms by bacteria and cyanobacteria that conduct **nitrogen fixation**. Plants that belong to the **legume** family play a very important role in nitrogen fixation. Bacteria that can fix nitrogen live in the roots of legumes. Thus, plants such as soy beans and clover are often planted to replace nitrogen in nutrient-poor soils. After cycling through plants and animals, nitrogen is returned to the atmosphere by **denitrifying bacteria**.

Water is also considered to be a nutrient. It, too, cycles in the **hydrologic cycle**. The ocean serves as the major reservoir for water. Water is evaporated from the oceans and returned to land as precipitation.

4) What Is Causing Acid Rain and Global Warming?

Acid rain, or **acid deposition**, occurs when nitrogen oxide and sulfur dioxide combine with water vapor in the atmosphere. Subsequently, nitric acid and sulfuric acid are formed. The acids then come down in the form of acid rain or as dry particles interacting with the structures and organisms they touch.

From the start of the Industrial Revolution, carbon, stored in fossil fuels, has been released into the atmosphere. This has increased the CO_2 content of the atmosphere 25 %, from 280 ppm or .028% preindustrial revolution to over 360 ppm or .036 %. Carbon dioxide in the atmosphere traps heat from the sun and radiating from Earth. This provides a natural **greenhouse effect**, keeping our atmosphere warm to sustain life. However, as the CO_2 levels increase, the warming effect increases also, leading to warmer temperatures globally, or **global warming**.

KEY TERMS AND CONCEPTS

Fill-In: From the following list of key terms, fill in the blanks in the following statements.

abiotic	detritis feeders	legumes
acid deposition	energy pyramid	macronutrients
autotrophs	food web	micronutrients
biodegradable	food chain	net primary productivity
biodiversity	fossil fuels	nitrogen fixation
biological magnification	global warming	primary consumers
carnivores	greenhouse effect	producers
consumer	greenhouse gas	reservoir
decomposer	herbivore	secondary consumers
deforestation	heterotrophs	tertiary consumer
denitrifying bacteria	hydrologic cycle	trophic level

(1) _____ are organisms that produce their own organic material using inorganic materials as their energy source, while organisms that use other living organisms as a source of energy are called (2) _____, and are also known as (3) _____.

Plants are (4) _____, using sunlight as their source of energy to make organic material, creating the first trophic level. The amount of energy captured by these organisms is measured as (5) _____. If the amount of energy captured is measured in dry weight of the organisms, the (6) _____ has been determined

Plants as producers form the first (7) _____. Organisms that ingest plant material are the (8) _____ and are also called (9) _____. Animals that eat only other animals are called (10) _____. These animals are the (11) _____, making up the third trophic level, and the (12) _____, the fourth trophic level.

Animals such as earthworms, millipedes, and termites, which feed on dead organic matter, are referred to as (13) _____. Fungi and bacteria make up the (14) _____, releasing the final nutrients.

Approximately 10 % of biomass energy will be passed on to the next energy level forming an (15) _____.

Some human-made chemicals are not readily broken down in the environment. These chemicals are not (16) _____. Therefore, their concentration may increase in the organisms as you go up the trophic levels in the environment. This increase is called (17) _____.

Nutrients that are required by organisms in large amounts are (18) _____, while those needed only in trace amounts are (19) _____. Within the ecosystem, nutrients are recycled in the (20) _____, also called the (21) _____.

In each cycle, large storage areas serve as (22) _____. These storage areas are often nonliving or (23) _____. During the carboniferous period, the remains of ancient plants and animals were covered with sediment. Today, these carbon stores are (24) _____, burned for electricity production.

Unusable nitrogen gas is converted to usable ammonia by bacteria living in the roots of plants in the (25) _____ family. This process is called (26) _____. After passing through organisms in the ecosystem, nitrogen is returned to the atmosphere by (27) _____.

Water evaporating from the oceans and falling back to land as rain is part of the (28) _____. When evaporated water in the atmosphere mixes with nitrogen oxide or sulfur dioxide, nitric acid and sulfuric acid are formed. With their formation, (29) _____ follows, damaging ecosystems and human-made structures.

Atmospheric CO_2 absorbs heat creating a natural (30) _____ around Earth. As CO_2 levels increase due to deforestation and fossil fuel burning, the average temperature of Earth is expected to increase in an event known as (31) _____.

The "richness" of an ecosystem, measured as the total number of different species in that ecosystem, is referred to as (32) _____.

Key Terms and Definitions

abiotic (a-bi-ah´-tik): nonliving; the abiotic portion of an ecosystem includes soil, rock, water, and the atmosphere.

acid deposition: the deposition of nitric or sulfuric acid, either in dissolved in rain (acid rain) or in the form of dry particles, as a result of the production of nitrogen oxides or sulfur dioxide through burning, primarily of fossil fuels.

autotroph: a "self-feeder," usually meaning a photosynthetic organism.

biodegradable: able to be broken down into harmless substances by decomposers.

biological magnification: the increasing accumulation of a toxic substance in progressively higher trophic levels.

biomass: the dry weight of organic material in an ecosystem.

carnivore: literally "meat eater," a predatory organism feeding on other heterotrophs; a secondary (or higher) consumer.

consumer: an organisms that eats other organisms; a heterotroph.

decomposers: a group of organisms, mainly fungi and bacteria, that digest organic material by secreting digestive enzymes into the environment. In the process they liberate nutrients into the environment.

deforestation: the excessive cutting of forests, primarily rain forests in the tropics to clear space for agriculture.

denitrifying bacteria: bacteria that break down nitrates, releasing nitrogen gas to the atmosphere.

detritus feeders: a diverse group of organisms ranging from worms to vultures that live off the wastes and dead remains of other organisms.

energy pyramid: a graphical representation of the energy contained in succeeding trophic levels, with maximum energy at the base (primary producers) and steadily diminishing amounts at higher levels.

food chain: an illustration of feeding relationships in an ecosystem using a single representative from each of the trophic levels.

food web: a relatively accurate representation of the complex feeding relationships within an ecosystem, including many organisms at various trophic levels, with many of the consumers occupying more than one level simultaneously.

fossil fuels: fuels such as coal, oil, and natural gas, derived from the bodies of prehistoric organisms.

global warming: a gradual rise in global atmospheric temperature as a result of an amplification of the natural greenhouse effect due to human activities.

greenhouse effect: the ability of certain gases such as carbon dioxide and methane to trap sunlight energy in the atmosphere as heat. The glass in a greenhouse does the same, hence the name. This warming of the atmosphere is being enhanced by human production of these gases.

greenhouse gas: a gas, such as carbon dioxide or methane, that traps sunlight energy in a planet's atmosphere as heat; a gas that participates in the greenhouse effect.

herbivore: literally "plant-eater," an organism that eats food directly and exclusively on producers; a primary consumer.

heterotroph: literally "other-feeder," meaning an organism that eats other organisms.

hydrologic cycle: the water cycle, driven by solar energy; a nutrient cycle in which the main reservoir of water is the ocean and most of the water remains in the form of water throughout the cycle (rather than being used in the synthesis of new molecules).

legumes: a group of plants (including alfalfa, peas, soybeans and clover) that harbor colonies of nitrogen-fixing bacteria in special swellings on their roots.

macronutrients: molecules required by organisms in relatively large quantities.

micronutrients: molecules required by organisms in trace quantities.

net primary productivity: the energy stored in the primary producers of an ecosystem over a given time period.

nitrogen fixation: the process of combining atmospheric nitrogen with hydrogen to form ammonia.

nutrient cycle: a description of the movement of a specific nutrient (carbon, nitrogen, phosphorus, water, etc.) through the living and nonliving portions of an ecosystem.

primary consumer: an organism that feeds on producers; an herbivore.

producer: a photosynthetic organism; an autotroph.

reservoir: the major source of any particular nutrient in an ecosystem, usually in the abiotic portion.

secondary consumer: an organism that feeds on primary consumers; a carnivore.

tertiary consumer (ter´-she-ar-e): a carnivore that feeds on other carnivores (secondary consumers).

trophic level: literally, "feeding level"; the categories of organisms in a community, and the position of an organism in a food chain, defined by the organism's source of energy; producers, primary consumers, secondary consumers, and so on.

THINKING THROUGH THE CONCEPTS

True or False: Determine if the statement given is true or false. If it is false, change the underlined word so that the statement reads true.

33. _____ <u>Almost all</u> of the energy produced from the sun reaches the Earth.

34. _____ Ecosystems that have a low producer biomass will have <u>low productivity</u>.

35. _____ Herbivores constitute the <u>first</u> trophic level.

36. _____ <u>Omnivores</u> consume both plant and animal material.

37. _____ A <u>food chain</u> can be very complex since it considers all the feeding relationships between organisms.

38. _____ Vultures and hyenas are <u>detritus feeders</u>.

39. _____ The transfer of energy between trophic levels is extremely <u>efficient</u>.

40. _____ The most infamous chemical related to biomagnification is <u>DDT</u>.

41. _____ Biogeochemical cycles connect the <u>biotic and abiotic</u> portions of the ecosystem.

42. _____ Plants of the <u>rose</u> family are important because they house nitrogen-fixing bacteria in root nodules that can directly absorb and use nitrogen gas from the air, and directly convert atmospheric nitrogen into fertilizer.

43. _____ Human activities are <u>stabilizing</u> the balance of biogeochemical cycles.

44. _____ In the hydrologic cycle, some of the water enters the <u>living community</u> before it evaporates back into the atmosphere.

45. _____ Acid particles can fall from the atmosphere in <u>dry form</u>.

46. _____ Twenty-five percent of the Adirondack lakes are <u>dead</u>.

47. _____ Acid deposition <u>increases</u> lead and other heavy-metal poisoning in animals.

48. _____ Climate prediction for the future is <u>certain</u> in the face of global warming.

Matching: Members of the ecosystem and their trophic level:

49. _____ bracket fungus   Choices:
50. _____ squirrel   a. producer
51. _____ sheep   b. primary consumer
52. _____ dead leaves   c. detritus feeder
53. _____ oak tree   d. detritus
54. _____ hawk   e. decomposer
55. _____ mushrooms   f. secondary consumer
56. _____ algae
57. _____ shark
58. _____ bacteria
59. _____ earthworm
60. _____ snake skin

True or False Determine if the following statements regarding the phosphorus (P) cycle:

61. _____ The main source of P is from rock.
62. _____ The P cycle is greatly accelerated by transport through the atmosphere.
63. _____ Organic P is transported through the food chain.
64. _____ P is the limiting nutrient in many ecosystems.
65. _____ Humans have affected the phosphorus cycle by inadvertent and undesirable fertilization of waterways, accumulation of phosphorus-rich pollutants in the atmosphere, and depletion of soil supplies by overcutting and erosion.

Short answer:

66. Define the term abiotic and provide two examples of abiotic reservoirs each for the carbon, nitrogen, and phosphorous cycles.

_____
_____
_____
_____
_____
_____
_____
_____

67. Which of the following derive energy for growth <u>directly</u> from light?

autotrophs   heterotrophs   tertiary consumers

decomposers   primary consumers   producers

68. Which of the following derive energy for growth <u>indirectly</u> from light?

autotrophs   heterotrophs   tertiary consumers

decomposers   primary consumers   producers

69. Identify six trophic levels. Explain how they interact with one another.

_____

_____

_____

_____

_____

_____

_____

_____

_____

CLUES TO APPLYING THE CONCEPTS

These practice questions are intended to sharpen your ability to apply critical thinking and analysis to the biological concepts covered in this chapter.

70. Assuming typical efficiency of energy transfer from one trophic level to the next, describe how 1000 calories in producer biomass might be converted to carnivores. How many calories will a carnivore receive from the original 1000 calories?

_____

_____

_____

_____

_____

_____

71. Why, do you suppose, there has been no mention in this chapter of quaternary or fourth-level consumers? Energetically, is a fifth trophic level possible? Why or why not?

_____

_____

_____

_____

_____

_____

72. Using the following list, carefully trace the path of a carbon atom through the carbon cycle.

secondary consumer	fecal matter	atmosphere
sugar molecule in plant	fungus	detritus
primary consumer	plant stem	cell respiration
green leafy plant	detritus feeder	decomposer

_____

_____

_____

_____

_____

_____

_____

_____

_____

73. Briefly describe three ways in which humans have affected the nutrient cycles of carbon, nitrogen, and phosphorus. **Do not** state three examples for each cycle--give examples of how humans have affected the cycling of nutrients as a whole.

_____

_____

_____

_____

_____

_____

ANSWERS TO EXERCISES

1. Autotrophs
2. heterotrophs
3. consumers
4. producers
5. net primary productivity
6. biomass
7. trophic level
8. herbivores
9. primary consumers
10. carnivores
11. secondary consumers
12. tertiary consumers
13. detritis feeders
14. decomposers
15. energy pyramid
16. biodegradable
17. biomagnification
18. macronutrients
19. micronutrients
20. nutrient cycles
21. biogeochemical cycles
22. reservoirs
23. abiotic
24. fossil fuels
25. legume
26. nitrogen fixation
27. denitrifying bacteria
28. hydrologic cycle
29. acid deposition
30. greenhouse effect
31. global warming
32. biodiversity
33. false, relatively little
34. true
35. false, second
36. true
37. false, food web
38. true
39. false, inefficient
40. true
41. true
42. false, legume
43. false, threatening
44. true
45. true
46. true
47. true
48. false, uncertain
49. e
50. b
51. b
52. d
53. a
54. f
55. e
56. a
57. f
58. e
59. c
60. d
61. true
62. false
63. true
64. true
65. true

66. Abiotic means nonliving. Carbon reservoirs include gas in the atmosphere, dissolved carbon in the oceans, and fossil fuels. Nitrogen reservoirs include gas in the atmosphere, nitrogen-containing molecules, and wetlands. Phosphorous reservoirs include phosphates in rock, bird guano, and animal teeth and skeletons.

67. autotrophs, producers

68. heterotrophs, primary consumers, decomposers, tertiary consumers

69. Producers capture sunlight energy. Primary consumers feed on producers. Secondary consumers feed on primary consumers. Tertiary consumers feed on secondary consumers. Detritis feeders feed on dead producers, and on waste material from or dead primary, secondary, and tertiary consumers. Decomposers externally digest what remains of producers, and of primary, secondary, and tertiary consumers, and release the remaining nutrients to the environment.

70. 1000 calories exists in producer biomass. If 10 % of the energy is lost with each trophic level, 100 calories will be converted to herbivore (primary consumer) biomass. As a carnivore (secondary consumer) feeds on the herbivore, it will receive only 10 calories from the original 1000 in the producer biomass.

71. The existence of a quaternary consumer would indeed be very rare. Using exercise 70 and continuing to assume 10 percent energy is lost with each trophic level, a tertiary consumer would only receive 1 calorie from the original 1000. If a fourth-level consumer existed, it would receive 0.1 calories from the original 1000. In order to maintain energy demands, this organism would need to consume very large quantities of food.

72. Possible scenario: Atmosphere ------->green leafy plant ------>sugar molecule in plant ------->cell respiration -------> atmosphere ------->green leafy plant ------>plant stem ------->detritus ------->detritus feeder ------->decomposer ------->atmosphere ------->green leafy plant ------>primary consumer -------> secondary consumer ------->cell respiration -------> atmosphere ------->green leafy plant ------>primary consumer -------> secondary consumer ------->fecal matter ------->decomposer ------->atmosphere.
Note* the cycle always starts with and comes back to the atmosphere.

73. The carbon cycle has been affected by the burning of fossil fuels. Massive quantities of carbon were stored in plant material during the carboniferous period, a period that lasted approximately 150 million years. In the roughly 150 years since the beginning of the industrial revolution, humans have returned much of this stored carbon to the atmosphere at a rate that is orders of magnitude greater than the rate at which it was stored. In the meantime, deforestation and the burning of the tropical rainforests have added additional carbon to the atmosphere while at the same time removing areas that could absorb CO_2 from the atmosphere.

The nitrogen cycle has been affected by human-made nitrogen fixation: the production and overuse of fertilizer. Fertilizer runoff from farms into streams and rivers has led to massive die-offs of aquatic ecosystem life. Nitrogen oxides put into the atmosphere by the burning of fossil fuels mixes with water vapor forming nitric acid. Nitric acid is, in part, responsible for damage to crops, aquatic systems, and human-made structures.

The phosphate cycle has been affected, again, by the formation and overuse of fertilizer. Phosphorous-rich soil runoff into waterways stimulates the growth of producers such as algae. When the algae die, their decomposition uses oxygen at a high rate, suffocating fish and other aquatic organisms; ultimately disrupting the aquatic balance.

Chapter 41: Earth's Diverse Ecosystem

OVERVIEW

This chapter looks at how weather patterns and climate determine how the organisms on Earth are distributed. In terrestrial ecosystems, climate determines which plants inhabit an area. The plants present, in turn, determine the animal life existing there. This chapter outlines the major biomes and their characteristic plant communities. The impact humans have on each biome is addressed. Additionally, since the oceans cover 71% of Earth, aquatic life, both freshwater and marine, are discussed. And again, the impact humans have on aquatic ecosystems is considered.

1) What Factors Influence Earth's Climate?

Earth's **climate** is determined by the patterns of **weather** that exist in a specific region. The climate in a region is influenced by **latitude**, air currents, ocean currents, the presence of continents, and elevation. Latitude is measured as the distance north or south of the equator. The sunlight that heats Earth's surface hits Earth at an angle. The degree of the angle determines the overall temperature for a region. Because of Earth's rotation and temperature differences in air masses, air currents are generated. As warm, moist air rises, it is cooled causing precipitation. At the same time, ocean currents are formed from the Earth's rotation, wind, and the sun's heating of the water. The presence of continents causes the ocean currents to circulate forming **gyres**. The circular patterns generate moderate weather patterns on the coastal regions of a land mass. The presence of differing elevations within continents also generate climate variations. As moist air approaches a mountain, it rises, cools, and the moisture condenses causing precipitation. The now dry air passes over the mountain, absorbing moisture as it flows. This causes a **rain shadow** phenomenon on the far side of the mountain.

2) What Are the Requirements of Life?

Four basic resources are required for life. Nutrients are needed to build living tissue, energy is needed to produce the tissues, and in order for metabolic reactions to occur, liquid water must be present and temperature must be appropriate. Within diverse ecosystems, the organisms have adapted the mechanisms necessary to acquire resources and to survive.

How is life on land distributed?

Basically, temperature and availability of water determine the distribution of terrestrial organisms. Both temperature and water are unevenly distributed, and organisms have adapted various mechanisms to tolerate unfavorable temperatures and drought or flood conditions.

Communities on land are defined by the plants dominant there. In the tropics, the **tropical rain forest** is dominated by a diverse population of huge, broadleaf, evergreen trees that grow well along the equator where temperatures are invariant and 250 to 400 cm of rain falls. The soil in this biome is infertile since all the nutrients are tied up in the lush vegetation. Still, humans are deforesting the area for agriculture, inefficient in the infertile soil, causing the loss of countless species and unmeasurable potential. **Tropical deciduous forests** are located slightly north and south of the equator where rainfall is not as predictable. During the dry season, trees drop their leaves to reduce water loss. Farther away from the equator, the grasses of the **savanna** dominate. Any trees here are scrubby with thorns protecting their sparse foliage. The root systems of the grasses can withstand the severe droughts common to the savanna regions. Human settlements have spread into the savanna threatening the black rhino and African elephant populations through poaching and habitat conversion to cattle pastures.

Where rainfall decreases below 50 cm a year, the savanna gives way to **desert**. Plants that survive in the deserts have extended root systems to quickly absorb any rain. The absorbed water is stored in fleshy stems covered with wax to prevent evaporation. The area of Earth covered by deserts is increasing as human activities cause **desertification** of fragile habitats. Desertification is caused by deforestation, soil destruction, and over use of once productive land. Deserts that merge with a coastal region give rise to **chaparral** ecosystems. These unique ecosystems are found along the Mediterranean Sea and along the southern California coast. In the centers of Eurasia and temperate North America, **prairies**, or **grasslands**, dominate. Limited rainfall and relatively frequent fires restrict the growth of trees to river banks while promoting the growth of the prairie grasses. Humans have used the fertile soil created by the dense grass growth for intense cereal agriculture or pasture land. Overgrazing on western prairies in the United States has lead to desertification while agriculture has left only a few, small, remnants of original prairie habitat.

As the amount of precipitation increases trees can take root, shading out the grasses. This gives rise to the **temperate deciduous forests**. Water is unobtainable to the trees during winter, thus they drop their leaves to reduce moisture loss. Spring brings short-lived wildflowers to bloom before new leaves form on the trees, shading the forest floor. Along the southeastern coast of Australia, the southwestern coast of New Zealand, and the northwestern coast of North America lie very wet ecosystems called **temperate rain forests**. These rare habitats receive upward of 400 cm of rain a year, giving rise to lush plant growth of evergreen conifers, ferns, and mosses.

In the interior of northern North America and northern Eurasia, harsh temperatures and a short growing season produce the **northern coniferous forests**, or the **taiga**. Humans have clear-cut vast areas of the taiga forests for timber; however, due to its harsh climate, much has remained undisturbed. Farther north the climate becomes even more extreme. Winter temperatures of the **tundra** may drop below -55 C. A permafrost layer is impervious to water when temperatures do allow for the top meter or so to thaw. Thus a wet marshy expanse is formed. During the brief summer thaw, small flowering plants quickly grow, bloom, and die back until the next thaw.

4) How Is Life in Water Distributed?

Water presents unique characteristics that limit where life can or does exist in aquatic habitats. Water is a great insulator, thus it warms and cools slowly. The sunlight that penetrates into water is quickly absorbed so that depths below 200 m will not receive enough light energy to conduct photosynthesis. Any suspended sediment will reduce the depth of light penetration. Available nutrients settle to the bottom of aquatic ecosystems where light often does not reach, limiting life.

Large freshwater lake ecosystems have distinct life zones. The **littoral zone** occurs where water is shallow, light is abundant, and plants can root and find nutrients in the bottom sediments. Among the rooted plants, **plankton** drift. **Phytoplankton**, such as algae and photosynthetic bacteria, and **zooplankton**, such as protozoa and minute crustaceans, add to the diversity of this ecosystem. In deeper water, light is limited to the **limnetic zone** where the aquatic food web is supported by abundant phytoplankton. Light does not reach the deeper depths of the **profundal zone** and the detritus feeders and decomposers living on the lake floor receive nutrients from detritis from the limnetic zone. Lakes that are extremely nutrient poor are referred to as being **oligotrophic**. These lakes are often formed from deep depressions from glaciers and fed by mountain streams. These lakes are often clear, deep, and oxygen rich. Lakes that receive nutrients and sediments in high amounts as run off from surrounding areas are **eutrophic**. Eutrophic lakes are clouded with suspended sediment and phytoplankton. Oxygen content in the profundal zone is often limited as decomposers use any available oxygen feeding on dead phytoplankton.

Marine ecosystems are also divided into life zones based on depth and availability of light energy. The **photic zone** is the upper region where light energy supports photosynthesis. Below the photic zone is the **aphotic zone** where nutrients settle. Nutrients from the aphotic zone rise to the photic zone during **upwelling** events caused by surface winds displacing surface water. Closer to shore, organisms of the **intertidal zone** experience alternating dehydration, as the tides recede, and rehydration, as the tides rise again. Bays and coastal wetlands make up the **near-shore zone**, which is constantly submerged. Both of these ecosystems support diverse and abundant life. Human activities have greatly impacted the fragile coastal ecosystems. By filling in or dredging, humans have removed almost all of the world's wetlands. Equally fragile are **coral reef** communities. Minute animals, the corals, live symbiotically with algae, building great structures from calcium carbonate secreted from their bodies. These structures serve as habitat for unique and specialized organisms. As sediment and silt from nearby land areas cloud the water around a coral reef, the algae are not able to photosynthesize and provide the energy necessary for the organisms to survive. Ultimately, the reef will die due to human activities.

The open ocean supports free swimming or **pelagic** life forms in the upper photic zone. Here, the food web is supported by phytoplankton. The aphotic zone, however, supports unique life forms of its own. Even the open ocean has been subjected to misuse by humans. Trash, dumped by ships or blown in from shore, endangers the lives of many animals in the sea as they mistake it for food. Radioactive wastes have also been dumped in the open ocean under the pretenses that it could "do no harm out there." However, unique life forms may live in undiscovered areas of the deep ocean. **Vent communities** represent one such example. Deep in the ocean, cracks in Earth's crust heat surrounding water, supporting an ecosystem containing 284 new species. These organisms do not use the sun's energy for life support. Instead, they use hydrogen sulfide from Earth's crust by a process called chemosynthesis.

KEY TERMS AND CONCEPTS

Fill-In: From the following list of key terms, fill in the blanks in the following statements.

near-shore zone
northern coniferous forest
phytoplankton
temperate rain forest
temperate deciduous forest
tropical rain forest
tropical deciduous forests
vent community

desertification
intertidal zone
limnetic zone
oligotrophic
photic zone
profundal zone
zooplankton

coral reef
grassland
littoral zone
ozone layer
permafrost
rainshadow
upwelling

chaparral
eutrophic
latitude
plankton
prairie
savanna
weather

aphotic
climate
desert
gyres
pelagic
taiga
tundra

Short term changes in temperature, precipitation, or cloud cover determines the (1) _____ of an area. However, temperature or precipitation patterns over the long term determine the (2) _____ of a region. The equator at zero degrees (3) _____ has consistently warm temperatures. Ocean currents occur in circular patterns called (4) _____ as determined by the Earth's rotation, wind, warming of the water, and continents.

Mountains modify precipitation patterns. As air passes over a mountain it releases precipitation. On the far side of the mountain, then, the air is dry creating a (5) _____.

The plant life in a terrestrial biome may be determined by the amount of rainfall and the frequency with which it occurs. In the (6) _____ temperature and rainfall averages are consistent from year to year and the organisms found there are the most diverse. However, if rainfall were not so consistent so that there were distinct wet and dry seasons, the plants would drop their leaves during the dry season. This is more characteristic of the (7) _____. Where grasses are dominant and very tolerant of the severe dry season, the (8) _____ exists. The use of this biome for cattle grazing is threatening its wildlife. When rainfall averages less than 25 - 50 cm, water-storing cacti live, making up the flora of the (9) _____. Over use of land compromised by drought, deforestation, and soil destruction has lead to irreversible (10) _____ in many regions.

Coastal areas on the margins of deserts result in the unique biome of the (11) _____. In temperate regions, as rain fall increases away from the deserts, grasses and wildflowers grow. This is the (12) _____ biome, more commonly referred to as the (13) _____.

In areas that can support tree growth, grasses are shaded out and forests dominate. Regions that have cold winters with below freezing weather, trees drop their leaves to conserve water. This describes the (14) _____. However, if cold temperatures are moderated and heavy rain fall occurs, as along the Olympic peninsula, then a (15) _____ exists.

Long, hard, cold winters and short springs of Southern Canada result in heavy growth of evergreen conifers in the (16) _____ or (17) _____ biome. As temperature becomes even colder, the fragile (18) _____ exists as the last biome before the polar ice caps.

Aquatic ecosystems are also defined by their unique characteristics. Areas of freshwater lakes are divided into three zones. The (19) _____ has shallow water with diverse communities. Drifting among the plants in this zone are the microscopic organisms, the (20) _____. If these organisms are photosynthetic bacteria or protists such as algae, then they are called (21) _____. If, instead, the organisms consist of protozoans and minute crustaceans, they are called (22) _____.

Deeper open water that does not allow plants to be rooted to the bottom consists of the upper (23) _____ which allows photosynthesis to occur, and the lower (24) _____. Here, light does not penetrate to allow photosynthesis to occur.

Lakes are also classified according to their nutrient content. (25) _____ lakes are very low in nutrients, having been formed by glaciers and fed by mountain streams. On the other hand, (26) _____ lakes have high sediment rates and high deposition rates of organic and inorganic material from their surroundings.

Ocean ecosystems consist of the upper layer that allows light penetration for photosynthesis, the (27) _____, and the deeper, nonphotosynthetic region, (28) _____. Nutrients from the lower oceanic regions may be brought to the surface as (29) _____ occurs.

Aquatic areas along the coast may alternately be wet and dry, this occurs at the (30) _____ as the tide rises and falls. Shallow, submerged bays and wetlands make up the (31) _____.

Off shore areas with all the right parameters allow corals and algae to build structures of calcium carbonate. The diverse habitats of these (32) _____ are extremely fragile and sensitive.

Life in the open ocean primarily exists in the upper photic zone. The life forms here are free swimming or (33) _____. However, new and unusual life forms have been discovered far from the ocean surface. Deep in the ocean, extremely hot water, heated by Earth's core, supports the (34) _____.

Key Terms and Definitions

aphotic zone: the region of the ocean below 200 m, where sunlight does not penetrate.

biodiversity: the total number of species within an ecosystem and the resulting complexity of interactions among them.

biome: a terrestrial region with a characteristic climate and vegetation. Regions with similar climates will have similar vegetation and belong to the same biome (grassland, desert) even though widely separated and inhabited by different species.

chaparral: a temperate coastal biome with hot dry summers and cool, somewhat rainy winters with frequent fogs. The typical vegetation consists of small trees or large bushes that are drought and fire resistant.

climate: prevailing weather patterns over a particular region as determined by availability of sunlight, water, and temperature.

coral reef: the most diverse marine ecosystem, formed by the bodies and calcium carbonate skeletons of corals and algae in warm, relatively shallow water.

desert: a biome in which potential evaporation greatly exceeds rainfall. Perennial vegetation is widely spaced and has drought-resistant adaptations, such as waxy or spiny leaves.

desertification: the conversion of productive land in warm, dry areas into unproductive desert land through loss of soil and vegetation, often as a result of overgrazing or other improper farming methods.

estuary: a wetland formed where a river meets the ocean; the salinity there is quite variable but lower than in seawater and higher than in freshwater.

eutrophic lake: a lake that is rich in nutrients and supports dense communities of organisms.

grassland: a biome often found at the centers of continents characterized by a relatively continuous ground cover of grasses, with few or no trees except along rivers. Typically, grassland biomes have a prolonged dry season and frequent fires, both of which favor grasses over trees.

gyre: a massive, roughly circular ocean current. Gyres flow clockwise in the northern hemisphere and counterclockwise in the southern hemisphere.

hydrothermal vent community: a community of unusual organisms, living in the deep ocean near hydrothermal vents, that depends on the chemosynthetic activities of sulfur bacteria.

intertidal zone: the area along the edges of land masses that is alternately covered and uncovered by the rising and falling of the tides.

limnetic zone: the upper, lighted, open-water region of a lake, where phytoplankton can carry out photosynthesis.

littoral zone: the near-shore, shallow-water region of a lake where anchored plants grow, and light and nutrients are abundant.

near-shore zone: shallow-water areas of the ocean just below the low-tide line. Because nutrients are abundant and sunlight strong, the near-shore zone often has the greatest concentration of life in the ocean.

northern coniferous forest: a northern biome dominated by conifers; taiga.

oligotrophic lake: a lake that is low in nutrients and supports sparse communities of organisms.

ozone layer: layer of atmosphere between 10 and 50 kilometers above the earth, which contains relatively high concentrations of ozone (O_3). Ozone absorbs harmful ultraviolet radiation.

pelagic: referring to organisms that are found free-swimming or floating in open water.

permafrost: soil layer no more than a half a meter below the surface of the tundra that stays frozen year round.

photic zone: the surface waters of the ocean, into which sunlight penetrates well enough to support photosynthesis.

phytoplankton: microscopic photosynthetic protists, bacteria, and algae that form the basis of freshwater and marine food webs.

plankton: microscopic pelagic protists and animals found in the photic zone of the oceans, whose movement is determined primarily by movement of the water. Phytoplankton are the main producers in the open ocean.

prairie: a biome, located in the centers of continents, that supports grasses; also called *grassland.*

profundal zone: the depths of a lake where light is inadequate for photosynthesis.

rain shadow: an area of low rainfall on the side of a mountain facing away from the prevailing winds.

savanna: a transition biome between the tropical deciduous forest and tropical grassland, characterized by a nearly continuous cover of grasses with occasional drought-resistant trees.

taiga: the cold coniferous forest biome, characterized by evergreen trees with small, waxy needles. These needles reduce water loss by evaporation during winter.

temperate deciduous forest: a biome having cold winters but a fairly long, moist growing season. The dominant vegetation is broadleaf deciduous trees, which shed their leaves in winter as a protection against water loss.

temperate rain forest: a biome of very limited distribution, along certain coasts where the climate is fairly mild year round and rainfall is very high, characterized by coniferous evergreen trees, often of enormous size.

tropical deciduous forest: a broadleaf deciduous forest biome found in areas with a warm climate year round but a pronounced wet and dry season. The trees drop their leaves in the dry season.

tropical rain forest: a broadleaf evergreen forest biome found in the tropics, where rainfall is adequate and temperatures are warm year round. The tropical rain forest has the greatest diversity of both plants and animals to be found anywhere on land.

tundra: a treeless biome characterized by permafrost and an extremely short growing season. Most of the plants are low-growing perennial shrubs, grasses, and wildflowers.

upwelling: movement of cold, nutrient-rich water from the ocean depths to the surface.

weather: fluctuations in temperature, humidity, cloud cover, wind, and precipitation occurring over a period of hours or days.

zooplankton: a diverse group of small protists and animals, such as tiny crustaceans, that serve as food for larger freshwater and marine invertebrates.

THINKING THROUGH THE CONCEPTS

True or False: Determine if the statement given is true or false. If it is false, change the underlined word so that the statement reads true.

35. _____ The tilt of Earth's axis influences air and ocean currents.

36. _____ The presence of continents affects ocean currents, producing gyres.

37. _____ Deserts are often found in the rainshadow of a mountain range.

38. _____ The specific requirements of life for each organism are the same.

39. _____ Since plants are specifically adapted to their environment, terrestrial communities are defined by the plant life found there.

40. _____ It is possible to harvest products from the tropical rain forests without damaging the ecosystem.

41. _____ Grazing cattle on the range land of the savanna has had little impact on the wildlife there.

42. _____ The total area of land occupied by deserts is <u>decreasing</u> due to human activities.

43. _____ The occurrence of <u>fire</u> plays an important role in the maintenance of healthy prairies.

44. _____ In the temperate rain forest, seedlings often take root in a newly fallen tree, which serves as a <u>nurse log</u>, providing nutrients and protection for the developing plant.

45. _____ Clearcutting of forests in the <u>tundra</u> has destroyed habitat in large regions.

46. _____ The major factors that determine the quantity and type of life in aquatic ecosystems are <u>energy and nutrients</u>.

47. _____ Wastes from sewer treatment facilities and over use of fertilizers may lead to <u>excessive amounts of nutrients</u> being washed into aquatic ecosystems.

48. _____ Coral reef ecosystems are <u>ultimately resilient</u> and, therefore, are <u>unaffected</u> by human activities.

49. _____ Fish populations in the ocean have <u>declined dramatically</u> due to over fishing practices.

50. Fill in the table below with characteristics of the biomes listed.

Biome	Precipitation range	Temperature range	Typical plants present	Typical animals present
Tropical rain forest				
Tropical deciduous forest	NA	NA		NA
Savanna		NA		
Desert				
Chaparral		NA		NA
Prairie; Grassland				
Temperate deciduous forest				
Temperate rain forest				NA
Taiga; Northern coniferous forest				
Tundra				

Identify the following: Determine if the following characteristics belong to **eutrophic** lakes or **oligotrophic** lakes.

51. _____ receive high inorganic material from area runoff

52. _____ contain very little nutrients

53. _____ are often clear

54. _____ profundal zones often low in oxygen

55. _____ tend to be good trout fishing lakes

56. _____ receive high organic material from area runoff

57. _____ are oxygen rich

58. _____ are murky with a dense phytoplankton population

59. _____ experience seasonal algal blooms

60. _____ limnetic zone may extend to the bottom

Identify the following: Determine if the following statements are representative of **human** impacted ecosystems or of **undisturbed** ecosystems.

61. _____ animals in the highest trophic levels are rare.

62. _____ species diversity is high.

63. _____ fueled by sunlight.

64. _____ nutrients are recycled.

65. _____ fueled by fossil fuels.

66. _____ fertilizers, pesticides, and topsoil pollute streams and rivers.

67. _____ water is stored.

68. _____ water is polluted.

69. _____ water is filtered and purified.

70. _____ natural predators control population growth.

71. _____ crops planted close together encourages pest outbreaks.

72. _____ population is expanding exponentially.

73. _____ populations are relatively stable.

74. _____ ecosystems are simple.

Short answer:

75. Identify the four fundamental resources necessary for life.

_____

_____

_____

_____

_____

CLUES TO APPLYING THE CONCEPTS

These practice questions are intended to sharpen your ability to apply critical thinking and analysis to the biological concepts covered in this chapter.

76. Discuss how the limnetic zone and profundal zone of fresh water lakes relate to the photic zone and aphotic zone of oceans.

_____

_____

_____

_____

_____

77. Discuss how the Ozone Layer is formed in the stratosphere and its function. How is the layer being destroyed and what are the possible results of this destruction?

_____

_____

_____

_____

_____

_____

_____

78. How can humans reverse the destructive trends of our activities?

_____

_____

_____

_____

_____

_____

_____

_____

ANSWERS TO EXERCISES

1. weather
2. climate
3. latitude
4. gyres
5. rainshadow
6. tropical rain forest
7. tropical deciduous forest
8. savanna
9. desert
10. desertification
11. chaparral
12. grassland
13. prairie
14. temperate deciduous forest
15. temperate rain forest
16. northern coniferous forest

17. taiga
18. tundra
19. Littoral zone
20. plankton
21. phytoplankton
22. zooplankton
23. limnetic zone
24. profundal zone
25. oligotrophic
26. eutrophic
27. photic
28. aphotic
29. upwelling
30. intertidal zone
31. near-shore zone
32. coral reefs
33. pelagic
34. vent communities

35. false, rotation of Earth
36. true
37. true
38. false, general requirements
39. true
40. true
41. false, life threatening impact
42. false, increasing
43. true
44. true
45. false, taiga
46. true
47. true
48. false, extremely fragile, severely affected
49. true

50. Biomes table

Biome	precipitation range	temperature range	typical plants present	typical animals present
Tropical rain forest	250 cm - 400 cm	25 - 20 C	broadleaf evergreens	arboreal monkeys, birds, and insects
Tropical deciduous forest	NA	NA	deciduous trees	NA
Savanna	30 cm	NA	scrubby trees, grasses	antelope, buffalo, lions, wildebeest, elephants
Desert	< 25 cm	ave 20 C	cacti, succulents	lizards, snakes, kangaroo rats
Chaparral	< 25 cm	NA	small evergreen trees, bushes	NA
Prairie; Grassland	> 25 cm	ave 7 C; hot summers	grasses	bison, antelope
Temperate deciduous forest	75 cm - 150 cm	periods of subfreezing temperatures	deciduous trees, wildflowers	black bear, deer, wolves, shrews, raccoons, birds
Temperate rain forest	400 cm	moderate	evergreen conifers	NA
Taiga; Northern coniferous forest	25 cm - 30 cm	ave -4 C	evergreen conifers	bison, grizzly bear, fox, moose, wolves
Tundra	< 25 cm	below -55 C	perennial wildflowers, dwarf willows	caribou, lemmings, birds mosquitos, snowy owls

ANSWERS TO EXERCISES

1. weather
2. climate
3. latitude
4. gyres
5. rainshadow
6. tropical rain forest
7. tropical deciduous forest
8. savanna
9. desert
10. desertification
11. chaparral
12. grassland
13. prairie
14. temperate deciduous forest
15. temperate rain forest
16. northern coniferous forest

17. taiga
18. tundra
19. littoral zone
20. plankton
21. phytoplankton
22. zooplankton
23. limnetic zone
24. profundal zone
25. Oligotrophic
26. eutrophic
27. photic zone
28. aphotic
29. upwelling
30. intertidal zone
31. near-shore zone
32. coral reefs
33. pelagic
34. vent communities

35. false, rotation of Earth
36. true
37. true
38. false, general requirements
39. true
40. true
41. false, life threatening impact
42. false, increasing
43. true
44. true
45. false, taiga
46. true
47. true
48. false, extremely fragile, severely affected
49. true

50. Biomes table

Biome	precipitation range	temperature range	typical plants present	typical animals present
Tropical rain forest	250 cm - 400 cm	25 - 20 C	broadleaf evergreens	arboreal monkeys, birds, and insects
Tropical deciduous forest	NA	NA	deciduous trees	NA
Savanna	30 cm	NA	scrubby trees, grasses	antelope, buffalo, lions, wildebeest, elephants
Desert	< 25 cm	ave 20 C	cacti, succulents	lizards, snakes, kangaroo rats
Chaparral	< 25 cm	NA	small evergreen trees, bushes	NA
Prairie; Grassland	> 25 cm	ave 7 C; hot summers	grasses	bison, antelope
Temperate deciduous forest	75 cm - 150 cm	periods of subfreezing temperatures	deciduous trees, wildflowers	black bear, deer, wolves, shrews, raccoons, birds
Temperate rain forest	400 cm	moderate	evergreen conifers	NA
Taiga; Northern coniferous forest	25 cm - 30 cm	ave -4 C	evergreen conifers	bison, grizzly bear, fox, moose, wolves
Tundra	< 25 cm	below -55 C	perennial wildflowers, dwarf willows	caribou, lemmings, birds mosquitos, snowy owls

51. eutrophic	59. eutrophic	67. undisturbed
52. oligotrophic	60. oligotrophic	68. human
53. oligotrophic	61. human	69. undisturbed
54. eutrophic	62. undisturbed	70. undisturbed
55. oligotrophic	63. undisturbed	71. human
56. eutrophic	64. undisturbed	72. human
57. oligotrophic	65. human	73. undisturbed
58. eutrophic	66. human	74. human

75. Nutrients to construct living tissue; energy to support the construction of living tissue; liquid water in which metabolic reactions can occur; appropriate temperature such that the metabolic reactions can occur.

76. In both the limnetic zone of a lake and the photic zone of the ocean, enough sunlight energy penetrates to support photosynthesis. Both areas may provide the energy to support the profundal and the aphotic zones respectively. The profundal and the aphotic zones do not receive enough light energy to support photosynthesis and rely on the limnetic zone or the photic zone for energy.

77. The stratospheric ozone layer is formed by ultraviolet light striking oxygen, forming O_3. Ozone functions to filter damaging UV radiation, greatly reducing the amount that reaches life on Earth. This layer of ozone is destroyed when CFC's (chlorofluorocarbons) are degraded by UV radiation releasing chlorine atoms. Chlorine reacts with the O_3 molecules, breaking it into O_2 molecules. The chlorine stays in the stratosphere reacting with ozone molecule after ozone molecule. The result of the destruction of the stratospheric ozone layer is increased UV radiation reaching Earth. This will ultimately result in increased health concerns, such as reduced immune function and increased skin cancer in humans. Globally, it will result in reduced net primary productivity and increased damage to DNA molecules in most organisms.

78. Humans can reverse the destructive trends our activities have on the ecosystem through understanding how healthy ecosystems function, educating ourselves and others about the destruction that is occurring and how it can and should be reversed, and by making a commitment, globally, to the reversal of the destruction. Through the appropriate use of technology, water and air do not have to be polluted, soil does not have to be depleted of nutrients so that artificial fertilizers are needed, and we do not have to rely on fossil fuels to provide energy for our needs. Finally, humans need to stabilize population growth to reduce the expansion of human-dominated ecosystems.